DATE DUE			

TECHNOLOGY
AND THE RISE OF
THE NETWORKED CITY
IN EUROPE
AND AMERICA

TECHNOLOGY AND URBAN GROWTH

a series edited by

Blaine Brownell
Donald T. Critchlow
Mark S. Foster
Mark Rose
Joel A. Tarr

TECHNOLOGY
AND THE RISE
OF THE NETWORKED
CITY
IN EUROPE
AND AMERICA

Edited by
JOEL A. TARR *and*
GABRIEL DUPUY

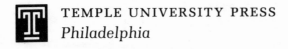

TEMPLE UNIVERSITY PRESS
Philadelphia

Temple University Press, Philadelphia 19122
Copyright © 1988 by Temple University. All rights reserved
Published 1988
Printed in the United States of America

Library of Congress Cataloging-in-Publication Data

Technology and the rise of the networked city in Europe and America /
edited by Joel A. Tarr and Gabriel Dupuy.
p. cm. — (Technology and urban growth)
ISBN 0-87722-540-0 (alk. paper)
1. Infrastructure (Economics)—Europe—History—Case studies.
2. Urban policy—Europe—History—Case studies. 3. Technological
innovations—Economic aspects—Europe—History—Case studies.
4. Infrastructure (Economics)—United States—History—Case studies.
5. Urban policy—United States—History—Case studies.
6. Technological innovations—Economic aspects—United States—
History—Case studies. I. Tarr, Joel A. (Joel Arthur), 1934–
II. Dupuy, Gabriel. III. Series.
HC240.9.C3T43 1988   87-27787
363'.09173'2—dc19   CIP

TO TOVA AND CLAUDINE—

who know something about international relations

Contents

TABLES   xi
ILLUSTRATIONS   xii
PREFACE   xiii
ACKNOWLEDGMENTS   xix

PART I   *Transportation*   1

    1   Comparative Perspectives on Transit in
           Europe and the United States, 1850–1914   3
           John P. McKay

    2   Street Transport in the Second Half of the
           Nineteenth Century: Mechanization Delayed?   22
           Anthony Sutcliffe

    3   Economic Aspects of Public Transit in the
           Parisian Area, 1855–1939   40
           Dominique Larroque

    4   Urban Pathways: The Street and Highway,
           1900–1940   67
           Clay McShane

PART II   *Water Systems*   89

    5   The Genesis of Water Supply, Distribution,
           and Sewerage Systems in France, 1800–1850   91
           André Guillerme

6   The Development of Water and Sewerage
     Systems in France, 1850–1950   116
     Jean-Pierre Goubert

7   Fire and Disease: The Development of Water
     Supply Systems in New England, 1870–1900   137
     Letty Anderson

PART III   Waste Disposal   157

8   Sewerage and the Development of the
     Networked City in the United States,
     1850–1930   159
     Joel A. Tarr

9   Historical Origins and Development of a
     Sewerage System in a German City:
     Bielefeld, 1850–1904   186
     Georges Knaebel

10   Technology Diffusion and Refuse Disposal:
     The Case of the British Destructor   207
     Martin V. Melosi

PART IV   Energy, Heat, and Power   227

11   Urban Gas and Electric Systems and Social
     Change, 1900–1940   229
     Mark H. Rose

12   City Lights: The Electrification of the Chicago
     Region, 1880–1930   246
     Harold L. Platt

13   Municipalities as Managers: Heat Networks
     in Germany   282
     Roselyne Messager

14   Utility Networks and Territory in the Paris
     Region: The Case of Andresy   295
     Gabriel Dupuy

PART V   *Communication*   307

15   Cities and Communication: The Limits of
      Community   309
      Seymour J. Mandelbaum

16   Telephone Networks in France and Great
      Britain   322
      Chantal de Gournay

THE CONTRIBUTORS   339

List of Tables

3.1	Number of passengers carried, Paris area, 1890 and 1913	58
4.1	The growth of the automotive industry, 1900–1940	68
4.2	Parks in major American cities, 1890–1916	73
4.3	Articles dealing largely with parkways, *American City Magazine*, 1909–1939	74
4.4	Motor vehicle fatalities in the United States	78
5.1	Water supply systems in France, selected cities	104
7.1	Water utilities, United States and New England	138
7.2	Size distribution of New England towns	151
8.1	Sewer mileage by type and population group, 1905, 1907, 1909	167
8.2	Total U.S. population, urban population, water treatment, sewers, and sewage treatment, 1880–1940	171
11.1	Indices of the cost of electricity, gas, and coal for household use in Denver and Kansas City, 1919–1936	235
13.1	Electric companies in Germany, by type of corporation and composition of capital, 1913	288
16.1	Telephone equipment, France, Great Britain, and Ireland, 1899–1908	326

Illustrations

PART

I	Transportation	45
II	Water Systems	123
III	Waste Disposal	197
IV	Energy, Heat, and Power	255

FIGURE

3.1	The CGO transportation network, 1861	42
3.2	Survey of CGO equipment for five operations, from 1879–1909 annual reports	52
12.1	Comparative growth of Chicago's homes and its gas, telephone, and electric utilities	269
14.1	Extensions of the Mallet factory network to supply water, gas, and electricity, 1895–1932	299
16.1	The French telephone network in 1891	324
16.2	The English telephone network in 1892	332

Preface

TECHNOLOGY is critical to the city-building process and the operation of cities, but historians have not paid serious attention to its vital role in shaping the urban environment until the last decade or so. Although technology and cities have always been interdependent, only since the advent of industrialism in the nineteenth century have urban technological networks evolved. Today, what we call the *urban infrastructure* provides the technological "sinews" of the modern metropolitan area: its road, bridge, and transit networks; its water and sewer lines and waste-disposal facilities; and its power and communication systems.

These "sinews" guide and facilitate urban functioning and urban life in a multitude of ways, some positive and others negative, some visible and others invisible. The infrastructure includes not only networks but also structures and machines, and it is located both above and below the ground. It is both publicly and privately constructed and operated, with variations not only from country to country but also from time period to time period within nations. Technological infrastructure makes possible the existence of the modern city and provides the means for its continuing operation, but it also increases the city's vulnerability to catastrophic events such as war or natural disaster. While technology may enhance the urban quality of life, it may also be a force for deterioration and destruction of neighborhoods, as well as a hindrance to humane and rational planning.

The 1970s and 1980s saw a growing interest by European and American scholars and professionals in the effects of technology on society. This interest extended to urban technologies and was reflected in an increase in the number of scholars exploring such questions. In recognition of this growing body of scholarship, an International Conference on the City and Technology took place on December 12–15, 1983, at the Ecole Nationale des Ponts et Chausses in Paris. The purpose of this conference, jointly sponsored by the Centre National de la Recherche Scientifique and the National Science Foundation, was to explore the role of technology in urban settings in the United States and in various European countries during the nineteenth and twentieth centuries. The goal of the conference was to provide insight into the evolution of urban technological systems and, whenever possible, to bring such knowledge to bear on our understanding of the contemporary urban environment.

The history of European cities is, of course, much older than that of cities in the United States and Canada. Western European cities such as Paris, Rome, and London, as well as those in other parts of Europe, experimented with various technologies to improve their functioning and their quality of life long before cities of any size existed in America. And yet, once one enters the nineteenth century, the transatlantic experience in regard to urban technology adoption, in spite of national variations, appears generally more similar than different. That is, by the first decade of the twentieth century, most large and medium-size cities in North America and Western Europe had equipped themselves with an infrastructure of pipes, tracks, and wires. The explanation for this similarity may lie in the fact that cities, regardless of the national context, had common needs for water, transport, energy, and communication as they expanded in the nineteenth century. And industrialism was able to provide the means to satisfy some of the needs by furnishing a range of materials and techniques not previously available. Technology transfer was important for cities on both continents, as information about various technological innovations flowed first from Europe to the United States and then reversed toward the end of the century.

In his study of the city-building process in the nineteenth and twentieth centuries, Josef W. Konvitz maintains that industrialism produced a new kind of city. These were cities whose patterns were almost solely shaped by economic factors rather than political and social institutions. City-building techniques that focused on economic criteria, he argues, "produced environments ill-prepared to adjust to many of the changes accompanying urban development."[1] Aside from a very few massive rebuilding and planning efforts, such as that of Baron Georges Haussmann in Paris in the 1850s and 1860s, most infrastructures were unevenly inserted into existing urban environments without drastically affecting the uses of space. As the urban historian Jon A. Peterson has observed, nineteenth-century city building was characterized by a "piecemeal, decentralized approach."[2] Even when opportunities for radical change in a rational direction existed, such as after great fires, alterations in the urban fabric were, as Christine Rosen has shown, much less than were desirable and theoretically possible.[3] Still, new technologies in transport, energy, and communication did strongly influence the character of urban life for some city dwellers, although the essential fabric was often slow to change.

The speed of change accelerated in the last decades of the century in both the United States and the nations of Western Europe, as demand and supply factors such as rapid urbanization, industrialization, new engineering knowledge, and capital availability interacted to stimulate a wave of infrastructure construction in cities. Some of the most striking advances included the development of electricity and its application to transport, lighting, and power; the growth of telegraph and telephone networks; a host of innovations in materials and building that led to new structures such as steel-frame skyscrapers and steel and concrete bridges; and the application of new scientific knowledge to water and sewage treatment.[4] By World War I, the major cities in the United States and Western Europe were rapidly progressing to the stage where they were fully networked with wires, pipes, and tracks. Increasingly, in the period after World War I, structures and networks to accommodate the automobile, such as paved streets, parkways, and highways, along with bridges and parking structures, also became part of the infrastructure.

The infrastructure constructed in the period after the 1880s had somewhat different effects from that built earlier. Konvitz has argued that these systems developed in a manner that reduced the importance of economic criteria in city building and replaced it by what he calls the "regulatory mode."[5] That is, because these infrastructures produced major alterations in street patterns and had effects throughout the urbanized areas in which they were constructed, political and social as well as economic considerations entered into infrastructure decision making. Increasingly, especially in Europe but also to a somewhat lesser extent in the United States, bureaucratic organizations, such as special authorities and regulatory bodies of various kinds, made many of the basic decisions that guided metropolitan growth. Such involvement, however, did not necessarily reduce the fragmentation of urban areas in the twentieth century. Specialized actors involved in the city-building process have continually constructed infrastructures without consideration of their relationship to the total urban fabric.

To a large extent, the contributors to this book deal with the critical period of technology adoption and infrastructure construction from the middle of the nineteenth century to the middle of the twentieth century. The specific subject areas covered include urban transportation, water supply, wastes, energy, and communication. Some chapters are specifically comparative; others deal with developments in one country. They make clear that while the essential transformation in the cities of both Western Europe and the United States was from the pedestrian to the networked city, the process of transformation differed between continents and nations and also within countries. The reasons for these differences vary from country to country, and have to do with a variety of cultural, economic, organizational, political, social, and technological factors that interacted over time. These reasons for the rise of the networked city in Western Europe and the United States, as well as for the variations in patterns, are explored in the chapters included in this volume.

The nineteenth-century Industrial Revolution produced the technologically networked city of the twentieth century. Today a revolution is occurring in communication technologies and information systems that is producing new urban forms and patterns. Some of these patterns involve further residential and industrial dispersal, as well as the revival of a form of cottage industry but with high levels of electronic interaction. Others include the revitalization of the central business districts of certain "world cities" and the gentrification of inner-city neighborhoods. Still others involve the decline of older industrial cities or their shift to service-based economies. To what extent the cities of the future will continue to depend on the infrastructure technologies of the nineteenth century, and to what extent they will incorporate new and more flexible technologies, is unclear, as is the prospective form and shape of the cities themselves. The contributors to this volume provide an important step toward understanding the effects of technology on cities during the last technology-driven period of urban transformation. We hope that these essays may also serve to provide insight in regard to our current transformation.

Notes

1. Josef W. Konvitz, *The Urban Millennium: The City-Building Process from the Early Middle Ages to the Present* (Carbondale, Ill., 1985), pp. 100–101, 129–130.

2. Jon A. Peterson, "The Impact of Sanitary Reform upon American Urban Planning," *Journal of Social History* 13 (Fall 1979): 84–89.

3. Christine Rosen, *The Limits of Power: Great Fires and the Process of City Growth in America* (New York, 1985).

4. The following studies include comparisons of urban technologies in cities of Western Europe and the United States: For the 1870–1900 period, see Jon C. Teaford, *The Unheralded Triumph: City Government in America, 1870–1900* (Baltimore, 1984), pp. 217–250; for transit, see John P. McKay, *Tramways and Trolleys: The Rise of Urban Mass Transport in Europe* (Princeton, N.J., 1976); for the development of electrical networks, see Thomas P. Hughes, *Networks of Power: Electrification in Western Society* (Baltimore, 1983); for the telephone, see Ithiel de Sola Pool, ed., *The Social Impact of the Telephone* (Boston, 1977); for materials and structures, see David P. Billington, *The Tower and the Bridge: The New Art of Structural Engineering* (Princeton, N.J., 1983); and for water and sewers, see Joel A. Tarr and Gabriel Dupuy, "Sewers and Cities: France and the U.S. Compared," *Journal of the Environmental Engineering Division, Proceedings of the American Society of Civil Engineers* 108 (April 1982): 327–338. An excellent study of the emergence and regulation of gas, water, electricity, and telephone systems in Canadian cities with some comparison with cities in the United States is Christopher Armstrong and H. V. Nelles, *Monopoly's Moment: The Organization and Regulation of Canadian Utilities, 1830–1930* (Philadelphia, 1986). In the twentieth century came the development of the automobile, with its far-reaching effects on American and then European cities. There is no study that compares the impact of the automobile on American and European cities, but see the essays in John Brotchie, Peter Newton, Peter Hall, and Peter Nijkamp, *The Future of Urban Form: The Impact of New Technology* (New York, 1985).

5. Konvitz, *Urban Millennium*, pp. 132–134.

Acknowledgments

THE conference upon which this book is based was originally supported by CNRS (Programme Science, Technique et Société) and the National Science Foundation. The English-language publication of the book was generously supported by the National Science Foundation, Grant No. INT-83-13397. Any opinions, findings, and conclusions or recommendations expressed in this publication are those of the authors and do not necessarily reflect the views of the National Science Foundation. The essays included here were originally published in French in *Les annales de la recherche urbaine*, Nos. 23–24, July–December 1984, and are published here with the journal's permission. A number of the essays have been revised for this English-language edition. The editors would like to thank Josef W. Konvitz for his helpful suggestions in regard to the organization of this volume and various colleagues and students who struggled to translate the French essays into English.

PART I

Transportation

TRANSPORT innovation played the key role in the transformation of the spatially limited pedestrian city to the much more extended networked metropolis. Transport technologies of the greatest importance in this change were the electrically powered streetcar (or tram) and the gasoline internal combustion engine–powered automobile, truck, and bus. In the larger cities, such as Paris, London, Philadelphia, and New York, other transport technologies, such as subways, elevated railways, and commuter steam railways, also played important roles. These transport technologies helped determine the direction and extent of urban areal expansion and the growth of specialized residential, commercial, and industrial districts.

The four essays in this part emphasize various aspects of transport development. In his comparative contribution, John P. McKay details the factors that caused Europe initially to lag behind the United States in the diffusion of street railways, especially electrified systems. By 1914, however, McKay finds that Europe's largely municipally owned systems were providing service that was equal or even superior to that of privately owned lines in the United States. In a study focusing on technological innovation, Anthony Sutcliffe explores the reasons for the delay in mechanization of urban street transport from approximately 1870, when the horsecar was near the peak of its technical development, to about 1890, when electric traction became available. He finds the answer in a variety of economic, institutional, and technological factors, not the least of which, especially in Great Britain, was a vested interest in horses. Dominique Larroque, in a case study of public transport in the Paris region, focuses on the relation between the development of the regional system and global economic forces. While transport policy shifted over time from private-company neglect of workers' needs to

1

an acceptance by a publicly owned system of its social and urban responsibilities, the system remained attuned to the profit and production
imperatives of corporations. Finally, to conclude this part, Clay Mc-
Shane explores the changes that the automobile prompted in highway
and street design, especially those arteries linking city and suburb. Early
parkway systems, argues McShane, served local recreational needs as
well as citywide transportation functions, therefore providing larger
urban benefits than the high-speed highways built in the post–World
War II period.

1 Comparative Perspectives on Transit in Europe and the United States, 1850–1914

John P. McKay

W<small>HEN</small> one considers the city in a broad historical perspective, it is clear that public transit is a recent phenomenon and that sophisticated transit technologies are even more recent. Medieval cities were unquestionably "walking cities," to use the graphic term coined by Sam Warner.[1] And although the baroque city of the centralizing European state expanded the urban horizon both spatially and functionally, the enthusiastic acceptance of the privately owned coach by nobles and then by wealthy bourgeois was the era's most notable change in urban passenger transportation. Hackney carriages for hire—forerunners of the modern taxi—also appeared in great cities such as London and Paris to serve the comfortable classes. But there is no evidence of a successful transport service carrying paying passengers along a predetermined itinerary within cities before the nineteenth century.[2]

Nor did the great transformations associated with the Industrial Revolution and accelerated urban growth in the nineteenth century have much immediate effect on urban public transportation. Railroads, of critical importance in moving goods and passengers between cities and promoting urban concentration, were of very little value in moving people within cities. Steam railroads also did little to develop suburbs, as John Kellett has demonstrated for provincial Great Britain.[3] Even in gigantic London, only 3 to 4 percent of those going to work in the 1850s rode the railroads.[4] Expensive, inflexible, and relatively inaccessible, early railways failed to provide public transit capable of alleviating the growing urban concentration and overcrowding they themselves encouraged.

In fact, the old transportation technology of animate power—the horse—proved more adaptable to emerging transit needs than the new

3

one of steam and iron. Thus horse-drawn omnibuses appeared as the first significant form of urban public transportation. A modification of existing intercity stagecoaches and the innovation of a retired French army officer, omnibus service quickly spread from Nantes and Paris to London (1829) and New York (1831). Other cities followed, and omnibus transit became fairly well established on both sides of the Atlantic by 1850. The possibilities for subsequent traffic growth were substantial in large cities. In Paris, for example, the number of omnibus passengers increased threefold, from 40 million in 1855 to 120 million in 1867, as the city grew and Baron Haussmann's new thoroughfares facilitated public transit. More generally, it appears that omnibus service played a rather significant role in the mid-century growth of new middle-class residential areas located 1 to 2 miles from the historic city center in both Europe and the United States.[5]

Yet the factors limiting the significance of the omnibus were real, as a brief examination of supply and demand makes clear. Above all, the supply of omnibus transit was limited by inadequate motive power, rough roadways, and crowded streets. Omnibus service also responded poorly to wide fluctuations in hourly and seasonal demand, for holding extra coaches, teams, and drivers in reserve for peak-load periods was uneconomical. Demand was also limited. With an average speed of perhaps 5 miles per hour, the omnibus beat walking—but not by much. Moreover, although fares gradually trended down as companies did such things as add top decks to increase capacity and attract economy-minded passengers, they remained relatively high and restricted the regular clientele to the reasonably well-to-do. As the historian of public transit in Dresden writes, "Omnibus travel had many inconveniences attached to it which limited its use by the wealthy, and it was not cheap enough to be used to even the same extent by the poorer classes."[6]

Within the comparative framework of this study, these systematic shortcomings were certainly felt more keenly in the United States than in Europe. Thus whereas early omnibus service proceeded at roughly the same pace in both areas, Americans undoubtedly took the lead in developing the next major transit innovation: horse-drawn street railways.

Street railways, or tramways in English, French, and Russian terminology, grew out of experiences in coal mines and embodied a simple idea: Use iron rails to reduce surface friction and dramatically increase the efficiency of horse traction. First applied in a suburban environment in 1832, on a horse-drawn railroad line connecting New York and the proposed terminus on the Harlem River, serious street railway construction dated from 1852. In that year, a French engineer living in New York, Alphonse Loubat, constructed a brand new line using grooved rails that lay flush with the pavement. This was a significant innovation because the step rails used previously on New York and Harlem streets (and elsewhere) protruded above the roadway, interfering with other vehicles and creating hostility that prevented acceptance. Loubat's grooved rail proved much less of a nuisance.[7]

By 1860, horsecars were replacing omnibuses in many American cities, and the trend accelerated during the Civil War. Once established, per capita usage grew rapidly. In New York, for example, horse railway traffic almost tripled in the 1860s, reaching 115 million passengers by 1870—or about a hundred rides per year per capita.[8] In Europe, in contrast, only a few large cities had established the new form of transit by 1869, and the 1870s marked only the beginning of serious street railway development in most British, French, and German cities. Moreover, whereas streetcars quickly replaced omnibuses in most American cities, their European counterparts often used streetcars to complement omnibuses, which remained in service into the twentieth century in some instances.

Although Europe's lag and America's lead in adopting and developing the new transit technology may seem a minor or perhaps accidental matter, the early lack of sychronization in increasing the supply of urban public transportation reflected significant differences between the two urban environments. Indeed, a comparative study of urban transportation enables us to isolate forces and institutions that have shaped purely technical developments—so often the primary interest of transportation historians—and to analyze the consequences of alternative technologies for economy and society.[9]

First, the rate of urban growth was generally more rapid in the United States, and this growth apparently created a more intense demand for urban transit at an earlier date. More certainly, the difference in timing reflected contrasting physical environments. Fast-growing American cities, with their easily extended grid pattern, had fewer paved streets than their Western and Central European counterparts, and the gain in efficiency and profit from replacing omnibuses with streetcars was correspondingly greater.[10] Moreover, although the larger "walking cities" in the United States were perhaps as congested and overcrowded as their European counterparts—New York's population density surpassed London's in 1850[11]—the desire to move toward the fringes, toward the suburbs, was also stronger than in continental Europe (though perhaps not in Great Britain). On the Continent, centuries of city walls and intensive utilization of a carefully defined built-up area limited the appeal of an emerging suburban ideal directly dependent on improved transit.[12]

Transit technology was also embedded in different institutional structures. Cities on both sides of the Atlantic relied on privately owned street railway companies, but these private transit firms were more strictly regulated by national and municipal governments in Europe than in the United States. Franchises granted to private entrepreneurs in the United States had few specific provisions—paving between and beside the rails was often the most onerous obligation—and they were often perpetual in their duration. Even when municipalities stipulated a fixed term, what was to occur when the franchise expired was "a subject upon which almost all of the limited franchises granted in the United States are silent," according to an authoritative 1898 report.[13] Instead, with both the public and real estate speculators avid for more urban transit, and politicians susceptible to corruption, American cities generally relied on a multiplicity of competing franchises and market forces to regulate companies and provide good service.[14] Around 1880, New York had seventeen urban transit companies and Boston had seven; Philadelphia had more than twenty by 1865.[15]

In Europe, however, street railway companies, like their mainline railroad predecessors, were regulated by public authority from the beginning. This European regulatory framework placed restraints on transport entrepreneurs, scrutinizing plans to replace omnibuses with streetcars, for example, and often checking the spread of the American innovation.[16] Generally speaking, European states established a comprehensive juridical framework and well-defined procedures for granting "concessions" to petitioning entrepreneurs, who worked out specific terms with local governments and technical experts from the central administration. Concessions were normally for long periods—forty to fifty years—but on expiration, all the immovable property—mainly the tracks—reverted to the city free of charge. In addition to contractual provisions on fares, routes, paving obligations, and frequency of service, national and local authorities had extensive powers allowing them to regulate the speed and capacity of cars, or require or reject modifications as a matter of course. As a result, European cities were more disposed to accept one or a very few transit companies and generally avoided a proliferation of competing franchises.[17]

Here, as elsewhere, there were variations on European themes. Supposedly laissez-faire Great Britain laid down rigorous control in the Tramways Act of 1870. Municipal governments were permitted to build tracks and own lines, although they were required to lease them to a private operating company for not more than twenty-one years.[18] France, a model for much of Europe in street railway legislation, tightened up initial regulation in its tramways law of June 11, 1880. This law called on state engineers to verify the expenditures of tramway constructors and check the speculative inflation of the capital account noted in the tramway boom of the 1870s. German regulation was strict, but at the state rather than the imperial (that is, national) level.

All these factors continued to operate in the 1880s. And there can be little doubt that in 1890, America continued to lead in the development of urban transit. In the first place, the urban population in the United States simply used urban transit to a much greater degree than its European counterpart. In 1890, when rapid electrification had barely begun, per capita transit usage was three to four times as high in the United States as in Europe. For example, in large German cities, the arithmetical average of annual rides per capita was 39; in large American cities, it was 143. In 1890, the inhabitants of the four largest American cities averaged 195 rides annually on all modes of urban public transportation, while the inhabitants of Great Britain's four largest cities averaged but 56.[19]

This quantitative advantage in usage was probably the critical difference for most Americans who thought about the matter. Thus the 1890 American transportation census admitted that "comparisons much to our disadvantage are frequently made between street railway systems of our own and foreign cities, the grounds of conclusions favorable to the foreign service being that it is managed with more regard for public convenience and made to contribute more to municipal expense." But with evident pride in the American achievement, the writer concluded that in their "most important function"—that of carrying passengers— "our lines have attained a development unapproached in Europe. . . . The street railway system of Berlin, justly regarded as the most perfect in Europe, carried in proportion to the inhabitants of the city less than one-third as many passengers as the city of New York."[20]

The 1890 difference reflected substantial growth in per capita usage in the United States in the 1880s. In Europe, in contrast, the rapid growth of the 1870s was followed by stagnation in the 1880s in such cities as Paris, Leipzig, and Vienna.[21] Once again, there was apparently less demand for more transit in Europe than in the United States in these years. Part of the reason was macroeconomic—less robust economic growth in Europe. But more important was urbanizing America's greater areal dispersion, most strikingly evidenced by its "streetcar suburbs"— new, relatively low-density residential areas linked by daily commuting and a symbiotic relationship to a sharply defined central business district. To cite but one example, the well-studied growth of such suburbs in Boston coincided with a sharp rise in rides per capita, from 118 in 1880 to 175 in 1890.[22] In Europe, horse-drawn tramways had succeeded in tapping an expanded clientele with lower fares, but before 1890 they clearly contributed much less to spatial diffusion than their American counterpart.[23]

Finally, as is well known, the United States took the lead in applying mechanical traction to urban public transportation. What is less well known is that innumerable experiments first proceeded on both sides of the Atlantic and that, in addition to the rather obvious greater demand in the United States, the different institutional patterns we have noted and the contrasting aesthetic perceptions we have hinted at played significant roles in the differential development and diffusion of new transit technologies.

Steam technology was one line of attack. Steam engines suitable for urban transit purposes seemed to promise faster, more flexible, and potentially cheaper service than horse traction. Steam traction would also be more dependable—horses were subject to illness and disease—and would reduce the hygiene problem inherent in horse droppings. The challenge, of course, was to meet environmental objections to noisy, dangerous engines in busy public streets, either by building specially designed dummy locomotives for street railways or special, self-propelled steam streetcars. Yet, in spite of many efforts, such as ingeniously enclosed coke-fired boilers capable of reducing noxious smoke and cinders, steam-powered tramways never succeeded in meeting environmental requirements and simultaneously achieving a clear economic advantage over horse traction. Only on a few purely suburban lines did they attain real urban significance in Europe, and apparently in the United States as well.[24]

Yet there were important differences in steam-traction development. Americans appear to have experimented less and Europeans more with pollution-reducing steam engines and self-propelled cars. Moreover, steam street railways carried 287 million passengers in the United States in 1890, almost one-fourth as many as were carried by horse traction and a far larger portion than in European countries.[25] Above all, in a single but extremely significant case, New York City allowed massive smoke-belching, mainline railroad locomotives to run through the streets on elevated tracks, despite the obvious drawbacks of these "environmental monstrosities."[26] To be sure, as Charles Cheape has shown, New York's use of elevated railroads for mass transit grew incrementally out of an unsuccessful cable-powered franchise, and he has argued further that environmental damage was thus accepted because it was not immediately apparent.[27] Yet the fact that major elevated extensions were granted in 1875, after the initial elevated system was fully functioning,[28] suggests that transit needs took precedence over cultural–aesthetic–environmental considerations in New York's acceptance, however grudging, of new technology. By contrast, in the largest European cities, steam-powered elevated railways in city streets were never accepted. Although the similarity is not perfect, it seems clear that the same factors accounted for the Americans' greater use of steam for urban transit as for its earlier, more wholehearted acceptance of horse-railway tracks in city streets.

The extensive development of the cable car was another aspect of the decisive U.S. lead in mechanization. Developed in 1873 in San Francisco, where extremely steep grades combined with straight streets to provide strong incentives and ideal conditions for a continuously moving cable laid between the tracks, cable systems were installed in twenty-six American cities by 1891. Requiring heavy investment and therefore suitable only in high-density areas, cable cars ran twice as fast as horsecars and were nonpolluting.[29] By 1890, they were carrying 373 million passengers in the United States, as opposed to 1.23 billion on all horse-drawn street railways (and 287 million on all forms of steam-powered street railways).[30] In contrast, cable cars were used on only a few lines in Europe, mainly in London, Edinburgh, Birmingham, and Paris.[31] Differences in physical environments were probably the most critical factor. Moving well on straight, grid-set thoroughfares and turning poorly on curves, cable cars were ill suited for the sinuosities of Europe's historic urban landscape.

Efforts to apply electric traction to urban transit, which were eventually to prove highly successful, are particularly interesting from a comparative point of view. A whole series of pioneers and inventors in many countries made important technical contributions before the decisive breakthrough, a familiar pattern in the process of strategic innovation. The contributions depended in turn on basic scientific discoveries in electricity and the development of electrical engineering capability associated with the rapid emergence of the electric industry.[32] Without this sophisticated international effort, the development of electric street railways would have been impossible.

The fundamental step was taken in the 1860s and 1870s with the development of the dynamo, which was indispensable in that it provided a source of cheap current. As Harold Passer has noted in his fine study of the American electric manufacturers, power obtained from the primary storage battery was twenty times as costly as power from a steam engine; and "until a more economical source of current was available, there was no possibility of electric power's coming into general use."[33] Used at first primarily for lighting, cheap electricity from the dynamo reawakened interest in electric traction. The mechanical energy of the steam engine was to be converted by the electric generator—the dynamo—into electricity and then reconverted into mechanical energy by a motor on the streetcar. The great appeal was electricity's transmissibility and divisibility; generated efficiently at a single point, it would then be parceled out to vehicles all over town. The vision was simple and seductive, but it required a full decade of many-sided experiments to make it a reality.

The critical problem was finding an effective way to provide the car motor with current. The Siemens and Halske Company of Berlin, the Continent's preeminent electrical enterprise, developed several methods between 1879 and 1881, including a third-rail conductor, a live surface rail, and a slotted overhead conductor. Werner von Siemens also opened the first electric tramway in the United Kingdom in 1882, and within two years Magnus Volk and others had opened third-rail systems in amusement parks. French inventors focused simultaneously on self-contained battery cars. Until 1882, largely because of Siemens' work, Europe led in the development of electric traction.[34]

Yet Europeans failed to capitalize on their initial advance, in part because Siemens did not believe electricity would be permitted in public streets, and Americans achieved the decisive breakthrough between 1883 and 1888.[35] In those years, following Thomas Edison's momentary interest in 1880, several inventors and businessmen competed to create what became the standard American technology for urban street transit. That technology—first successfully synthesized in Richmond, Virginia, by Frank Sprague in February 1888—combined improved electric motors with an overhead conductor method. An under-running trolley on the streetcar picked up current for the motor from a single copper-wire conductor attached to poles installed for that purpose. The rails were used for the return circuit.

Overhead electric traction promised—and then delivered—substantially greater profits to street railway companies by cutting per unit operating costs on horsecar operations. Although there were severe difficulties in calculating exactly how much savings were involved, costs per rider declined by roughly 50 percent even as fixed investment per unit of line doubled.[36] Still more important than these cost reductions, fare reduction in general, and low-cost workman's tickets for morning and evening commuting in particular. Fare structures varied between individual cities, but a dramatic reduction in European fares actually paid was unmistakable and unmatched by a similar fare reduction in the United States—the second notable difference. In France, for example, where the carefully renegotiated contract with private industry reigned supreme, the average fare per tramway rider fell from 15.4 centimes in 1896 to 8.8 centimes in 1910, or from about 3 cents to about 1.7 U.S. cents.[51] As a confidential internal report by financial experts at France's largest deposit bank concluded in late 1900, "It is above all the public which has benefited from the application of electric traction. The companies have not lost by it, but the [financial] benefits have rarely been as high as could have been expected."[52]

Sharply reduced fares and reasonable profits in Europe are related to the third difference: the general lack of outrageous inflation of capital stock in Europe compared to the United States. Although financial manipulation and speculation were not unknown in Europe, the verification of actual expenditures by state engineers and municipal authorities limited unhealthy abuses.[53] Capital accounts reflecting genuine cash outlays also laid a solid foundation for the systematic (and very un-American) amortization of fixed investment in the long period before the prolonged concession expired and all immovable property reverted to the city. Finally, and less certainly, it would appear that Europe's regulated, renegotiated framework of tramway electrification short-circuited the hostility and sense of injustice that Americans seemed to develop toward their transit companies. This greater hostility in the United States may have encouraged the growth of a more general American dissatisfaction with public transportation per se, and combined with other well-known factors to contribute to a more rapid decline of transit systems in the United States than in Europe after 1920.[54]

This study has concentrated on the street railways found in every city and slighted the rapid transit of elevateds and undergrounds found only in a very few of the largest agglomerations before 1914; thus it has stressed American–European differences in transit development and downplayed the obvious similarities, which have seemed less significant in a comparative perspective. Nevertheless, two important and growing similarities dating from the electrification era should be noted, if only briefly.

First, whereas electric traction evolved out of ongoing mechanization in the United States, it produced a sharp, revolutionary break in Europe's public transportation history. And as a result of Europe's more rapid rate of change, European cities closed a very large part of the pre-viously existing transit gap, as measured by per capita usage and similar indicators. Thus per capita usage in the four largest British and German cities was almost four times as great in 1910 as in 1890. In 1910, Euro-pean per capita usage had come to represent about 70 percent of the level in comparable American cities, as opposed to less than 30 percent in 1890.[55] Moreover (and more work needs to be done here), electric traction facilitated areal expansion and suburbanization in Europe as it did in America, allowing a reduction in overcrowding at the center. The familiar American argument that electric street railways would en-courage generally desired suburban development found many echoes in Europe. It was important in overcoming opposition to overhead electric traction, and it played a part in the partial or total abandonment of the long-standing zone system in favor of American-style flat fares in many French and German cities.[56]

Second, there were growing similarities in terms of industrial struc-ture. In the United States, electrification allowed nimble entrepreneurs and street railway magnates to combine competing horsecar franchises into unified, monopolistic transit systems in many cities.[57] Thus the American pattern moved toward that in Europe, and in both areas a unified transit operation faced the riding public and regulators. A re-lated similarity was on the equipment producer side. In both the United States and leading European countries, the electrical industry quickly became highly concentrated, so electric installations on street railways and subsequent expansions were planned and executed by a small num-ber of firms locked in intense but clearly monopolistic competition.[58] The desire of such firms to sell their equipment diffused electric traction around the world. For example, Belgian firms led in carrying tramway electrification throughout the Russian empire.[59]

Our focus on European and American comparisons has mentioned differences between European countries only in passing. A few of these, like the contrasting patterns of electric traction adoption noted earlier, were important. Above all, Great Britain's slow acceptance of electric traction was tied up with the impending municipalization of private operating companies, as the twenty-one-year leases came due.[60] Secur-ing finally the authority to operate as well as own their systems, British cities then made rapid progress and emerged as the new leaders in pub-lic transportation in some eyes, particularly American eyes.

Yet while some American reformers, such as Frederic Howe, urged their countrymen to follow the municipal ownership model of almost all British (and some German) cities, at least one famous British expert

—James Dalrymple, general manager of Glasgow Tramways—disagreed. In 1906, called to study Chicago's street railway system by a reformist mayor, Dalrymple agreed that the system was "altogether out of date" and that "there is no wonder that the inhabitants are intensely dissatisfied with their transit facilities." But Dalrymple counseled against municipalization, accepting implicitly the American streetcar industry's contention that in the United States, politicians and cities were too dishonest and corrupt to establish efficient, successful municipal enterprises on the European model.[61] It was another indication of how, by 1914, Europeans had closed the earlier large gap in urban transit and perhaps even moved ahead of their American cousins in some important ways.

Notes

1. Sam B. Warner, Jr., Streetcar Suburbs: The Process of Growth in Boston, 1870–1900 (Cambridge, Mass., 1962), pp. 15–21.

2. The great Blaise Pascal is generally credited with establishing just such an urban service in Paris in 1662. But the venture soon failed, no doubt partly because only "bourgeois and people of merit" were allowed to use Pascal's coaches. See Alfred Martin, Etude historique et statistique sur les moyens de transport dans Paris (Paris, 1894), pp. 27–36, 68–79; and Octave Uzanne, La locomotion à travers le temps, les moeurs et l'espace (Paris, n.d.), pp. 67–86.

3. John R. Kellett, The Impact of Railways on Victorian Cities (London, 1969), pp. 354–365. Judging on the basis of Boston, railroads may have played a somewhat more positive role in the development of early suburbs and commuter traffic in the United States. See Charles J. Kennedy, "Commuter Services in the Boston Area, 1835–1860," Business History Review 36 (1962): 153–170.

4. T. C. Barker and Michael Robbins, A History of London Transport: Passenger Travel and the Development of the Metropolis, vol. 1, The Nineteenth Century (London, 1963), p. 53.

5. For further discussion of omnibus development and citation of sources, see John P. McKay, Tramways and Trolleys: The Rise of Urban Mass Transport in Europe (Princeton, N.J., 1976), pp. 9–13. The omnibus has received little attention from urban historians, although Kellett's consideration of their importance for middle-class housing in Great Britain is excellent.

6. Hermann Grossmann, Die kommunale Bedeutung des Strassenbahnwesens beleuchet am Werdegange der Dresdner Strassenbahnen (Dresden, 1903), p. 15.

7. On the early development of street railways, see George Rodgers Taylor, "The Beginnings of Mass Transportation in Urban America," parts I and II, Smithsonian Journal of History 1 (Summer and Autumn 1966): 35–50 and 39–

52. On European developments, see McKay, *Tramways and Trolleys*, pp. 13–18, and sources cited there.

8. Charles W. Cheape, *Moving the Masses: Urban Public Transit in New York, Boston, and Philadelphia, 1880–1912* (Cambridge, Mass., 1980), p. 25. Cheape has an excellent bibliography. Also see Arthur J. Krim, "The Innovation and Diffusion of the Street Railway in North America," MA thesis, University of Chicago, 1967, pp. 56–73.

9. While preparing this paper, I was struck by Raymond Grew's observation that historians often shy away from comparative studies because of the difficulty in mastering equally the countries being compared. This difficulty I cannot claim to have overcome, since my own research in urban transportation has focused on selected European nations. See Raymond Grew, "The Case for Comparing Histories," *American Historical Review* 85 (1980): 767. It should be noted that the literature on American and European street railways is large, but it is diverse and of very uneven quality. For some discussion of sources, see McKay, *Tramways and Trolleys*, pp. vii–ix, 247–255; and Glen E. Holt, "The Main Line and Side Tracks: Urban Transportation History," *Journal of Urban History* 5 (1979): 397–400.

10. D. Kinnear Clark, *Tramways: Their Construction and Working*, 2nd ed. (London, 1894), pp. 5–10; Alexander Easton, *A Practical Treatise on Street or Horse-Power Railways* (Philadelphia, 1859), pp. 4–8; Clay McShane, *Technology and Reform: Street Railways and the Growth of Milwaukee, 1887–1900* (Madison, Wis., 1974), pp. 3–4.

11. Cheape, *Moving the Masses*, pp. 21–22.

12. With an extensive bibliography, maps, and capsule histories, Robert E. Dickinson, *The West European City* (London, 1951), provides a wealth of material supporting these generalizations.

13. State of Massachusetts, *Report on the Relations between Cities and Towns and Street Railway Companies* (Boston, 1898), p. 74. The work of a special committee headed by Charles Francis Adams, this excellent report contains considerable comparative material on European and American practice. Also see McKay, *Tramways and Trolleys*, pp. 89–95, and sources cited there, particularly Delos F. Wilcox, *Municipal Franchises*, 2 vols. (New York, 1911), which contains a wealth of information on street railway franchises in the United States.

14. McShane, *Technology and Reform*, pp. 4–5; Wilcox, *Municipal Franchises*, 2:30–31, 99.

15. Cheape, *Moving the Masses*, pp. 40, 111, 159.

16. For early attempts in Great Britain and Paris, see Charles Klapper, *The Golden Age of Tramways* (London, 1961), pp. 16–26; and Martin, *Etude historique*, pp. 90–91, 148–149.

17. See McKay, *Tramways and Trolleys*, pp. 19–21, 89–95, and sources cited there, particularly C. Colson, *Abrégé de la législation des chemins de fer et tramways*, 2nd ed. (Paris, 1905), pp. 198ff., which has a good discussion of

the concession system on the Continent, as well as an exhaustive analysis of French practice.

18. In addition to note 17, see Klapper, Golden Age of Tramways, pp. 27–34.

19. For an extended comparison of European and American per capita usage at different points in time, see McKay, Tramways and Trolleys, pp. 192–198.

20. U.S. Bureau of the Census, Report on Transportation Business in the United States, 1890, part 1, Transportation by Land (Washington, D.C., 1895), p. 685.

21. McKay, Tramways and Trolleys, pp. 23–25. For Paris, in addition to Martin, Etude historique, see the major work by Pierre Merlin, Les transports parisiens (Paris, 1967), pp. 36–48.

22. Warner, Streetcar Suburbs; Cheape, Moving the Masses, p. 214. Joel A. Tarr, Transportation Innovation and Changing Spatial Patterns in Pittsburgh, 1850–1934 (Chicago, 1978), concludes that street railways (and commuter railroads) had gone far toward creating the shape of a modern metropolis in Pittsburgh, complete with considerable suburban expansion, by 1888.

23. The impact of horse tramways on European living patterns has been little studied, but see McKay, Tramways and Trolleys, pp. 22–25, 205ff.; and see the important article by David Ward, "A Comparative Historical Geography of Streetcar Suburbs in Boston, Massachusetts and Leeds, England: 1850–1920," Annals of the Association of American Geographers 54 (1964): 477–489.

24. H. A. Whitcombe, History of the Steam Tram (South Godstone, Surrey, England, n.d.), pp. 12–18; McShane, Technology and Reform, pp. 7–8; and McKay, Tramways and Trolleys, pp. 27–34, and sources cited there.

25. U.S. Bureau of the Census, Transportation, 1890, I:682.

26. McShane, Technology and Reform, p. 7.

27. Cheape, Moving the Masses, pp. 31, 38.

28. Ibid., pp. 31–36.

29. Edgar Kahn, Cable Car Days in San Francisco (Stanford, Calif., 1940), pp. 27–42; and George W. Hilton, The Cable Car in America (Berkeley, 1971), pp. 21–50. There is a good study on Melbourne, Australia, which had the largest cable system outside the United States: John D. Keating, Mind the Curve! A History of the Cable Trams (Melbourne, 1970).

30. U.S. Bureau of the Census, Transportation, 1890, 1:682.

31. Hilton, Cable Car, p. 476.

32. On the growth of the electrical industry, see the articles and bibliography in Charles Singer et al., A History of Technology, 5 vols. (Oxford, 1958), 5:177–234. See also Malcolm MacLaren, The Rise of the Electrical Industry (Princeton, N.J., 1943); and David S. Landes, The Unbound Prometheus: Technological Change and Industrial Development in Western Europe from 1750 to the Present (Cambridge, 1968), pp. 281–290.

33. Harold C. Passer, The Electrical Manufacturers, 1875–1900: A Study in Competition, Entrepreneurship, Technical Change, and Economic Growth (Cambridge, Mass., 1953), pp. 211–212.

34. For more extended treatment and citation of relevant materials, see McKay, *Tramways and Trolleys*, pp. 35–41.

35. Ibid., pp. 43–44. The American story has often been told. The most authoritative account is Passer, *Electrical Manufacturers*, pp. 213–255; but see Frank Rowsome, Jr., *Trolley Car Treasury: A Century of American Streetcars-Horsecars, Cablecars, Interurbans, and Trolleys* (New York, 1956), pp. 54–69; and John A. Miller, *Fares, Please!: From Horse-Cars to Streamliners* (New York, 1941), pp. 54–69. Both of these popular histories are also useful for earlier transit developments.

36. For discussion of this problem and relevant sources, see McKay, *Tramways and Trolleys*, pp. 58–67.

37. See, for example, Eric Lampard, "The Social Impact of the Industrial Revolution," in *Technology in Western Civilization*, ed. Melvin Kranzberg and Carroll Pursell, Jr., 2 vols. (New York, 1967), I:305–306; and E. A. Wrigley, "The Supply of Raw Materials in the Industrial Revolution," *Economic History Review* 15, 2nd series (1962): 1–16.

38. McKay, *Tramways and Trolleys*, pp. 51–58.

39. Ibid., pp. 50–51, for full citation of sources.

40. George W. Hilton, "Transport Technology and the Urban Pattern," *Journal of Contemporary History* 4 (1969): 126.

41. *Electrical Engineer* (London) 9 (1892): 196.

42. For documentation and extended discussion of the points in this paragraph, see McKay, *Tramways and Trolleys*, pp. 67–83.

43. Report of Birmingham Subcommittee on Tramways, as quoted in *Electrical Engineer* (London) 19 (1897): 569–570. See McKay, *Tramways and Trolleys*, pp. 84–88, for similar reactions.

44. "Remarks of Mr. F. J. Sprague on Electricity as a Motive Power," American Street Railway Association, *Proceedings* 6 (1887–88): 60–68. also see Passer, *Electrical Manufacturers*, p. 216.

45. McShane, *Technology and Reform*, pp. 23–24. In Fort Wayne, Indiana, and elsewhere, the company simply announced it was switching to electric traction and did so. George K. Bradley, *Fort Wayne's Trolleys* (Chicago, 1963), p. 17.

46. McShane, *Technology and Reform*, pp. 24–26.

47. Ibid., pp. 33–35; McKay, *Tramways and Trolleys*, pp. 94–95; Wilcox, *Municipal Franchises*, 2:25–30.

48. For analysis of these efforts and relevant source citation, see McKay, *Tramways and Trolleys*, pp. 95–101.

49. Ibid., pp. 101–106.

50. Ibid., pp. 107–124, for elaboration, based mainly on French and German case studies and sources.

51. France. Ministère des Travaux Publics. Direction des Chemins de fer. *Statistique des chemins de fer français* (1906), II:655, 667; idem (1912), II:895, 909.

electric traction broke the long-standing bottleneck of inadequate supply of urban transit. It permitted companies to increase enormously their capacity to carry passengers on "saturated" lines and networks. Increased supply capacity, so critical for economic growth in general and not merely urban transport development,[37] resulted principally from increased car speed, larger cars (and/or double-deckers and trailers), and much better performance on hills and in snow. Moreover, the increased quantity of transport was of better quality because electric trams rode more smoothly, stopped more easily, and did not lurch from side to side.[38] In short, the gains from technology were very substantial, however they might be distributed between passengers, operating companies, equipment producers, and taxing authorities.

As we might expect at this point, the initial responses of American and European cities to practicable electric traction were quite different. Electrification shot through the American street railway industry like current through a copper wire. Within two short years after Sprague's Richmond installation, one-sixth of all street railway track in the United States was electrified. By the end of 1893, total street railway track had doubled from the 1890 level to 12,000 miles; ten years later, it had reached 30,000 miles. Of this greatly increased total, fully 60 percent was electric by the end of 1893, and 98 percent was electric by the end of 1903.[39] In the words of a leading transportation historian, the electric streetcar in the United States "was one of the most rapidly accepted innovations in the history of technology."[40]

In Europe, however, electric traction proceeded slowly amid great discussion for several years. As one of many European observers put it in 1892, the development of electric street railways had been prodigious in the United States, since it became a "practical commercial success in 1888, while Europe has shown scarcely any improvement."[41] Then, from the end of 1893, Europe increased its electric line about tenfold, to almost 1,900 miles at the end of 1898, when an electric tramway boom was in full swing. By the time the depression of 1901 arrived, electric systems were either in operation or under construction in almost all European cities. As might be suspected, there were national differences. Germany clearly led the way in European street railway electrification, and Great Britain lagged behind Germany by about a decade. France and indeed most continental countries occupied the middle ground, leaning more toward the German than the British extreme. More surprisingly, although the standard trolley-wire conductor used almost exclusively in the United States clearly predominated in Europe, European cities also used a complex variety of electric traction systems well into the twentieth century. The extent to which these different electric systems were used in different countries may be determined with some precision through 1898, thanks to the annual tramway censuses of the French journal *L'Industrie Electrique*.[42]

In explaining contrasting patterns of European and American street railway electrification, factors evident in the horsecar era continued to exert a strong influence. Quite clearly, the greater preference for spatial diffusion and central-business-district concentration in the United States combined with more rapid urban growth to create a stronger demand for urban transit. Yet continuing greater pressure from demand and a more vigorous supply response in the form of more rapid mechanization and technical change—the kind of argument preferred in economic analysis of technological improvement—is only part of the story. Equally significant was the way cultural–aesthetic judgments regarding the new technology continued to combine with the previously noted institutional differences to determine the course of diffusion and modification of American innovation.

Judging by a wealth of reports in the technical press, municipal debates, and scattered monographic studies, there is no doubt that large numbers of influential Europeans thought that overhead-wire conductors and their support poles—the very essence of the American innovation in surface transit—were extremely ugly and aesthetically unacceptable. Better, they argued, to send inventors and engineers back to the drawing boards than accept such "visual pollution." This cultural reaction was most pronounced in connection with putting poles and wires along broad boulevards and through historic squares, and was less pronounced regarding many suburban lines and lines through working-class and industrial districts. And, unlike early objections to electric traction that were related to things like safety for horses or telephone reception, the aesthetic critique rested on subjective values that could not be refuted by technical experts.

An 1897 report by Birmingham's Subcommittee on Tramways of its special tour of various continental systems fairly suggests the scope of antitrolley feeling.

> Your subcommittee would point out that the unsightliness and otherwise
> objectionable features of the overhead wire are recognized in many of
> the places they have visited. The municipal authorities at Paris, Vienna,
> and Berlin stated that in no case would they permit overhead wires in the
> central portion of those cities. At Brussels several miles of conduit are
> being laid to avoid them; at Dresden and Budapest they are not allowed in
> the principal streets.

And the list of cities went on and on, leading the subcommittee to "rec-ommend strongly that no consent be given for the erection of overhead wires in any part of the city."[43]

Interestingly, there was also initial hostility toward overhead con-ductors on aesthetic and safety grounds in the United States. While in-stalling his pioneering Richmond operation, Sprague himself admitted to street railway operators that the overhead system "may be unsightly" and "somewhat in the way," so it "perhaps would not be tolerated in twenty large cities in the country."[44] Yet only in New York and Wash-ington did such considerations delay the installation of the trolley for any considerable length of time.[45]

Although aesthetic opposition to overhead electric traction was by every indication less in the United States than in Europe, differing in-stitutional arrangements were the decisive factor. American cities in-herited from the horsecar era a lack of effective control, and street railway operators could ignore or easily outflank antitrolley opposition. Thus most companies in the United States were able to electrify their lines almost as quickly as they saw that it paid, and without even having to modify their existing franchises.[46]

Technical freedom in the United States had consequences that went well beyond more rapid diffusion and aesthetic controversies. First, companies in the United States were usually able to maintain a flat 5-cent fare, which had already emerged as the standard in most fran-chises during the horsecar era, and they were not required to share the cost-reducing gains from technology through fare reduction. Or, more accurately, companies reduced fares by extending lines and pro-viding longer rides for the same price, as electric traction subsidized suburbanization on a massive scale. Second, with their perpetual or quasi-perpetual franchises intact, American companies did not need to worry much about amortization in a given time period of the very large capital increases that electrification necessarily required and that were swollen further by manipulation and watered stock.[47]

In European cities, where technological departures in urban transit customarily required regulatory permission, opponents of poles and wires could call on existing institutional arrangements to assess the "environmental impact." Thus one result of time-consuming opposition to overhead conductors was a strong stimulation of European efforts to develop technological alternatives to meet aesthetic–environmental objections all through the 1890s, long after the technical issue was a closed book in the United States.[48]

While continuing earlier efforts with battery-powered cars, Europeans in the 1890s focused primarily on developing various surface-conduit systems, as well as highly imaginative "mixed" systems—systems that "mixed" an overhead conductor and an alternative such as the conduit conductor on the same car and same line. One variant, perfected by Siemens and Halske, was used in Berlin, Prague, Budapest, and Vienna, for example. Another, developed by General Electric's French affiliate, Société Thomson-Houston, was used in Bordeaux, Marseilles, and especially Paris. In both cases, cars used a more expensive, less dependable surface conduit—an electric wire laid in a special tube between the tracks in the street—to traverse historic or very fashionable areas. The same car then switched to an overhead-wire conductor in parts of the city where aesthetic objections could be overcome. In addition, European constructors modified early American practice to improve the visual effect, reducing especially the unsightliness of overhead wires and ornamenting their poles.[49]

These experiments and modifications allowed European street railway constructors, regulators, and the general public to explore the possibilities and tradeoffs of different modes of electric traction, the installation of which was then negotiated between the parties and spelled out between them in a new concession. In spite of important variations, a fairly common pattern may be discerned for most continental cities.[50] Basically, the street railway companies and the electrical equipment producers that financed them promised lower fares, more frequent service, extension of existing horsecar lines, somewhat higher municipal taxes, and some improvement in working conditions. In return, they required permission to introduce electric traction in general, a delineation of areas where trolley wires were unacceptable, and a substantial prolongation of the concession. The extra expense of antitrolley limitations to preserve urban beauty had to compete therefore with other desirable goals—lower fares, expanded service, and the rest.

This pattern of careful regulation and renegotiation, so different from that in the United States, produced some significantly different results. First, in Europe, aesthetic–environmental concerns—probably part of a broader cultural commitment to a harmonious urban landscape—were incorporated into street railway electrification to a larger extent than in the United States. Many a European street railway had to make do grudgingly with more expensive alternatives, like conduit traction, on at least some of their lines for many years, although the number of such lines declined with time and continued reevaluation stemming from company complaints and proposals.

Use of more expensive technologies declined to permit maximum fare reduction in general, and low-cost workman's tickets for morning and evening commuting in particular. Fare structures varied between individual cities, but a dramatic reduction in European fares actually paid was unmistakable and unmatched by a similar fare reduction in the United States—the second notable difference. In France, for example, where the carefully renegotiated contract with private industry reigned supreme, the average fare per tramway rider fell from 15.4 centimes in 1896 to 8.8 centimes in 1910, or from about 3 cents to about 1.7 U.S. cents.[51] As a confidential internal report by financial experts at France's largest deposit bank concluded in late 1900, "It is above all the public which has benefited from the application of electric traction. The companies have not lost by it, but the [financial] benefits have rarely been as high as could have been expected."[52]

Sharply reduced fares and reasonable profits in Europe are related to the third difference: the general lack of outrageous inflation of capital stock in Europe compared to the United States. Although financial manipulation and speculation were not unknown in Europe, the verification of actual expenditures by state engineers and municipal authorities limited unhealthy abuses.[53] Capital accounts reflecting genuine cash outlays also laid a solid foundation for the systematic (and very un-American) amortization of fixed investment in the long period before the prolonged concession expired and all immovable property reverted to the city. Finally, and less certainly, it would appear that Europe's regulated, renegotiated framework of tramway electrification short-circuited the hostility and sense of injustice that Americans seemed to develop toward their transit companies. This greater hostility in the United States may have encouraged the growth of a more general American dissatisfaction with public transportation per se, and combined with other well-known factors to contribute to a more rapid decline of transit systems in the United States than in Europe after 1920.[54]

This study has concentrated on the street railways found in every city and slighted the rapid transit of elevateds and undergrounds found only in a very few of the largest agglomerations before 1914; thus it has stressed American–European differences in transit development and downplayed the obvious similarities, which have seemed less significant in a comparative perspective. Nevertheless, two important and growing similarities dating from the electrification era should be noted, if only briefly.

First, whereas electric traction evolved out of ongoing mechanization in the United States, it produced a sharp, revolutionary break in Europe's public transportation history. And as a result of Europe's more

rapid rate of change, European cities closed a very large part of the previously existing transit gap, as measured by per capita usage and similar
indicators. Thus per capita usage in the four largest British and German
cities was almost four times as great in 1910 as in 1890. In 1910, European per capita usage had come to represent about 70 percent of the
level in comparable American cities, as opposed to less than 30 percent
in 1890.[55] Moreover (and more work needs to be done here), electric
traction facilitated areal expansion and suburbanization in Europe as
it did in America, allowing a reduction in overcrowding at the center.
The familiar American argument that electric street railways would encourage generally desired suburban development found many echoes in
Europe. It was important in overcoming opposition to overhead electric
traction, and it played a part in the partial or total abandonment of the
long-standing zone system in favor of American-style flat fares in many
French and German cities.[56]

Second, there were growing similarities in terms of industrial structure. In the United States, electrification allowed nimble entrepreneurs
and street railway magnates to combine competing horsecar franchises
into unified, monopolistic transit systems in many cities.[57] Thus the
American pattern moved toward that in Europe, and in both areas a
unified transit operation faced the riding public and regulators. A related similarity was on the equipment producer side. In both the United
States and leading European countries, the electrical industry quickly
became highly concentrated, so electric installations on street railways
and subsequent expansions were planned and executed by a small number of firms locked in intense but clearly monopolistic competition.[58]
The desire of such firms to sell their equipment diffused electric traction
around the world. For example, Belgian firms led in carrying tramway
electrification throughout the Russian empire.[59]

Our focus on European and American comparisons has mentioned
differences between European countries only in passing. A few of these,
like the contrasting patterns of electric traction adoption noted earlier,
were important. Above all, Great Britain's slow acceptance of electric
traction was tied up with the impending municipalization of private
operating companies, as the twenty-one-year leases came due.[60] Secur-
ing finally the authority to operate as well as own their systems, British
cities then made rapid progress and emerged as the new leaders in public transportation in some eyes, particularly American eyes.

Yet while some American reformers, such as Frederic Howe, urged
their countrymen to follow the municipal ownership model of almost
all British (and some German) cities, at least one famous British expert

—James Dalrymple, general manager of Glasgow Tramways—disagreed. In 1906, called to study Chicago's street railway system by a reformist mayor, Dalrymple agreed that the system was "altogether out of date" and that "there is no wonder that the inhabitants are intensely dissatisfied with their transit facilities." But Dalrymple counseled against municipalization, accepting implicitly the American streetcar industry's contention that in the United States, politicians and cities were too dishonest and corrupt to establish efficient, successful municipal enterprises on the European model.[61] It was another indication of how, by 1914, Europeans had closed the earlier large gap in urban transit and perhaps even moved ahead of their American cousins in some important ways.

Notes

1. Sam B. Warner, Jr., *Streetcar Suburbs: The Process of Growth in Boston, 1870–1900* (Cambridge, Mass., 1962), pp. 15–21.

2. The great Blaise Pascal is generally credited with establishing just such an urban service in Paris in 1662. But the venture soon failed, no doubt partly because only "bourgeois and people of merit" were allowed to use Pascal's coaches. See Alfred Martin, *Etude historique et statistique sur les moyens de transport dans Paris* (Paris, 1894), pp. 27–36, 68–79; and Octave Uzanne, *La locomotion à travers le temps, les moeurs et l'espace* (Paris, n.d.), pp. 67–86.

3. John R. Kellett, *The Impact of Railways on Victorian Cities* (London, 1969), pp. 354–365. Judging on the basis of Boston, railroads may have played a somewhat more positive role in the development of early suburbs and commuter traffic in the United States. See Charles J. Kennedy, "Commuter Services in the Boston Area, 1835–1860," *Business History Review* 36 (1962): 153–170.

4. T. C. Barker and Michael Robbins, *A History of London Transport: Passenger Travel and the Development of the Metropolis*, vol. 1, *The Nineteenth Century* (London, 1963), p. 53.

5. For further discussion of omnibus development and citation of sources, see John P. McKay, *Tramways and Trolleys: The Rise of Urban Mass Transport in Europe* (Princeton, N.J., 1976), pp. 9–13. The omnibus has received little attention from urban historians, although Kellett's consideration of their importance for middle-class housing in Great Britain is excellent.

6. Hermann Grossmann, *Die kommunale Bedeutung des Strassenbahnwesens beleuchet am Werdegange der Dresdner Strassenbahnen* (Dresden, 1903), p. 15.

7. On the early development of street railways, see George Rodgers Taylor, "The Beginnings of Mass Transportation in Urban America," parts I and II, *Smithsonian Journal of History* 1 (Summer and Autumn 1966): 35–50 and 39–

52. On European developments, see McKay, *Tramways and Trolleys*, pp. 13–18, and sources cited there.

8. Charles W. Cheape, *Moving the Masses: Urban Public Transit in New York, Boston, and Philadelphia, 1880–1912* (Cambridge, Mass., 1980), p. 25. Cheape has an excellent bibliography. Also see Arthur J. Krim, "The Innovation and Diffusion of the Street Railway in North America," MA thesis, University of Chicago, 1967, pp. 56–73.

9. While preparing this paper, I was struck by Raymond Grew's observation that historians often shy away from comparative studies because of the difficulty in mastering equally the countries being compared. This difficulty I cannot claim to have overcome, since my own research in urban transportation has focused on selected European nations. See Raymond Grew, "The Case for Comparing Histories," *American Historical Review* 85 (1980): 767. It should be noted that the literature on American and European street railways is large, but it is diverse and of very uneven quality. For some discussion of sources, see McKay, *Tramways and Trolleys*, pp. vii–ix, 247–255; and Glen E. Holt, "The Main Line and Side Tracks: Urban Transportation History," *Journal of Urban History* 5 (1979): 397–400.

10. D. Kinnear Clark, *Tramways: Their Construction and Working*, 2nd ed. (London, 1894), pp. 5–10; Alexander Easton, *A Practical Treatise on Street or Horse-Power Railways* (Philadelphia, 1859), pp. 4–8; Clay McShane, *Technology and Reform: Street Railways and the Growth of Milwaukee, 1887–1900* (Madison, Wis., 1974), pp. 3–4.

11. Cheape, *Moving the Masses*, pp. 21–22.

12. With an extensive bibliography, maps, and capsule histories, Robert E. Dickinson, *The West European City* (London, 1951), provides a wealth of material supporting these generalizations.

13. State of Massachusetts, *Report on the Relations between Cities and Towns and Street Railway Companies* (Boston, 1898), p. 74. The work of a special committee headed by Charles Francis Adams, this excellent report contains considerable comparative material on European and American practice. Also see McKay, *Tramways and Trolleys*, pp. 89–95, and sources cited there, particularly Delos F. Wilcox, *Municipal Franchises*, 2 vols. (New York, 1911), which contains a wealth of information on street railway franchises in the United States.

14. McShane, *Technology and Reform*, pp. 4–5; Wilcox, *Municipal Franchises*, 2:30–31, 99.

15. Cheape, *Moving the Masses*, pp. 40, 111, 159.

16. For early attempts in Great Britain and Paris, see Charles Klapper, *The Golden Age of Tramways* (London, 1961), pp. 16–26; and Martin, *Etude historique*, pp. 90–91, 148–149.

17. See McKay, *Tramways and Trolleys*, pp. 19–21, 89–95, and sources cited there, particularly C. Colson, *Abrégé de la législation des chemins de fer et tramways*, 2nd ed. (Paris, 1905), pp. 198ff., which has a good discussion of

the concession system on the Continent, as well as an exhaustive analysis of French practice.

18. In addition to note 17, see Klapper, *Golden Age of Tramways*, pp. 27–34.

19. For an extended comparison of European and American per capita usage at different points in time, see McKay, *Tramways and Trolleys*, pp. 192–198.

20. U.S. Bureau of the Census, *Report on Transportation Business in the United States, 1890*, part 1, *Transportation by Land* (Washington, D.C., 1895), p. 685.

21. McKay, *Tramways and Trolleys*, pp. 23–25. For Paris, in addition to Martin, *Etude historique*, see the major work by Pierre Merlin, *Les transports parisiens* (Paris, 1967), pp. 36–48.

22. Warner, *Streetcar Suburbs*; Cheape, *Moving the Masses*, p. 214. Joel A. Tarr, *Transportation Innovation and Changing Spatial Patterns in Pittsburgh, 1850–1934* (Chicago, 1978), concludes that street railways (and commuter railroads) had gone far toward creating the shape of a modern metropolis in Pittsburgh, complete with considerable suburban expansion, by 1888.

23. The impact of horse tramways on European living patterns has been little studied, but see McKay, *Tramways and Trolleys*, pp. 22–25, 205ff.; and see the important article by David Ward, "A Comparative Historical Geography of Streetcar Suburbs in Boston, Massachusetts and Leeds, England: 1850–1920," *Annals of the Association of American Geographers* 54 (1964): 477–489.

24. H. A. Whitcombe, *History of the Steam Tram* (South Godstone, Surrey, England, n.d.), pp. 12–18; McShane, *Technology and Reform*, pp. 7–8; and McKay, *Tramways and Trolleys*, pp. 27–34, and sources cited there.

25. U.S. Bureau of the Census, *Transportation, 1890*, I:682.

26. McShane, *Technology and Reform*, p. 7.

27. Cheape, *Moving the Masses*, pp. 31, 38.

28. Ibid., pp. 31–36.

29. Edgar Kahn, *Cable Car Days in San Francisco* (Stanford, Calif., 1940), pp. 27–42; and George W. Hilton, *The Cable Car in America* (Berkeley, 1971), pp. 21–50. There is a good study on Melbourne, Australia, which had the largest cable system outside the United States: John D. Keating, *Mind the Curve! A History of the Cable Trams* (Melbourne, 1970).

30. U.S. Bureau of the Census, *Transportation, 1890*, 1:682.

31. Hilton, *Cable Car*, p. 476.

32. On the growth of the electrical industry, see the articles and bibliography in Charles Singer et al., *A History of Technology*, 5 vols. (Oxford, 1958), 5:177–234. See also Malcolm MacLaren, *The Rise of the Electrical Industry* (Princeton, N.J., 1943); and David S. Landes, *The Unbound Prometheus: Technological Change and Industrial Development in Western Europe from 1750 to the Present* (Cambridge, 1968), pp. 281–290.

33. Harold C. Passer, *The Electrical Manufacturers, 1875–1900: A Study in Competition, Entrepreneurship, Technical Change, and Economic Growth* (Cambridge, Mass., 1953), pp. 211–212.

34. For more extended treatment and citation of relevant materials, see McKay, *Tramways and Trolleys*, pp. 35–41.

35. Ibid., pp. 43–44. The American story has often been told. The most authoritative account is Passer, *Electrical Manufacturers*, pp. 213–255; but see Frank Rowsome, Jr., *Trolley Car Treasury: A Century of American Streetcars-Horsecars, Cablecars, Interurbans, and Trolleys* (New York, 1956), pp. 54–69; and John A. Miller, *Fares, Please!: From Horse-Cars to Streamliners* (New York, 1941), pp. 54–69. Both of these popular histories are also useful for earlier transit developments.

36. For discussion of this problem and relevant sources, see McKay, *Tramways and Trolleys*, pp. 58–67.

37. See, for example, Eric Lampard, "The Social Impact of the Industrial Revolution," in *Technology in Western Civilization*, ed. Melvin Kranzberg and Carroll Pursell, Jr., 2 vols. (New York, 1967), I:305–306; and E. A. Wrigley, "The Supply of Raw Materials in the Industrial Revolution," *Economic History Review 15*, 2nd series (1962): 1–16.

38. McKay, *Tramways and Trolleys*, pp. 51–58.

39. Ibid., pp. 50–51, for full citation of sources.

40. George W. Hilton, "Transport Technology and the Urban Pattern," *Journal of Contemporary History* 4 (1969): 126.

41. *Electrical Engineer* (London) 9 (1892): 196.

42. For documentation and extended discussion of the points in this paragraph, see McKay, *Tramways and Trolleys*, pp. 67–83.

43. Report of Birmingham Subcommittee on Tramways, as quoted in *Electrical Engineer* (London) 19 (1897): 569–570. See McKay, *Tramways and Trolleys*, pp. 84–88, for similar reactions.

44. "Remarks of Mr. F. J. Sprague on Electricity as a Motive Power," American Street Railway Association, *Proceedings* 6 (1887–88): 60–68. also see Passer, *Electrical Manufacturers*, p. 216.

45. McShane, *Technology and Reform*, pp. 23–24. In Fort Wayne, Indiana, and elsewhere, the company simply announced it was switching to electric traction and did so. George K. Bradley, *Fort Wayne's Trolleys* (Chicago, 1963), p. 17.

46. McShane, *Technology and Reform*, pp. 24–26.

47. Ibid., pp. 33–35; McKay, *Tramways and Trolleys*, pp. 94–95; Wilcox, *Municipal Franchises*, 2:25–30.

48. For analysis of these efforts and relevant source citation, see McKay, *Tramways and Trolleys*, pp. 95–101.

49. Ibid., pp. 101–106.

50. Ibid., pp. 107–124, for elaboration, based mainly on French and German case studies and sources.

51. France. Ministère des Travaux Publics. Direction des Chemins de fer. *Statistique des chemins de fer français* (1906), II:655, 667; idem (1912), II:895, 909.

52. Archives du Crédit Lyonnais, Paris. Etudes Financières. "Etude sur les résultats de la traction électrique . . . en France," December 1900.

53. The interplay between private entrepreneurship, speculation, and government regulation has been examined in some detail for France in McKay, Tramways and Trolleys, pp. 125–162.

54. See the suggestive comments in Edward S. Mason, The Street Railway in Massachusetts: The Rise and Decline of an Industry (Cambridge, Mass., 1932), pp. 185–186.

55. McKay, Tramways and Trolleys, pp. 192–198. Also see Holt, "Main Line and Side Tracks," pp. 401–402.

56. McKay, Tramways and Trolleys, pp. 113–114, 205–225.

57. This is a primary theme of Cheape's excellent study, Moving the Masses. On consolidation in Massachusetts, see Mason, Street Railway in Massachusetts, pp. 41–70. Mason's study has an abundance of economic information.

58. McKay, Tramways and Trolleys, pp. 125–126ff.; Passer, Electrical Manufacturers, pp. 249–275, 335–348; V. S. Diakin, Germanskie kapitaly v Rossii: elektroindustriia i elektricheskii transport (Leningrad, 1971).

59. John P. McKay, Pioneers for Profit: Foreign Entrepreneurship and Russian Industrialization, 1885–1913 (Chicago, 1970), pp. 100–102; Diakin, Germanskie kapitaly v Rossii, p. 117; McKay, Tramways and Trolleys, pp. 144–145.

60. For an interpretation of the British pattern of tramway electrification and municipal ownership, see McKay, Tramways and Trolleys, pp. 163–191.

61. Ibid., pp. 198–201.

2 Street Transport in the Second Half of the Nineteenth Century: Mechanization Delayed?

Anthony Sutcliffe

Most historical analyses of the rise of urban transport allot an important place to the turn of the nineteenth century.[1] In a suggestive and convincing study, John P. McKay has identified a veritable urban transport revolution at that time.[2] McKay's periodization is based on his work on the electrification of street railways in Europe, but its main outlines have been followed by Charles Cheape in his authoritative study of street railway development in the United States.[3] Of course, this transformation of the street railways was followed by the rapid diffusion of the gasoline-driven motor vehicle that, in the long term, came to exercise an even more important influence on urban spatial structures. Nevertheless, Robert Fogelson's work on Los Angeles has suggested that the electric trolley car, even in Los Angeles, prompted the initial spatial expansion that the motor vehicle merely confirmed and accelerated.[4]

Seen in the context of the long-term development of the world economy, the distinction between the electric motor and the internal combustion engine is in any case of little significance. The theory of economic fluctuations remains an area of great controversy, but our experience in the 1980s perhaps makes us unusually susceptible to the long-swing hypotheses of Nikolai D. Kondratieff, Joseph Schumpeter, Simon Kuznets, and others.[5] Although the idea of a "Great Depression" between the early 1870s and about 1895 does not convince certain economists and historians,[6] no one would deny that the later 1890s and the early years of this century were marked by unusually rapid economic growth. Mechanical engineering played a central part in this expansion. Seen in this perspective, the rapid development of electricity and the internal combustion engine form part of a single movement, symbolized by the meteoric careers of Thomas Edison and Henry Ford, who became close

22

friends. This acceleration of economic growth was associated with a new phase in the urbanization of the industrialized world after about 1890, as I have suggested in a recent study.[7] This phase was associated with the development of the *theory* of modern urban planning and, to a large extent, the *practice* of planning. The creators of this new phenomenon of planning saw decentralization as one of their prime objectives; planning may thus be seen as, to some degree, a product of the urban transport revolution.[8]

In this discussion, however, I propose to concentrate, not on the urban transport revolution itself, but on a question that has been partly inspired by the American "new economic history." Why was urban street transport not mechanized earlier? The counterfactual character of this question is less disturbing if we recognize that the absence or the deficiency of a given technical network often exercises a positive influence on other networks. A comprehensive appreciation of the inter-relationships of a variety of technical networks within a city, like that essayed by Gabriel Dupuy in Part IV of this volume, becomes possible only if one takes into account what is *not yet* present as well as what has *already* been created. In the field of transport, an improvement or an extension of the existing networks tends to increase the accessible extent of the urban area, resulting in a fall in urban land rent and a tendency toward a decline in residential densities and sometimes in employment densities.[9] The distribution of building across the urban area thus in itself constitutes a technical network. High-rise building, for instance, with its very important technological elements, such as frame construction, elevators, and fire prevention, is intimately linked to the arrangement of urban transport networks.[10] We are only too well aware of the decentralizing tendency of the twentieth-century urban phenomenon. In many instances it has encountered the inertia gener-ated by the nineteenth-century legacy of a massive accumulation of very dense building. For most of the nineteenth century, in contrast, when urban populations were growing at an unprecedented rate, certain im-portant means of transport were not available. Any analysis of Victorian urban patterns thus has to take their absence into account.

The statistical analyses of McKay, J. R. Hume, and others have shown an increase in per capita usage of public transport in cities when their tramways were electrified.[11] This increase was partly the result of fare reductions, which often became possible thanks to the important social saving effected by electric traction. In most instances, however, elec-trification was accompanied by an improvement in service. Frequency was increased, the average speed was doubled, and many lines were extended.[12] To some extent, therefore, the increase in usage suggests that a latent demand existed for the improvement of urban transport before electrification was carried out. This latent demand would have accumulated over time. The longer this time period, the more we are tempted to suspect some kind of technological delay or blockage. This possibility forms the subject of what follows.

For the economic historian, the theory of technological innovation has been transformed by the lifelong researches of Jacob Schmookler.[13] For Schmookler, innovation and even invention are not a *deus ex machina*, as the traditional historians of technology sometimes used to imagine. On the contrary, they are the direct product of supply–demand relationships. From this point of view, there can be no such thing as technological retardation. Nevertheless, to stimulate a given innovation, demand conditions have to be positive, and here institutions and even mentalities have their part to play. In this broader context, the possibility of technological retardation cannot be ruled out.

If a technological deficiency existed in urban transport, its beginnings would presumably be found in the 1870s when the horsecar was approaching the peak of its technical development.[14] Subsequently encountering diminishing marginal returns, improvements in the horsecar stagnated. Meanwhile, the horse-drawn omnibus, which the street railway had partly replaced, generated even less technical improvement. Urban street transport usage consequently increased slowly in the 1880s despite the increase in real income enjoyed by most urbanites, thanks in part to the fall in world primary prices that had begun in the 1870s. This deflation, which brought such substantial benefits to wage earners, was principally the product of the extension of the world transport system and an associated redistribution of labor that resulted from the spread of steam railways and steamships from the 1840s, which produced a spectacular expansion of the exploited land area, particularly in the Americas.[15] From about 1850 onward, this new stage of world industrialization associated with the railways also brought about an acceleration of urbanization. The main beneficiaries were the largest cities, whose accessibility gain within the railway networks was greater than that enjoyed by small and medium-size cities. We may therefore suppose that in the large cities, there was an increased potential for the mechanization of intraurban transport from the middle of the century, in that the continued growth of urban areas was stimulated by an innovation, the steam railway, that was susceptible *a priori* of being adapted to the needs of movement *within* cities.

Indeed, we find that in the biggest cities, a debate springs up from the earliest years of railway construction about the intraurban role of the railway. In Europe, the terminal stations were normally built at the very edge of the built-up area. In London, however, the railway companies at first tried to drive their lines right into the heart of the commercial center in the City.[16] Parliament was largely indifferent to the fate of the poorer districts to the east of the City, and south of the Thames, but beginning in 1846, it refused to authorize the fragmentation of the upper- and middle-class West End by railway construction. The lines that approached London from the north and west thus had to terminate a mile or more away from the City. The most remote company, the Great Western Railway, rallied the others to build an underground line, the Metropolitan Railway, which opened in 1863 and extended their networks to the very gates of the City. This innovation appeared to herald a new era in which urbanites would travel more and more by rail. But such hopes were to be disappointed until the end of the century. Underground railways proved to be expensive to build and operate, and even in London they were not consistently profitable, although the deep boring of tube railways introduced around 1890 produced a degree of improvement.[17] In New York, an overhead solution was adopted beginning in 1870, but the results were unsatisfactory, partly because of the low speeds to which steam trains traveling on flimsy viaducts had to be restricted.[18]

In the suburbs, railways did not face the same disadvantages, mainly because the accelerated growth of the urban area after midcentury tended to incorporate railway installations that had been created on the surface on land acquired at or near agricultural value. In London, railway companies built a number of lines that were always intended to carry suburban traffic, and in the 1870s and 1880s, the central underground system was extended into the suburbs at or above the surface. In the provincial cities of Britain, however, the companies did little to encourage suburban traffic. Unlike London, these cities were too small, and their per capita incomes were too low, to generate an effective demand large enough to stimulate the necessary investment by the companies. So, whereas the railways played an important part in the growth of the London built-up area, they exercised little influence elsewhere in Britain and on the Continent.[19] Indeed, even in London, the main efforts to build up a big suburban traffic by attractive timetables and low fares were made by only two companies—the Great Eastern Railway, serving the eastern suburbs, and the Great Northern Railway, serving part of the northeastern suburbs. Elsewhere in the London suburbs, low-paid white-collar workers made less use of the trains, and manual workers hardly used them at all. Railway transport thus re-created in the suburbs the traditional distinction between East End and West End in the central districts. Still, there is no doubt that the railways contributed significantly to the articulation of a London area that enjoyed the paradoxical distinction of being both the world's largest city and the least densely populated of the giant cities of the later nineteenth century.[20]

There is a striking contrast between the steam railway and the other main application of railway technology within urban areas. This alternative technology was the street railway. In contrast to the railway, the construction cost of the street railway was very low even in the central districts because it used the existing public thoroughfare. The highway, already in public ownership and satisfying a multitude of public and private needs, could incorporate the street railway, so little additional land had to be acquired. Indeed, part of the construction cost of the street railway was in theory attributable to the public authority, which provided the foundation and the drainage of the thoroughfare. Naturally, the public authority endeavored to negotiate an agreement with the street railway company that would maximize the rent the company paid for the use of the highway (and of the track, where this was built by the public authority and leased to the company), and in many instances the company was obliged to take responsibility for street paving in the vicinity of the rails. Nevertheless, the community derived social advantages from the streetcar lines, and its stipulations were not usually exaggerated to the point where the tramway became less profitable than its main competitor, the omnibus.[21] In contrast, in many cases the authorities kept streetcars out of certain streets and districts so that they could retain a social exclusiveness unthreatened by the omnibus and its well-heeled clientele. In Britain, London was the main victim of this discrimination, and so the street railway played a limited role there, being excluded from most of the central districts north of the Thames.[22] In other British cities, however, and in the United States, the horsecar, with its fares substantially undercutting those of the omnibus, took over from the omnibus to provide the main means of urban mass transport, beginning in the 1850s in the United States, and in the 1870s in Britain and elsewhere in Europe.[23]

The disadvantages of the streetcar lines arose not so much from the conditions of their initial construction as from the obstacles to their subsequent mechanization. The mechanization of railway traction had taken place without serious disappointments from the early years of the nineteenth century, thanks to the efforts of English entrepreneurs, inventors, and engineers, whose initial aim was to improve the transport of coal from the pithead to the sea or waterway. The efforts of a more disparate assemblage of groups, departments of public administration, and individuals to mechanize road transport were much less successful, even though they began in the later eighteenth century. They did not achieve their goal until the 1890s, with the diffusion of electric street railway traction. This delay of roughly a century requires some explanation.

Before the development of steam power from its origins in draining mines and driving factory machinery, there was no obvious source of inanimate power that would have permitted a mechanization of road transport. As soon as high-pressure steam engines became available, however, thought was given to their application to the traction of heavy loads on the public highway by means of locomotives. In the late eighteenth century, the main interest came from the military, the most striking result being Joseph Cugnot's steam artillery tractor.[24] In the nineteenth century, the demand was broadened and reinforced by the growing scale of exploitation in agriculture. When these heavy locomotives used metaled roads, however, they encountered a fundamental contradiction. To generate the necessary friction, they had to be either very heavy or equipped with wheels or tracks that penetrated the road surface. In neither event were macadamized surfaces able to stand up to the wear.[25] Between about 1800 and 1835, engineers, in Britain at any rate, made a degree of progress in this area, and some of them were optimistic about an eventual solution.[26] The most practical solution, however, was to provide a hard and smooth road surface comparable to the metal rail. Surfacing with paving stones or asphalt could meet this need, but it was expensive; it made sense to restrict the high-quality surface to what was needed by the locomotives, in the form of narrow, parallel strips. But the most economical means of creating this more limited installation was to lay metal rails. So, from the 1830s, now that the railways in the north of England, such as the Stockton–Darlington and Liverpool–Manchester lines, had demonstrated that the new technology was suitable to the transport of a variety of freight and passengers, the search for an efficient mechanization of highway transport began to concentrate on railed solutions.[27] A demand for heavy locomotives for agricultural use remained, and some engineers described a potential demand for light steam cars. This latter was, however, clearly only a luxury demand, and it was particularly restricted in England by a horse-oriented culture of the rich that seems to have been reinforced in the nineteenth century.[28] There was a place perhaps for small steam omnibuses, which seem to have gone through a continuous but inconspicuous evolution between the 1830s and the triumph of the internal combustion engine after 1900.[29] But steam omnibuses, like the other road locomotives, were hamstrung by the severe speed limits imposed by the authorities beginning in the middle of the nineteenth century, mainly to control the new generation of big agricultural locomotives (traction engines). In Britain, these limits remained in force until the authorities began to make exceptions in favor of vehicles driven by internal combustion engines at the turn of the century. Reduced to a walking pace, the road locomotives were not in a position to attract the substantial amounts of development capital that would have been needed to overcome their defects.

To discuss the ways in which promoters of the internal combustion engine succeeded in obtaining a modification of the speed limits lies outside the scope of this essay. In Britain, the key statutes were the Locomotives on Highways Act of 1896 and the Motor Car Act of 1903. Successively, they increased the speed limit imposed on "light locomotives" to 12 miles and to 20 miles an hour.[30] The previous restrictions, which had been established by the Locomotives Acts of 1861 and 1865,[31] had been tolerated by the supporters of steam, no doubt because the proliferation of street railways after 1870 had attracted the main traction development effort, and partly because the speed limits applied to street railways were more generous. To the legislator's eye, the advantage of the rail was that a runaway vehicle could not normally cause mayhem on the sidewalks. Therefore, in 1879, the Use of Mechanical Power on Tramways Act set a limit of 10 miles per hour, much higher than the limit of 4 miles per hour that applied to locomotives at that time.[32] With the problem of the contact between the wheel and the surface of the highway still unresolved, the rail retained its attractiveness to the promoters of mechanization. Consequently, between the 1870s and the 1890s, efforts concentrated on street railway mechanization. We have now reached the point where we can seriously consider the possibility of a divergence between the growth of demand (whether real or latent) and the progress of technology, at any rate in Britain. This divergence would have been reflected in an increase in per capita urban land rent, for which, as we shall see, there is some evidence.

The efforts to mechanize the streetcar lines from 1870 to 1890 can be classified in three categories. First, there were attempts to adapt steam locomotive technology. These efforts were not a complete failure, but serious problems remained unresolved when electric traction took up the running in about 1890. Second, attempts were made to develop nonlocomotive forms of traction and locomotive traction using power sources other than steam. These methods were not very successful, with the exception of cable haulage, which proved expensive to install. Third, there was the development of electric traction, which, beginning in 1890, established a clear advantage over all competing technologies and permitted a rapid and economical mechanization of the street railways. This electrification was virtually complete, both in North America and in Europe, by 1905.

Innovations in all three categories had to face a means of traction that had reached almost its full development: the horse. This simple technology persisted almost universally until the arrival of electric traction. So, before we look at the new technologies, we should give a moment's consideration to the horse.

The advantages of horse traction on street railways, and indeed in other forms of street transport, have been summarized by F. M. L. Thompson in a widely read article.[33] Harnessed to a railed vehicle with a load of up to thirty people, the horse could stop and start on numerous occasions during a day's work without seriously depleting its strength, while maintaining an overall speed of about 4 miles an hour. Breakdowns on horsecar lines were very rare, in contrast to the mechanical ones. The public, including children and the elderly, was accustomed to horses, and horses did not frighten other horses, as steam engines were said to do. According to some economic historians, it was expensive to maintain horses, but comparisons with rival technologies are difficult to make because of their frequent breakdowns and the fact that most were never developed sufficiently to generate a normal operating cost.[34] Horse traction enjoyed the additional advantage after about 1870 of a fall in the long-run price of fodder, which was partly the product of the creation of a more efficient world sytem of cereal production.[35] This development, as I have noted, was largely the result of the railway revolution and the mechanization of shipping. Paradoxically, therefore, the application of steam to interurban transport contributed to the survival of horse traction within the cities. At the same time, the fall in primary-product prices contributed to a general deflationary tendency that discouraged investment, particularly on the technological margin. This conjuncture is no doubt the origin of the delay in the development of urban street transport noted by McKay.[36] Nevertheless, the idea of a generalized "Great Depression" is disputed by many historians because some sectors, such as retail commerce, consumer goods manufacture, and leisure were stimulated by the fall in primary-product prices. At first sight, urban transport should have been one of the growth sectors because of its close links with mass consumption. To some extent, of course, growing demand could be met by a proliferation of the existing technology, but there were two main disadvantages. First, street congestion produced growing internal and external diseconomies, particularly in central districts. Toward 1890, this congestion was seriously worrying both public and private interests within cities.[37] The congestion sprang partly from the second disadvantage, which was that the area of the largest cities, with the possible exception of London and certain big American cities with very advanced transport systems, was probably beginning to be restricted by the maximum effective speed of the horse-drawn vehicle. This restriction of the aggregate area of urban land was reflected in a rise in the share of urban land rent in national income.

Thanks to the brilliant researches of Avner Offer, and Charles Feinstein's persuasive national income estimates from which they spring, we are now in a position to descry the possible dimensions of this rise in urban land rent in England and Wales.[38] We can note, first, that total urban residential rent (i.e., land and buildings) constitutes a growing share of domestic income between 1855 and 1910.[39] Pure urban land rent appears to have risen from about 20 percent of total urban residential rent in 1876 to 40 percent in 1896. Thereafter it fell slightly, coinciding with streetcar electrification, to around 30 percent in 1900. It then fluctuated, but in 1910, at the end of Offer's series, it had risen to 35 percent.[40] The second rising trend was due in part to a general movement of the population toward larger urban centers. The rapid rise in the 1870s and 1880s, however, followed by a sharp fall at the time of the mechanization of street transport, seems to be the product of a different causation, which is, of course, our concern in this study. The part of the increase that was not due to a displacement of population in favor of larger urban centers would represent, in general terms, the latent demand for urban transport. Why, then, was this demand not satisfied between the 1870s and the early 1890s?

I have already noted that the streetcars were the key sector in this shadowy confrontation of a latent demand and a potential technology. Following Schmookler, we can suppose that an adequate concentration of development capital would have removed the obstacles to successful mechanization. But, as time passed, the greater was the chance of a spinoff from technological sectors better endowed with development capital. Such a spinoff could, in theory, permit street railway mechanization without a major increase in the level of capitalization. Contemporary decision makers were not in a position to perceive the options so clearly, however.

Instead of plunging into the detail of a series of prolonged and partially fruitless efforts, I am going to take advantage of hindsight and begin at the moment when, in 1888, Frank Sprague set up an electric trolley system at Richmond, Virginia. This success, which prompted a general electrification of street railways, first in North America and subsequently in Europe, crowned half a century of experiments with electric motors. But it was not until the late 1860s that the generation of large quantities of cheap electricity became possible, with the development of the dynamo. Moreover, at that time, promoters were attracted principally by the possibility of applying electricity to lighting.[41] Nevertheless, the diffusion of the street railway from around 1870 soon offered an attractive opportunity to develop electric traction. The advantage of the street railway was that the metal rails could provide a circuit, while the small size of the vehicles and the restricted length of the lines created fewer problems for the engineer than did railway electrification.[42] Lighting, however, raised even fewer problems, and in the 1870s and 1880s the early electrical engineering companies, such as Edison and Siemens, concentrated thereon. Werner von Siemens was nevertheless more interested in traction than was Thomas Edison, and in the 1880s the German company looked more likely than any of its competitors to make the crucial breakthrough. Siemens, however, concentrated on separate track systems because he wanted to circulate current through the rails alone.[43] His experimental line at the Berlin industrial exhibition in 1879 attracted great attention, and in 1881 he built a line, more than 1.5 miles long, between the edge of the Berlin built-up area and the isolated villa district of Lichterfelde.[44] In 1882, Siemens was involved in an underground railway project in London that his company would have equipped, but the scheme failed to attract sufficient capital and collapsed.[45] A number of experiments were made in Germany in the early 1880s with overhead distribution systems,[46] but the progress made merely contributed, together with certain Belgian experiments, to the solution implemented by Sprague in 1888.

This uneven progress was principally a reflection of the difficulty of pursuing sustained experiments with a dangerous power source on the public highway. On short lines carrying only one or two vehicles, the tension could be so low that the rails were not dangerous to touch. The potential demand for electrification was greatest in big cities, however, and the lines there had to be longer and more heavily trafficked. In these circumstances, overhead distribution was the most convenient solution, but the authorities and public opinion were reluctant to tolerate its visual intrusiveness as long as its superiority over other distribution systems remained unproven. This was why the proliferation of electrification had to await Sprague's successful experiment in an insignificant city.

This story cannot fail to suggest that, with a bigger concentration of capital, an effective technology could have been developed earlier, perhaps around 1880. One of the main obstacles to such a concentration, in addition to those discussed above, was the equivocal relationship between street railways and big, capital cities. The largest cities tended to generate the greatest demand for mechanization, and technology was diffused from them more quickly than from smaller places. But it was precisely in the large cities that street railways were most restricted. Competing railway interests were more influential there than in smaller cities, and the residential interests of the rich, in alliance with the commercial interests that served them, limited horsecar access to central districts. So, in the London of T. C. Barker and M. Robbins, and the Paris described in the following article by Dominique Larroque, streetcars were allowed a free run only in outer districts. There, low fares attracted a clientele of moderate means in sufficient numbers to secure a profit on their total operation, but they discouraged speculative capital.[47] In the absence of a coherent, citywide transport strategy within which the street railways would have radiated around suburban railway stations, streetcars tended to compete directly with railways on their suburban runs, to the detriment of both. In these circumstances, the omnibus was in a dominant position, thanks to its fuller access to the city center, which was sustained by a tacit alliance with real estate interests among the rich. As I have noted, however, it was highly unlikely for technical reasons that omnibus mechanization would precede that of street railways. From this point of view, we can see that the development of an efficient technology of mechanical traction had to await the time when a city of moderate size would offer favorable conditions—including a tolerant attitude on the part of the authorities—for an initiative by an unusually bold entrepreneur. Richmond, Virginia, eventually provided this conjuncture, and we should not be surprised that this pioneer was an American city, in view of the combination that existed in the United States of high per capita earnings, low residential densities, long streetcar franchises, links between transport interests and land speculation, and, finally, a certain indifference toward the quality of the public environment.

Given that the application of electricity to streetcar lines was delayed in this manner, for a number of years there must have been an enhanced opportunity for steam and other nonelectrical traction systems. Indeed, we can detect an intensification of effort by engineers, inventors, and entrepreneurs starting about 1870. In the steam area, spinoff effects were to be expected given that railway technology had already reached an advanced level by the 1860s when the main national networks were already in existence, at any rate in Europe, and capital was turning away from railways to seek new applications.[48] Unfortunately, an economical adaptation of steam traction to streetcar needs remained impracticable until the success of electric traction deprived the exercise of all interest. The problem was to develop a light locomotive that would not seriously damage the track and that would accelerate without an excessive consumption of energy. This technical problem was serious enough in itself, but it was aggravated by the antipollution restrictions imposed by the authorities. Indeed, D. H. Aldcroft sees these restrictions as the main obstacle to the development of an economical street railway locomotive.[49] The use of steam locomotives therefore remained very limited, especially in the central areas of towns. According to McKay, there were only 700 steam locomotives in the United States, and 500 in Britain, on the eve of general electrification, and most of them worked peripheral or rural lines with infrequent stops.[50] According to Clay McShane, opposition by frontage owners was the fundamental obstacle to the greater use of steam,[51] a point I shall return to later in the context of the opposition to streetcars by real estate interests.

Cable railways encountered less opposition from the point of view of environmental impact. Developed in San Francisco by Hallidie in the early 1870s, they enjoyed a measure of success in the United States. Around 1890 there were 283 miles of cable-car line in the United States, carrying 373 million passengers a year.[52] In Europe, cable cars made little headway. The most important system was in Edinburgh, with a few lines in Birmingham and London. On the Continent, they do not appear to have been used at all. This pattern suggests that demand conditions were the decisive influence on cable-car diffusion. In Europe, with lower per capita incomes than in the United States, cable railways were apparently unprofitable, except in certain large British cities where streetcars were allowed to run through the central areas. Even in the United States, cable cars were profitable only in heavily trafficked central streets, and they made little contribution to the spread of the built-up area.

So, despite the availability of cable technology, efforts were made to find a more economical solution to street railway mechanization in the nonsteam era. These experiments, which included clockwork and compressed-air systems, gas engines, and an internal combustion engine—all of them tried in London between the 1870s and 1890[53]—were marginal in the sense that the development capital applied to them was clearly insufficient to overcome the technical problems of adaptation or the creation of a new technology. Thus general street railway mechanization had to wait patiently for the application of a leading branch of technology: electrification.

Consequently, I join McKay in detecting a retardation of the development of urban public transport between the 1870s and about 1890.[54] In McKay's opinion, the fact that the European urban public was tolerant of high residential densities contributed to low demand for transport. In England, however, there was a tradition of low residential densities combined with high personal incomes. English income levels were exceeded by the United States during the nineteenth century, but the English conjuncture remained more favorable to an improvement of urban transport than did that of the rest of Europe. For this reason, I attach more importance than McKay appears to do to institutions, particularly in the area of relationships between transport and real estate, as these were moderated by the local authorities. In the nineteenth century, these institutions were dominated by property interests.[55] Public transport, in modifying the structure of urban rent, also took a share of urban rent. Any transport improvement thus had a potential income transfer effect. To bring about such a transfer by improvements in street transport needed the approval of the local authority that owned the thoroughfare. Thus, behind what might appear at first sight to be a technological problem, we can discern a confrontation of interests in the area of political economy. This essay ends with a consideration of certain aspects of this confrontation.

First, we cannot fail to be aware of the influence of vested interests. The mechanization of street transport clearly favored tenants and property owners on the periphery of the urban area and beyond. It tended to threaten the interests of other owners, except insofar as it increased the total income of the city by allowing it to function more efficiently. Normally, the municipal council was more sensitive to this latter viewpoint than were individual owners, who often opposed mechanization, at any rate in the short term. As mechanized transport used the public thoroughfare and so involved little or no appropriation of private property, opposing owners had to enter the political arena. They did so by emphasizing negative externalities that arose from the imposition of an additional function on the street. Indeed, this kind of intervention dated from the beginnings of the street railways, long before mechanization. An alliance emerged between property owners, who stressed the environmental impact of mechanization on frontage owners, and owners of the existing means of transport. The interests of the latter were closely related in a variety of ways to *horses*. This association of land and the horse was highly influential because it reflected an urban power structure still dominated by landed interests that possessed property in both city and country. The actual users of the streets remained on the fringes of this debate; opinions were constantly ascribed to them, but they were rarely consulted.

Second, the development process of technical innovations merits attention. The main promoters of this development, the transport companies, had to share the gains that resulted from an improved transport system with the peripheral property owners. In Europe, the companies tended to be short of capital, partly because their leases on the highway were generally very short. Only in the United States, where very long franchises were the norm, and where the transport companies could take part more easily in land speculation, were conditions favorable to an energetic promotion of mechanization by the companies themselves. Consequently, in Europe the development costs tended to fall on the inventors and on companies producing tramway equipment that sought to increase their sales. Such actors certainly existed, but their application of important concentrations of development funds was discouraged by the limitations placed on railed street transport already discussed. There was always the possibility that this discouraging conjuncture would be transformed by some kind of spinoff. Indeed, electric traction may be regarded as to some extent a spinoff from the growth of a large and diverse electrical industry. Nevertheless, the spinoff was delayed by the electrical engineering industry's initial preference for lighting, which was partly the product of the disadvantages under which street transport labored. Opposition to electric lighting, which did not extend greatly beyond the polluting gas companies themselves, was a much less serious threat than opposition to the mechanization of urban street transport because the main landed interests, the mine owners, benefited indifferently from the use of electricity and gas.

Third, we ought to be aware of what all this implies for the role of the state, both centrally and locally. In the long run, scientific and technical progress tended to increase the perceived potential social saving of a mechanization of urban street transport. Where the state, as in North America, had conceded an important share in urban real estate to the transport companies in the form of the streets and such private land as they purchased, directly or indirectly, to engage in land speculation, no further intervention by the state was normally necessary. But where, as was generally true in Europe, transport interests were not able to share so fully in urban property, the local state was inclined to extend its property interest to the means of transport themselves and subsequently to peripheral land and even buildings. The municipal ownership of peripheral building land, which became widespread in Germany, and the construction of peripheral rental dwellings by British municipalities, are predictable results of this tendency. Such ambitious initiatives could not be countenanced, however, until "municipal socialism" reached maturity in the 1890s.

Finally, as I have noted, urban technical networks modify the structure of rental values within the city. These structures are based on social formations. A new technology is often associated in the first instance with social interests that are marginal to property interests. Eventually, new knowledge can be transformed into property rights, but only through a struggle between the existing owners of property and those excluded from it. In this struggle, a compromise solution is most likely at the time when, and in the place where, the new technology is likely to produce a net increase in the social product. Conditions were thus favorable during the period of accelerated investment between the 1890s and World War I. During the "Great Depression," in contrast, little progress toward this compromise was to be expected. The general conclusion to be drawn is that the discussion of structures that has dominated this essay should not be allowed to obscure the cyclical elements. Nevertheless, in the fields of urban policy and urban investment, these cyclical effects correspond more closely to the long cycles discerned by Kondratieff and others than to the shorter fluctuations that preoccupied the classical economists until the events of the later nineteenth century suggested the existence of the longer economic swings that seem still to be shaping our world today.

Notes

1. See, for instance, the recent study by J. R. Hume, "Transport and Towns in Victorian Scotland," in *Scottish Urban History*, ed. George Gordon and Brian Dicks (Aberdeen, 1983), pp. 197–232.

2. John P. McKay, *Tramways and Trolleys: The Rise of Urban Mass Transport in Europe* (Princeton, N.J., 1976).

3. Charles W. Cheape, *Moving the Masses: Urban Public Transport in New York, Boston and Philadelphia, 1880–1912* (Cambridge, Mass., 1980).

4. Robert W. Fogelson, *The Fragmented Metropolis: Los Angeles, 1850–1930* (Cambridge, Mass., 1967).

5. For an excellent summary of these debates, see J. J. Van Duijn, *The Long Wave in Economic Life* (London, 1983).

6. Such as S. B. Saul, *The Myth of the Great Depression, 1873–1896* (London, 1969).

7. A. Sutcliffe, *Towards the Planned City: Germany, Britain, the United States, and France* (Oxford, England, 1981).

8. This hypothesis has become one of the main themes of a collective volume: A. Sutcliffe, ed., *Metropolis, 1890–1940* (London, 1984).

9. These dynamic relationships are set out in the classic study by William Alonso, *Location and Land Use: Towards a General Theory of Land Rent* (London, 1964).

10. High-rise building is discussed, partly with regard to the theory of technological evolution in the context of the urban property market, in A. Sutcliffe, "La victoire de l'immeuble de rapport: un problème de l'histoire des grandes villes européennes au dix-neuvième siècle," *Histoire Sociale (Social History)* 13, no. 25 (1980): 215–224.

11. See, for instance, Hume, "Transport and Towns," p. 207.

12. See especially the studies of McKay and Cheape, cited in notes 2 and 3.

13. See, especially, J. Schmookler, *Invention and Economic Growth* (Cambridge, Mass., 1966), and idem, *Patents, Invention and Economic Change: Data and Selected Essays* (Cambridge, Mass, 1972).

14. The main outlines of the technical development of urban transport, with particular reference to Britain, are sketched out in T. C. Barker, "Towards an Historical Classification of Urban Transport Development Since the Later Eighteenth Century," *Journal of Transport History* 1, no. 1 (1980): 75–90.

15. It will be noted that this interpretation of world economic growth is inspired principally by W. Rostow, *The Stages of Economic Growth: A Non-Communist Manifesto* (Cambridge, England, 1960).

16. See T. C. Barker and M. Robbins, *A History of London Transport: Passenger Travel and the Development of the Metropolis*, vol. 1, *The Nineteenth Century* (London, 1963).

17. For construction costs, see John R. Kellett, *The Impact of Railways on Victorian Cities* (London, 1969), p. 289.

18. David C. Hammack, *Power and Society: Greater New York at the Turn of the Century* (New York, 1982), pp. 231–233, 237.

19. This fundamental difference between London and other British cities is the main conclusion of Kellett, *Impact of Railways*.

20. London's exceptionally low residential densities are particularly noted by S. E. Rasmussen, *London: The Unique City* (London, 1937).

21. This political economy of the streetcar is examined by A. D. Ochojna, "Lines of Class Distinction: An Economic and Social History of the British Tramcar with Special Reference to Edinburgh and Glasgow," Ph.D. dissertation, University of Glasgow, 1974.

22. Barker and Robbins, *History of London Transport*. 1:xx and *passim*.

23. See H. J. Dyos and D. H. Aldcroft, *British Transport: An Economic Survey from the Seventeenth Century to the Twentieth* (Leicester, England, 1969).

24. The general development of technology can be followed in Charles Singer, ed., *A History of Technology*, 5 vols. (Oxford, England, 1954–1958).

25. See, for instance, House of Commons, *Minutes of Evidence Taken Before the Select Committee on the Locomotive Bill with the Proceedings of the Committee*, sess. 2, Sessional Papers, 5:351, 1859. This committee, which studied the implications of a new generation of locomotives for roads, tolls, speed limits, and public order in general, was particularly interested in the impact of locomotives on road surfaces. See, especially, the evidence of William M'Adam, General Surveyor of Turnpike Roads, July 19, 1859, pp. 1–18.

26. R. P. Mahaffy, *Highway and Road-Traffic Law* (London, 1935).

27. See R. W. Fogel, "Railways as an Analogy to the Space Effort: Some Economic Aspects," in *Applied Historical Studies: An Introductory Reader*, ed. M. Drake (London, 1973), p. 152.

28. See F. M. L. Thompson, *English Landed Society in the Nineteenth Century* (London, 1963).

29. In London and Edinburgh, a limited diffusion of steam omnibuses seems to have occurred around 1870. See Barker and Robbins, *History of London Transport*, 1:293.

30. Mahaffy, *Highway and Road-Traffic Law*, pp. 56–57. The positive influence of this relaxation on the development of the motor omnibus in London is studied by Barker and Robbins, *History of London Transport*, 2:119.

31. Mahaffy, *Highway and Road-Traffic Law*, p. 27.

32. McKay, *Tramways and Trolleys*, p. 30.

33. F. M. L. Thompson, "Nineteenth-Century Horse Sense," *Economic History Review* 29, no. 1, 2nd series (1976): 60–81.

34. The heavy costs of horses are emphasised by Barker and Robbins, *History of London Transport*, 1:5–6.

35. Ibid., p. 176.

36. McKay, *Tramways and Trolleys*, pp. 6, 23.

37. This disadvantage is recognized even by Thompson, *English Landed Society*, p. 77.

38. A. Offer, "Ricardo's Paradox and the Movement of Rents in England, c. 1870–1910," *Economic History Review*, 33, no. 2, 2nd series (1980): 236–252.

39. Ibid., pp. 240, 250.

40. Ibid., p. 241, fig. 2.

41. Barker and Robbins, *History of London Transport*, 1:297–298.

42. Ibid., 2:21.

43. McKay, *Tramways and Trolleys*, pp. 37–39.

44. Ibid., pp. 37–38.

45. Barker and Robbins, *History of London Transport*, 1:304–305.

46. McKay, *Tramways and Trolleys*, p. 39.

47. For the evolution of this conjuncture in London, see Barker and Robbins, *History of London Transport*, 1:182–197.

48. See T. R. Gourvish, *Railways and the British Economy, 1830–1914* (London, 1980), p. 15.

49. D. H. Aldcroft, "Urban Transport Problems in Historical Perspective," in *Business, Banking and Urban History: Essays in Honour of S. G. Checkland*, ed. A. Slaven and D. H. Aldcroft (Edinburgh, 1982), p. 224. See also McKay, *Tramways and Trolleys*, pp. 30–31.

50. McKay, *Tramways and Trolleys*, pp. 30, 40–41.

51. C. McShane, "Transforming the Use of Urban Space: A Look at the Revolution in Street Pavements, 1880–1924," *Journal of Urban History* 5, no. 3 (1979): 289–290; and *Technology and Reform: Street Railways and the Growth of Milwaukee, 1887–1900* (Madison, Wis., 1974), pp. 1–9.

52. McKay, *Tramways and Trolleys*, p. 40.

53. Barker and Robbins, *History of London Transport*, 1:294.

54. McKay, *Tramways and Trolleys*, pp. 6–7.

55. See the brilliant thesis of A. Offer, *Property and Politics, 1870–1914: Landownership, Law, Ideology and Urban Development in England* (Cambridge, England, 1981).

3　Economic Aspects of Public Transit in the Parisian Area, 1855–1939

Dominique Larroque

FRENCH history commonly relegates economic considerations to a rather marginal role in any analysis of policies concerning the development of urban networks. Similar to the interpretation given to other urban networks, the Parisian transportation systems have been explained exclusively in technical terms, as a simple response to the needs of urban development. These distorted explanations reflect the lack of an adequate model for that historic period. Some analyses of transportation systems mainly refer to political and institutional factors. The direction of development is perceived as in the hands of the authorities, and phases of its evolution are explained by the political and institutional diversity of decision makers. Finally, recent studies on the role of sociopolitical determinants also tend, indirectly, to minimize the role of economic factors in the production of transportation systems. Actually, the transportation service is integrated in a complete system of urban policies whose main objective is a full social-control system to preserve threatened dominant values.[1]

The lack of attention paid to the role of economic factors in the development of urban transportation networks is unjustified because, even before the rise of the public network concept, transportation systems were considered as products that had to return a profit on private capital invested at risk in the company. In 1920, limited state control was introduced (régie intéressée) over private companies without ending the role of the private sector. Economics is still, however, an essential part of transportation policy, even if only internally specific to each company. Although nationalization decreases internal economic profit constraints, the external economic environment still exists. In this essay I consider the goals and strategies of industrial groups in the area of

40

urban transportation systems. But economic power is not monolithic; it has its own internal tensions. Its components, methods, and objectives sometimes differ radically, following restructuring movements of the capital invested in the Parisian transportation companies. To argue that economic factors alone explain the nature of urban networks would oversimplify the complexity of the city.

While the analysis is concerned with the transportation network, the events and the chronology as a whole coincide with other kinds of networks. This coincidence is not accidental in the history of Paris, where stages in the evolution of urban networks do not follow a horizontal line linked to the typology of equipment, but a vertical line reflecting deep common tendencies. The transportation system also gives us a very clear vision of the economic implications of an urban network. It is not simply a public service (at a time when the concept of public service was developing very slowly); in the nineteenth century it is a leading industry, attracting capital and stimulating innovation.

Contradictions about transport as both a public service and a profit-oriented industry appear especially in Paris, where the system was particularly in demand. Paris also offers an advantage for analysis because it is a prestigious city, a political, ideological, and cultural melting pot where other factors may counterbalance economic issues in urban decisions. In order to understand the importance of economic factors, they must be considered over a long period: from 1855, when public transportation became a monopoly, to 1939, when conditions for nationalization were present.

The major change in the economy during this period occurred immediately after the great depression of the 1880s. This chronological change is particularly clear in the evolution of the Parisian transportation systems; the overall economic situation cadenced a network highly involved in the process of industrialization. But before submitting to an industrial logic and to "production" imperatives, urban transportation systems followed other, more speculative impulses.

Urban Speculations and Urban Mobility, 1855–1890

Haussmann's Leadership

When the imperial administration ratified the creation of the Compagnie Générale des Omnibus, or CGO (General Company for Transportation by Omnibus) on February 22, 1855, and granted it a monopoly concession of public transportation in Paris for a fifty-year period, the

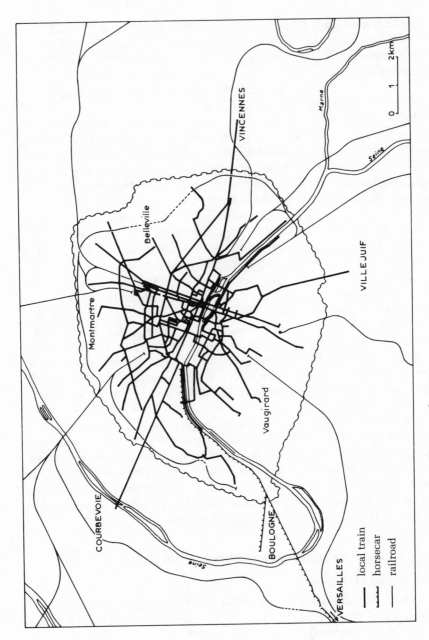

Figure 3.1 The CGO transportation network, 1861

COURBEVOIE

Montmartre

Belleville

VINCENNES

VILLE JUIF

Vaugirard

BOULOGNE

VERSAILLES

Seine

Seine

Marne

0 1 2km

— local train
···· horsecar
— railroad

legal communal territory stopped at the city's toll walls, almost following the actual "exterior" ring. But the population had passed these borders before the annexation in 1859–1860 by the public powers of the "nearby suburb" (see Figure 3.1). Modern Paris was born. Baron Georges Haussmann's administration divided the city into twenty wards and, by a new contract, expanded the territory of the CGO to that of today's capital. Urbanization, which had spread to what has since been called "old Paris" in 1880, far from covered the total annexed zone. Settlement was dense near the old gates and sparse at the peripheries between the main axes of communication. Annexation gave this territory an urban policy, which the first great achievements showed to be a voluntary one.

In this territory, Haussmann's urbanism did not encounter any problems. A municipal commission replaced the council, but gave only consultative advice; compulsory purchase of private property was extended at the expense of the right of ownership, and the procedure for it was simplified; under the rule of a chairman of the urban council, the city of Paris went into debt. Haussmann "tore Paris apart, hewed cut, sliced." The terminology used in many publications concerning the work of the Second Empire showed the extent of his voluntarism in urban politics.

It is not surprising that the concession contract with the CGO was extremely detailed. In itself, monopoly was already an authoritarian measure, imposed by ten former Parisian omnibus companies, which had sought the arbitration of public powers. Competition and anarchistic operation stopped. As a counterpart to the exclusivity and the time period granted to the CGO through the monopoly, the state could request a moderation in rate policy, demand the operation of unprofitable routes (compensated for by profitable ones), and impose a policy of regular expansion, guaranteed to pay off because of the length of the concession. But the contract went much further in defining the right of control and intervention by the state. Everything was regulated: routes, definition of equipment, rates, timetables, stations, technical progress—even dress and manners codes for drivers. The same concern for detail was seen in other urban facilities, such as water and gas.

The imperial administration gave itself the means of maintaining a strong urban policy, the orientation of which, if not the methods, would survive the fall of the regime. But despite a strong literal formulation, a contract has only the value that political powers can or wish to grant. In this light, the achievements of the concession contracts do not reflect their intentions.

Selective Mobility

The new routes were essentially the same as those formerly run by the ten separate companies: a concentric radius network, dense in the center—the business area—and tighter in the residential east than in the west and the northern populated areas. The omnibus network remained nearly unchanged, despite the extension of Paris in 1860 and tighter routing of suburban areas. It was not until after concessions by tramway lines in 1874 that the CGO accepted some scheduling corrections in favor of the suburbs. But more tramway service farther from the center, or increased capacity, did not actually modify the initial selective character of the operation.

There were two main reasons for this limitation. Rates on the tramways went unchanged, despite improved performance, and they were poorly adjusted to meet local needs. Charging 30 centimes for a simple roundtrip ticket (60 centimes with a change ticket) was prohibitive, since the daily average salary for a worker was from 1.50 to 2 francs for men and notably less for women and children; even this daily income was unreliable. The CGO did not attempt to serve this type of client. The working day lasted thirteen to fifteen hours (twelve for children, since the proclamation of May 19, 1874) and started long before the departure of the first omnibuses or tramways at 6:45 or 7:00 every morning. So working-class groups moved between home and work by foot. According to L. Lazare, one of the rare contemporary authors interested in these daily migrations, workers came down from Belleville every day to the gates of the factories.[2] Only a modest number of clerks walked the daily route to the business center; workers continued on to the industrial areas northeast of Paris, segregating activities geographically based on social status. It is impossible to quantify these movements now, since statistics are unavailable. But even in the 1890s, almost three decades after Lazare's description, walking was still the only means of reaching the factory, the workshop, and the small shops for most of the poor. This "social" inadequacy persisted in the CGO network of southern and eastern Paris (11th to 15th and 18th to 20th wards), which together housed 60 percent of the Parisian working population. The CGO mainly serviced the World Fairs; it resisted pressure from the City Council for a more social orientation, and maintained its rates despite general price deflation.

Correlation, if not causality, can be established between this selective organization of transportation systems and the urbanization methods characterizing the urban evolution of Paris. Employees lived in areas near the center, and factories were located on the high-density

Paris, 1913. This self-propelled Purrey steamcar satisfied many environmental concerns, but it was expensive and undependable.

Stuttgart, circa 1898. An electric motorcar pulling two trailers responds to the rush-hour demand for transit.

Canal Street, New Orleans, 1902. A thick network of overhead wires spans a major thoroughfare.

Bronx River Parkway before construction, in 1913, and after completion, in 1922, from *Engineering News-Record*, March 1, 1927.

Southern State Parkway, Long Island, in 1950. The Southern State was Robert Moses's first parkway. Photo courtesy of Long Island State Park Commission.

fringes; workers had to live near the factories in order to reduce travel-ling time, which became unbearable because of the long working hours. This mechanism, in which transportation systems—or better, lack of transportation—seems to be a main link, did not apply in the residen-tial West where there was early mobility, and its diffusion adapted to extended urbanized areas. The CGO only followed the trend that had arisen since the introduction of the first railway lines west of Paris; at that time, the ability to dissociate work and residence, settlement in sub-urbs and daily migrations, became a privilege of social groups whose working timetables and incomes gave them access to transportation sys-tems—to mobility.

Forty years after starting the first railway line between Paris and Saint Germain in 1837, this mode of transportation was sufficiently established to consider a more social application. Paris, the pivot of the whole national railway system, was overequipped. Six large Parisian railway stations were located inside the preannexation borders, toward which main national railway axes converged, connected by two circular lines: the small belt inside Paris, finished in 1867, and the large belt in the suburbs, built from 1877 to 1886. The inherent paradox in this system was that, with the exception of the western residential suburbs, its main goal was to avoid Paris rather than serve its population. In the other suburbs, the trains did not even stop. Even inside the fortifica-tions, small-belt traffic opened only reluctantly to labor transportation. Beginning in 1877, service started at 5:00 A.M. on the belt, and service became more frequent. In 1883, a special worker's rate, usable only on the first two morning trains and on evening trains, brought the price of a roundtrip ticket to 30 centimes regardless of the route. But the amount was still too high: In 1886, only 18 percent of small-belt riders used this special rate.

Networks and Speculation

This fact of partial inadequacy, of selective mobility, with its rami-fications for urban extension, simply reflected a hierarchy of values. Between the two cities that Lazare described in 1870, and not without exaggeration, as so "different and hostile, the city of luxury surrounded by the city of misery," transportation networks could not develop with-out open imbalances in a period when the concept of public service did not exist. In a system in which social values did not counterbalance unhindered economic liberalism, one must examine the internal logic of the concessionary companies in order to understand the evolution of the networks. Of course, treaties existed, and their extremely precise regulations suggested the omnipresence of the state. But the state neither abused nor used its power. Under the Second Empire, confusion between political powers and financiers is so well documented that there is no need to detail it in this chapter. Later, the regime changed, but not the economic power.

The lack of competition, forbidden by the monopoly, and of any real administrative control, gave considerable latitude to leaders of the CGO. Remarkably, the company continued to use the same strategy for over half a century. If interventionism, as written in the contract, was ambiguous about conceded activities, the policy of the CGO from the start left no doubt: Between the public service it had to assume and the return on capital it wanted, the company did not seek a balance, but intentionally shifted toward the latter. One may wonder if the production of transportation had, paradoxically, a relatively small place in a strategy geared mainly toward real estate operations and financial investment. This largely explains the orientation and strategy of the CGO's transportation policy.

When the CGO was established in 1855, the capital invested in the company was integrated with a chain of companies led by the Pereire brothers, founders of the Credit Mobilier and administrators of the East, West, and South Railway Companies. They invested in companies in order to control the water supply (a private service), lighting gas, running warehouses and general stores, and planning through their real estate company. This company invested in and supervised large projects in the center and west of the city and was granted every facility. Elsewhere, the configuration of urban networks showed the same defects as those of the transportation systems. The director of projects of the City of Paris, E. Belgrand, invoked a technical solidarity between the road system and the networks to justify the priorities granted by the real estate policy.[3] In fact, transportation systems during the Second Empire, like the networks in general, seemed to serve mainly as support for speculative urban real estate projects: Haussmann's "breakthrough" created new areas from the demolished spaces, areas where luxury buildings were later built. Real estate, as well as transportation and Haussmann's urbanism, addressed only a solvent clientele and acted within a limited area, beyond which working-class individuals were not included.

The financial integration linking urban networks to these speculative practices collapsed with the 1867 bankruptcy of the Credit Mobilier, which purportedly resulted from difficulties in real estate ventures. Nevertheless, if the CGO acquired a certain independence at this time, it did not alter the real estate objectives of the Pereire brothers. Their properties never suffered excessively from their bank's downfall. Of course, this is never clearly acknowledged in the annual reports of the Board of Administrators, but the "Buildings" section in the balance sheet of the company speaks for itself: Expressed in assessed value—without considering appreciation—with a stable franc, the value of the Pereire property increased from 2 million francs in 1858 to 74 million francs in 1903 (see Figure 3.2). The CGO invested in crisis as well as expansion periods, at the expense of transportation services, which the company did not hesitate to cut back whenever there were money problems. This strategy either prevented any extension of the network after the annexation of 1860 or maintained "social" gaps in the routing until the Great Exhibition of 1900. In 1905, according to the terms of the concession, only the equipment and the infrastructure, not the real estate properties, would come into the hands of the state; this did not encourage the CGO to change its policy.

The Parisian company persisted in its financial methods, uncommon in the nineteenth century, and its systematic practices paralleled the network's growth. The CGO did not designate any reserve fund for projects and did not use self-financing. It borrowed. Frequent use of the stock market forced the company to pay high dividends for its shares, explained its leaders, in order to maintain the credit-worthiness of the company. Since part of the loans was diverted to unproductive real estate acquisitions, the fare at the end of the chain included variables that extended beyond the transportation service. Loans weighed on fares; they also dictated the way the CGO was run. Without a reserve fund, extension of service was not progressive, even in a market that was growing steadily during the whole period. Instead, service was offered irregularly, when the pressure of demand was greatest or when a financial operation without risk was guaranteed. The World Exhibitions were an ideal occasion.

These large, international events gave new life to debate on the opportunity to build, or rather the nature of, a subway system. For about twenty years, there was conflict between the large railway companies and the City Council of Paris. The companies' strategy was, in a way, dictated by their national dimensions, and by the central position of Paris in the railway network. They had already connected the urban extension area between the small and large belts, which were operated by a syndicate. Between the two belts, the syndicate stopped a local railway project (the Brunfaut project) and filled the slots with new installations in order to oppose a final technical barrier.

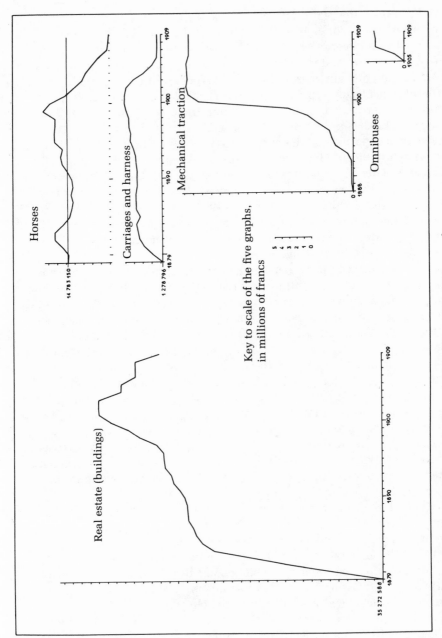

Figure 3.2 Survey of CGO equipment for five operations, from 1879–1909 annual reports

It was impossible for the companies to build a railway to support the urban growth; yet they wanted to protect national and international traffic from any local transformation attempt. In this spirit, they formed the idea, due less to absolute necessity than a concern about possible competition, of the Paris Metro. The main belt line (another belt!) connected networks at the Parisian railway stations for the national exchanges and branched off through a transverse line to the small belt, to confirm the classification of its lines as of general public interest, and to ensure a guaranteed state interest. The projects were not free of real estate implications. Systematically, and the Haag project is a clear example, the "overground" solution was recommended for some routes; this "Haussmannian" metro, associated with huge expropriations and the repurchase of properties, would doubtless have brought a large transfer of Parisian real estate and a large profit to the railway companies.

Finally, references to demographic pressure—so often cited as an unavoidable component and almost mandatory in analyses of urban phenomena—are rather useless here. Transportation systems during the Haussmannian period followed another logic, mostly speculative, rooted in daily life of Paris and the urban scene. The City Council, in one of its discussions, dared to speak of a boycott of Paris by the transportation companies. This was not completely unfounded, since transportation appealed only to a minority; it referred to product as well as to people's mobility. The speculative and financial roots of the Haussmannian system—one of its main features—generated, in fact, a conception of urbanism whose elite nature appears in many domains. The regime structured the Center and the West, but abandoned the periphery; Haussmann's "breakthrough" geared not only a flow of people but also of "quality" goods to commerce/businesses/banks/administration/recreation.[4] In contrast, industrial production, the factory itself and its labor, had no place inside the urban sphere of influence. Factories settled on space opened to them by the 1860 annexation, preferably near waterways, because railway companies were reticent about transporting workers. During the imperial administration, bathed in the industrial spirit envisioned by Saint-Simonians, productive space was not a priority or even a component of Haussmann's urbanism. Regulations were often restrictive or coercive concerning localizations (more rigorous application of the law on insanitary buildings) or access to urban equipment. Selective mobility also acted in this sense to exclude the transportation of workers from the activities of the CGO. Labor followed the factory to the periphery, obeying one-way legislation that forbade certain industry in the urban perimeter but kept silent on spontaneous urbanization near the factory. This was a necessary contradiction for a system that had simultaneously to reconcile endless working days and a lack of adequate transportation; it reveals a system that placed the industrial working class at the Parisian periphery and contradicts the logic of speculation that attracted, then abandoned, most of the inhabitants to individual solutions.

Economic Mutations and Generalized Mobility, 1890–1939

State Regulation

The CGO went through the great depression of the 1880s maintaining its position and methods, extending its real estate grip while distributing very respectable dividends to its shareholders. The difficulties encountered by the railway companies, intensified by the crisis, did not encourage them to expand their social services in Paris and its suburbs. But this time, the crisis raised a question to which economic recovery offered a positive answer. According to E. Beau de Lomenie, the demand for "transportation for all" found an echo and spokesman in the official local authority, breaking the long monologue of the "railway party" in the Chamber and the ministries.[5] Following the crisis, changes went very deep and affected areas other than transportation. The central problem for French society was to decide if, after half a century of economic growth within a limitless liberalism, the system had the capacity to incorporate social achievements. Elements of the answer would occur in successive strokes, but on the eve of World War II, prospects were slim.

The concessionary companies contributed to this process by widening the range of their clientele through rate modification; lowering rates even slightly in the context of social progress upset the data on urban transportation. Companies now had to think in terms of flow, automation, intensification, and a tighter network, thus offering a product of a very different nature. The standardization of urban transportation is thus impossible to dissociate from social transfer, of which it is both a consequence and a component. Local and national levels interact here, in a readjustment of the dominant system by social measures. Although this observation places the trend toward generalized mobility in a global evolutionary perspective, it does not explain its decision-making mechanisms or the speed and extent of change. Social transfer was slow in developing, and since the public powers had control over this activity, it was tempting to advocate state regulation as a central component of the system to standardize urban transportation. The time was ripe for an attempt at a larger compromise—led by the state—between owners and workers. But that was not its domain, and the results were mediocre.

Clearly, during the transition decade between the two centuries, the state radically modified its attitude concerning local demands, for which the City Council was the increasingly vocal spokesman. Since 1880, the local assembly had been in permanent conflict with the public service concessionary companies (CGO *et al.*) and the large railway companies. Year after year, relations deteriorated between this local institution, which was becoming increasingly radical, and the companies, which were using their monopolistic position to avoid or delay expanding services to meet social and urban needs. The powerless, if not purely consultative, Council of Paris needed to be backed by the state, which had shown until then a rather neutral attitude leaning toward the concessionary companies, or at least toward the large railway companies. Here we find explanations that reflect a sociopolitical dynamic:[6] In an explosive climate of protest at the turn of the century, the state decided to satisfy local desires so as not to concede anything regarding the social relationship underlying the dominant system. This schema justified, in effect, the shift in advancement of two levels of protest.

The major political event of the last years of the century was the rise of radicalism. In the majority for a decade in the City Council of Paris, the radical political party came to power and invaded all levels of the administration with its people and its ideas. In Paris, local and national authorities merged ideologically, and the urban sector was first to reflect this. We will see later what the result of an urban policy centered on those sociopolitical aspects is. For the moment, we can note only its relevance.

The local interest of the Parisian Metro was established in November 1895 by the short-lived radical government of Leon Bourgeois. Only a political change could have finally brought the state on the side of the City Council, after many years of loyal support of the large railway companies. The municipal commission of the Metro system feared a change of policy with the return of the right wing to power, so it injected into the project a technical interdict—with important consequences for the future—for the penetration of national or international trains in the future urban network. At first, the space between the rails had to be changed from 1.44 meters (normal railway) to 1 meter. After this proposal was rejected by the war ministry for defense reasons, the commission reduced only the size of the tunnels. Gearing the traffic to anything other than local service was thus impossible, whatever party was in power. This decision, which forbade any future connection between urban and interurban networks, suggests second thoughts about conducting an overly global analysis.

There is a profound analogy between the radical doctrine and political decisions concerning urban policy in Paris. Radical thought personified an ideal of social and economic moderation and balance. The minister of public works, Pierre Baudin, under whom the city assembly's transportation projects progressed decisively, did not hide his preferences for moderate productivity and concentration in France. There was a current of opinion, reaching further right than the radical political position in order to reach a fraction of the moderate voters, that in the transition years believed in a return to industrial "parceling out," after the dark nineteenth century of monopolies and excesses. They awaited a more harmonious form of growth, sidestepping the possible social consequences for the country of an abrupt economic change. This optimistic setup for the future was based on the emergence of new technologies, namely electricity and the automobile. Electrical energy was supplying many small Parisian workshops; it was still hard to conceive that automobiles could be produced outside small workshops. This was enough to revive discussion about technological progress that half a century of hard growth did not render unfashionable, even if hope was then centered on more social objectives. Analyzed from this political and doctrinal angle, the new orientation of urban network policy, in a framework of technical innovation, fits with the rise of radicalism.

Examine the facts: In 1887, the Parisian City Council, hostile to the monopolies, obtained from the public powers the suburb service concession for two competing companies (CGPT and TPOS), despite the fact that the CGO bid on it. The World Exhibitions—1889 in this case— had always been a lever for Parisian councillors because it indicated the multiplicity of concessions inside Paris for supplying electricity. The radicals considered—with reason—that the monopoly of the Parisian company guaranteed it an income that increased in proportion to the growing market, releasing it from any need to invest. For ten more years, the CGO resisted pressure from the local assembly, imposing old equipment and animal traction on the Parisian public on the eve of the "Great Exhibition" of 1900. Then the radical state took over, through a judicial stratagem allowing it to revoke the monopoly of the CGO: Ten new tramway companies were established for the Paris–suburbs lines. The target of this measure was the monopoly, and the large concentrations in the suburbs of outside companies, toward which the radical State had little complaint except their size. The ten new concessionary companies conformed to specifications imposing electric traction, reduced fares, and employment legislation then being prepared. The 1910 agreements finally extended these stipulations to all surface transportation, including a CGO promise of direct substitution of buses for the antique horse omnibuses. The agreement unified, but mainly organized and rationalized, the routing system, complementing the different modes of transportation. This touches on another aspect of radical doctrine, especially of the group of radicals in the government near business circles, such as Leon Bourgeois: The cult of organization, of rationalization, supplanted even the virtues of competitiveness that became an obstacle to the development of the company.

Reorganization of transportation at the turn of the century can be understood in the context of increased intervention of the state in a new political climate. Radicalism standardized transportation systems based on the performance characteristics of a new technique, gearing the system toward less concentrated but more rational forms, and financing a minimal program of social transfer. This state regulation appears much more clearly between the two wars. But circumstances changed, following a change in demand. With the growth of building in the suburbs, the product—transportation—had to cover longer routes within a more socially acceptable fare system. In this situation, the fiscal regulation restricting state power was lifted in 1917 after the imposition of progressive taxation on incomes and the transferral of old taxation to local collectivities. A decree of December 25, 1920, approved the leasing of the surface transportation network for a period of thirty years to a sole company, the Parisian Area Common Transportation Company (STCRP). In this agreement, the Department of the Seine, which had bought the networks of the old bankrupted companies after the war, was now in charge of infrastructural expenses and equipment. The leasing company insured only the funds for running the operation (personnel, upkeep, circulation), for a simple "payment." The transfer of the fixed assets of the transportation companies to the budget marked an important step in state interventionism, especially because this measure was then applied to other sectors, such as the large railway companies (except the West Company, which had been nationalized since 1907), the Parisian gas and electricity supply company, and the Metro.

Tighter routing and regionalization of transportation, its availability to lower social groups (via a change in fare policy on the whole network, in the suburbs as well as in Paris), the coordination of the activities of each company according to the routes, and technical innovation—after the World Exhibition of 1900, and especially from the reorganization of 1910, the system was deeply transformed, and it changed its nature by

TABLE 3.1.
Number of Passengers Carried, Paris Area, 1890 and 1913

Mode of transportation	1890	1913
Tramway	136,785,000	489,276,000
Omnibus	114,845,000	246,243,000
Subway	—	467,472,000

changing objectives. It was now a public service, participating in global production ensuring the transportation of workers. In other words, it mobilized the working force. Table 3.1 shows the number of passengers in 1890 and 1913.

After the war, the eight-hour working day, the demographic explosion, and the working-class exodus to the suburbs all caused the figures in the table to increase dramatically with the onset of commuting. But in the table, the most surprising fact is the large jump in the figures between 1890 and 1913: 251,630,000 compared to 1,202,991,000 in total passengers! Of course, a new infrastructure creates needs as well as satisfying them, in a two-way causal relation. It is nevertheless true that this jump reveals the extent of the distortion between need and demand maintained by a system centered only on a particular clientele.

There was also a change from the stringent economy used by CGO to run its network. In 1910, electric traction became standard, and the antique horse omnibus was replaced by the bus. In only five years, the STCRP and the Department rebuilt, modernized, and tightened the network. Then the General Council, whose majority became more radical over the years, addressed the social orientation of the transportation sector, as the City Council had done before the war. Fares stayed stable in a general price increase, and the railway and the urban network were modernized at the price of a growing deficit, written into the budget of the department since the leasing agreements of 1920.

Labor Transport and Production Imperatives

Beyond rhetoric citing the emergence of state regulation, one should question the economics underlying the sudden change. This is not clear in the reports and debates of the local assemblies or in the Archives of the Ministry of Public Works; they emphasize the political character of the measures. Nevertheless, at least for the period before 1914, another analysis may be proposed, one centered on purely economic mechanisms. After the war, the progress of state interventionism renders this kind of evaluation more problematic.

In my view, adapting the routes to social and urban realities came less from governmental initiative than from a profound internal economic evolution after the crisis of the 1880s. Of course, the public authorities had the ability to use certain pressure tactics from their agreements, and they used them in their conflicts with the CGO—without success for almost fifteen years. But high-level state executives, even those from the radical party, inherited with their charge an old bureaucratic tradition, and they were still quite attached to a very orthodox liberal group that fixed the limits of interventionism. In the long run, interventionism would have led to a profound change in the fiscal system, which the most influential part of society, even among the reformers, did not want. This follows because it was still private initiative that produced transportation during the transition to mass transport utilization. Thus, the maneuvering of the City Council and the state could only be tightly conditioned by the constraints and logic of companies still governed by market laws and free economic mechanisms. The chief engineer of Seine County, in his report of July 20, 1898, came to the same conclusion of powerlessness when he described the defeat of the public authorities in their quest for concessionaries of transport networks inside city boundaries. He ended, however, on a very significant note of hope: "Recently, a large movement of capital towards the electrical tramway companies occurred, which was translated into a large number of companies asking to become concessionaries. The Administration quickly accepted them."[7] This comment sums up the explanatory schema I think most convincingly explains the changes then occurring in Parisian transportation policy.

Down to the smallest detail, changes in the transportation industry mirror general economic evolution. After the crisis of 1880–1890, the economic system became more capitalistic, and the period of recovery saw new technologies around which capital was restructured, causing some economic historians to call this period the "second industrial revolution." Electricity replaced animal or mechanical traction in the Parisian transport system, and this sudden and radical change after years of stagnation demonstrates the accelerated pace of the economy via a new technique. The distribution of concessions in Paris desired by the radicals seemed to go against the general tendency of regrouping and more monopolistic forms. But this apparent concession to political currents, holding through the public services the keys to the national market, constituted in fact the establishment of a vast monopoly over

Paris: "For the public the campaign is directed only against the CGO with the slogan: tramways for 10 centimes," says a note addressed to the Ministry of Public Works in September 1898, "but in the shadows it is a great financial operation encompassing all the beneficiaries of the new concessions including the TPDS and the CGPT (the suburban companies)."[8] Both the Metro, and a few years later the CGO, were integrated into this new system, the origin of which lies in two companies: the French Thomson Houston Company, established in 1893 to operate the patents of the American Thomson Houston in France, Spain, and Portugal; and the General Traction Company, established in 1897 by the Belgian group Empain. In the entire advanced production sector, all the elements were present for a change in capitalism toward vertical and horizontal integration on an international level. It is interesting to note that American companies already controlled the future of the electrical industry, not only in their own country but also in Europe through their European branches, even before the real start of this new industrial era. According to the experienced observer R. L. Colson, integration not only affected the Parisian transportation systems but also provided a means, through standardization, for the "large companies" to conquer the national market.[9] In principle, the law of June 11, 1880, which standardized concessions nationally, should have helped the growth of many small local companies. In fact, it restricted the market to the few companies able to meet the specifications.

Winning the national market for electrical tools and equipment was the main goal, the sole imperative by which the Thomson and Traction companies drew their plans. We can thus suppose that integrating the Parisian transportation companies—now subsidiaries of Thomson and Traction after major capital investment by the two groups—was part of a global strategy and explains the change in routing. Supplying transportation to the capital, Paris, the showcase of World Exhibitions, was in fact a large plus in the advertising campaign for conquering the national market. Before reaching an agreement, the Thomson and Traction companies, in order to hold this position, competed vigorously through small transportation companies whose bids they controlled. It is an open question whether all the small Parisian transportation companies blooming just before the Exhibition of 1900 were concessions to the radical doctrine or whether they were only the aftereffects of the competition between the two companies. The Parisian area is also the main French (and possibly the main European) market for urban transportation. For the Thomson and Traction companies, whose goal was less one of profit from the transportation subsidiary companies than the enlargement of their commercial field of activity, the Parisian area was a very large part of the market. This marketing had social as well as urban consequences in a context of global productivity growth and a changing technological environment. Although railway transportation was not part of the "cutting edge," as electricity was, its productivity rose between 1890 and 1913 ($+$ 1.85 percent a year); part of this growth financed the general social rates and the extension of the suburban routes. These measures resulted more from existing demand than political actions.

When producers entered the transportation market, it changed the administration of the Parisian concessionary companies. Administration of a company was no longer centered on paying dividends to shareholders, but on one word: "Invest." The new leaders used the practice of retaining profits in order to justify the companies' administration—in this case, the Thomson and Traction companies—for which equipment was ordered by the subsidiary through a common administrator. Showing that times had definitely changed, CGO opened an extraordinary fund for its self-financing, and the "real estate" section of the balance sheet was halved between 1910 and 1914: The Parisian company got rid of unproductive elements. The agreements of 1910, unifying and rationalizing Parisian routing, should be analyzed as expressing judicially an existing organization of transportation, based on division of the area between two financial groups, Thomson and Traction: territorial division in the suburbs, functional division within Paris. Beyond the apparent diversity of the different concessions, it is less the radical state than the common administrators of these transportation companies— more "managers" than shareholders—that give unity and coordination to the system the 1910 Agreements ratify, after the competing phase in which each group delimited its territory. Should we accept the interpretation that the City Council imposed the urban Metro? One is tempted to answer affirmatively, for a number of reasons. In 1887, in a generally hostile climate toward the companies, thirty-seven votes were missing for the adoption by the Chamber of the railway companies project. After ten years, the local importance of the Metro was finally recognized, and the City Council presented as a victory what was due to the financial redistribution happening in surface transportation. As for the large railway companies, reasons linked to their internal administration, as well as to a change in Parisian demographics, pressed them to disengage from urban projects within Paris, with no encouragement from the City Council.

Parisian industry was decentralizing geographically; it became more dense at the periphery after the great depression of the 1880s. Transportation opened to direct production as well as to labor, as opposed to the former "Haussmannization," which had brought selective mobility. Factories were connected to railway networks, and the worker had access to mobility when the two large industrial companies colonized all of the Parisian transportation system.

A New Fact: The Automobile

Establishing restricted nationalization after the war changed relations between the state and private interests within the STCRP, the Metro, and the railway companies. Did it mean that, from now on, the very active Parisian transportation policy relieved the state of decisions regarding regulation and routing? This was the indirect argument of leaders of the STCRP when they tried to defend the usual deficits of the company by the "electoral routes" and the "social gifts" that the General Council of Seine County was offering to the people under its administration. But another analysis of the restricted nationalization is possible. It stresses the continuity of the system in place since before the war and follows the same line of overequipping and deficit, though now of private interests. Producers of goods (e.g., the Thomson and Traction companies) had forced their way into the public market by annexing the Parisian transportation companies, which they were supplying with equipment. But the system had its limits: After the war, the balance of the subsidiaries was challenged by structural changes in demand. With restricted nationalization, which transferred some fixed assets and expenses of the new infrastructures to the state, it was now the collectivity that had to assume the role the transportation companies once had. In the Chamber of Deputies in 1931, the socialist J. Moch exposed a veritable blank check given to the equipment industries for selling their production at the expense of the state.

Once more, the large railway companies were the preferred target; the STCRP and the Metro were just cited as examples. The evolution from tramways to buses in Paris and the suburbs illustrates another facet of these companies. The substitution was completed in 1938. It took six years, but the replaced equipment (tramways, rails, electrical installations), which had been totally modernized from 1920 to 1929, was still not fully paid for, the redemption period estimated at thirty years. Part of the General Council valued the equipment at the price of the steel, thanks to "common administrators," and sold to their private subsidiaries, who were reselling it for a high price in the provinces or the colonies. Let us note that substitution was justified in Paris, but certainly not in the suburbs. The tramway characteristics were well adapted to the Paris and suburban routes, along the major roads of the daily commute; the opening of tramway lines by other European capitals supports this. Also, the fixed infrastructure of the tramway orients the direction of urbanization, whereas automobiles are a factor in dispersion; due to the anarchistic urbanization tendencies of that time, this was not unimportant. The tramway disappeared from the day-to-day life of Parisians and suburbanites within six years. Urban transportation reached a second stage in the same manner: yesterday electric traction, today the automobile. The Thomson company still dominated the STCRP board of administration during the postwar reconstruction period. Then representation inside the leadership shifted toward new industries: automobiles and oil, with the International Oil Omnium, the French Fuel Company, the SITA, and the SCEMIA-Renault companies. Doubtless this new balance in the board of the STCRP was the main cause for the rapid and complete replacement of tramways by buses in the Paris area.

Let us not forget that restricted nationalization was still a great step forward in state interventionism. Even if the substitution of trams by buses was strongly suggested by the interested parties, the General Council could still oppose it, but at the risk of alienating voters. Different political and economic stakes forced technical innovation and tighter routing of the network while enlarging the deficit. But the problem now shifted from this seminationalized system that the transportation producers wanted to see disappear. The hypothesis can be offered that restricted nationalization was, in retrospect, a transitional judicial–economic substitution awaiting more efficient transportation techniques —namely, the automobile. When this technological change appeared in the 1930s, there was a significant change in transportation policy: railroad coordination, intended to open the market to a new industry.

The coordination problem was linked to the budget. From 1930 on, a very tight bus network developed in the Paris area. Not only did it take up the gaps in the public service, but it opened parallel lines, competing on the most profitable routes. The number of passengers dropped on state lines, and the deficit grew to become a real financial disaster for the state. The danger came less from small transportation companies offering a rebate than from Renault and Citroen, whose network systematically formed lines parallel to their routes. Coordination measures were one-way: In order to erase the deficit, the state raised fares by 74.5 percent within two years on the STCRP network, and made eighty route changes or cancellations. Ticket sales fell 20 percent in one year

(1938), and the forecast expected another 25 percent fall in 1939. Those measures came during a deflationist policy at the national level to try to solve the crisis, but they only facilitated the switch to private transporters, particularly Renault and Citroen. In widening the scope of the private automobile service, the measures may have been preparing the way for a reopening of transportation to recovered liberal orthodoxy. In 1942, the STCRP should have disappeared after balancing its deficit, but the war stopped the project. As with every advanced industry, competitive aspects of the automobile industry tended toward rejecting regulation. This industry wanted to enter a market free of any protectionist measures, where "every transportation mode should exist for its actual and not regulated value." This hope, expressed by the president of the National Transportation Federation of France in 1934, brings home the debate on liberalism: It is not a problem of doctrine, but, once again, of economic stakes in a recent technology, the innovating aspects of which open a new cycle in the evolution of transportation.

In the light of this economic logic, no contradictory proposal could hold sway, particularly not the planning projects of the General Council. These projects called for the Paris transportation network system to be widened and its social and urban role deepened, not via a restricted nationalization, but through municipal control. To this system the Regional Express Network was added, which corresponded almost to the actual network of the same name. These transportation infrastructures were to be the supports of a planned administration of the regional space around autonomous garden cities, associating all the functions—commercial and residential—of a city. In proposing a solution to city problems by an action on the periphery, the project broke away from Haussmann's centrality. It aimed at balancing the productivist mind-set in vogue from the beginning of the century to benefit a more social, or rather a more human, perspective. The Paris transportation systems were nationalized in the postwar period. The Directing Schema of the Parisian Area was established in 1964, the RER more recently; all these, with little change, were written in the projects of the General Council, and were reported at a later date. Meanwhile, the automobile imposed its own laws on Parisian urban space.

During the entire period when the urban transportation networks were structured, there were relations of dominance between the state, local collectivities, and private investors. These relationships were apparently opposed by the courts, but unsuccessfully. The involved interests never really lost control of a public service closely tied to the global production system. This is the context in which the units running the largest deficits, which the private sector did not want, had to be maintained because of their vital economic functions.

In this process of urban transportation production, the local companies and the state had a mutually dependent relationship, inducing them to modulate their attitude over changes they could not control. The state had only limited interventionist powers and was restricted to the administration of immediate problems. In this way, it imposed certain concessions. Nevertheless, attempts by the General Council to promulgate a coherent policy of urban space administration involving transportation failed.

Perhaps the elected representatives hid in a political framework that shielded them from aspects of a reality they could not control. During the last quarter of the nineteenth century, the Municipal Council of Paris waged an ideological battle against monopolies. It attributed progress to its own actions—actually more economic than social—when the progress was really in keeping with a mainly productive logic. Many "scandals" helped develop a general climate of hostility and, by extension, a somewhat Manichaean understanding of economic mechanisms. In fact, the large national railway companies and the Paris transportation companies established in the reorganization period from 1900 to 1910 were not "profit machines" but technocratic aggregates functioning within the global development process of the Parisian area. As with the automobile, there were both capital and industrial channels, of which the Parisian area was at the same time the support and the main beneficiary. This brings on a more global problem with state intervention in economic mechanisms after World War II. As an administrator of public services, the state had intervened in the sense of regulating a system whose complexity inevitably produced dysfunctions. The state also instigated industrial decentralization, and we are aware of its (minimal) consequences at the level of the region, but the negative results of it for the Parisian area have yet to be evaluated.

Notes

1. A. Cottereau, "Les origines de la planification urbaine en France," Tenth Conference on Urban Politics and City Planning, Dieppe, April 8–10, 1974.

2. L. Lazare, *Les quartiers de l'Est de Paris et les communes suburbaines* (1870).

3. E. Belgrand, *Des travaux souterrains de Paris*, 4 vols (Paris, 1875–1882).

4. On this theme, see M. Roncayolo, *La ville industrielle* (Paris, 1983).

5. E. Beau de Lomenie, *Les responsabilités des dynasties bourgeoises* (1947).

6. Cf. A. Cottereau, "l'Apparition de l'urbanisme comme action collective. L'agglomeration parisienne au début du siècle." Part I. "De Haussmann à la construction du métropolitain," *Sociologie du travail*, no. 4 (October–December 1969). Idem, Part II. "Le Mouvement municipal parisien," *Sociologic du travail*, no. 4 (October–December 1970).

7. National Archives, F14 15031.

8. Ibid.

9. L. Colson, member of the State Council and National Railway Council.

4

Urban Pathways:
The Street and Highway,
1900–1940

Clay McShane

THIS essay focuses on the evolution of urban streets and highways during the spread of the automobile in American cities. It is primarily concerned with the evolution of the urban–suburban highway, the main agent for changing the shape of twentieth-century cities. Attention also is paid to changes in street design, including pavements, and the rise of traffic engineering.

The years from 1900 to 1940 were tumultuous ones for American cities, marked by extraordinary growth, deconcentration to suburbs by residents and industry, and a great increase in heterogeneity as migrants from Europe and the rural South flocked to cities by the millions. Before the Great Depression of the 1930s, urbanites enjoyed an incredible economic prosperity. Middle-class Americans fled the tumult they saw downtown. They sought to spend their increased incomes and fulfill Jeffersonian fantasies by acquiring detached homes amid the ersatz rural scenery of the new, middle-class suburbs.

Facilitating such deconcentration was probably the primary goal of urban politics during this era. Progressive reformers warred constantly with transit companies over costly suburban extensions. The rise of zoning and the antiannexation movements of the 1920s facilitated the growth of class-segregated and land-use-restricted (no industry) suburbs for the growing middle class. Later, President Franklin D. Roosevelt's New Deal brought the first major federal intervention into American urban life. New Deal planners, while developing some public housing, preferred to focus on suburban "greenbelt" towns.

These were also the years in which automobile ownership spread throughout American society. In 1900, there were perhaps 8,000 cars in the United States, 1 for every 9,500 Americans. In 1940, there were

TABLE 4.1.
The Growth of the Automotive Industry, 1900–1940

	Automobiles Registered in U.S. (in 1,000s)	Persons Per Automobile
1900	8.0	9,499.0
1910	458.3	201.0
1920	8,131.5	13.0
1930	23,034.7	5.3
1940	27,465.8	4.8

Source: U.S. Bureau of the Census, Historical Statistics of the United States, Colonial Times to 1970 (Washington, D.C., 1975), pp. 716, 719.

27 million automobiles registered, 1 for every 4.8 Americans (see Table 4.1).[1] By 1923, Los Angeles, the premier city of the auto age, had 1 motor vehicle for every 3 residents.[2] Even in Chicago, an older, transit-dependent city, there was 1 auto for every 11 residents, and some observers were claiming that the death of the Loop, Chicago's central business district, was near.[3]

The spread of the auto was unpredictable. It seems to have occurred in four distinct stages, and only a few visionaries, never urban planners, foresaw the next. Even transit companies, which had the most to fear from the new technology, believed into the late 1920s that the auto would fade away. The first stage of the automotive age began about 1894 when the famous Chicago Times-Herald race introduced the American public to the internal combustion auto. At this time, the auto was primarily a plaything of the very rich. It appeared to be just another passing craze, like the bicycle, in the mid-1890s. As late as 1905, there were only 77,000 cars in the United States, fewer than 1 for every 100 Americans. Except for a very few speed and safety regulations, public policy took little note of the new technology. Woodrow Wilson, then president of Princeton University, did suggest banning the auto, but only because this new status symbol promised to exacerbate tensions between the rich and the poor.[4]

Beginning about 1905 or 1906 and lasting until 1914, a second stage of urban automobility appeared. Led by Henry Ford, auto manufacturers reduced prices for reliable vehicles below $1,000 for the first time, putting them within reach of upper-middle-class families. Cross-country tours, the vital role of autos in bringing aid to San Francisco after the devastating 1906 earthquake, and the development of motorized taxicab and delivery services suggested utilitarian applications of the automobile. Photos of downtown streets of the time show a few autos parked by the curb. Some urbanites, especially doctors, who still made house calls, began to use the auto for making their rounds. A few individuals started to use cars to commute downtown, although most urban drivers probably used their cars purely for pleasure. With increased auto usage, a rapidly growing accident rate developed, and safety regulations became common; there were tests for driver licenses, road-speed limits, anti-drunk-driving laws, and ordinances requiring headlights. Traffic police began to appear at busy intersections.

In the summer of 1914, frequent traffic jams occurred in major cities for the first time, another sign of the advent of the auto age. That year, Henry Ford opened his famous assembly-line plant at Highland Park, Detroit. By the early 1920s, he would knock the price of a new Model-T below $300, well within the reach of most American families. Traffic jams in the summer of 1914 reflected not only increased pleasure ridership but also the appearance of a new urban travel mode: the jitney. These vehicles, often Model-Ts whose owners had been laid off by that summer's recession, plied urban streets on quasi-fixed routes, charging passengers the same 5-cent fare as trolleys. Jitneys offered more comfort: riders got seats, the vehicles attained higher speeds because they could dodge through traffic, and there was more frequent service. Jitneys represented the first mass application of the automobile to the urban journey to work. Although streetcar companies got municipalities and even state governments to ban the jitneys, they were a harbinger of things to come.[5]

The period from 1914 to 1923 saw the first massive traffic jams, downtown streets narrowed by parked cars, and a surge in traffic fatalities. But most urban residents still believed that the changes were temporary and that the auto would be purely a recreational vehicle. As late as 1919, Henry Ford contemplated leaving the auto industry for street railways.[6]

By 1923, it was clear that the auto was becoming the dominant mode of urban transportation. In that year, for the first time, auto manufacturers sold more closed than open cars, a sign that purchasers no longer wanted vehicles primarily for pleasure driving. Moreover, across the nation, transit ridership declined for the first time since electrification.[7] Shortsighted streetcar corporations believed the decline was temporary,[8] but by that date perceptive urban planners, municipal engineers, and elected officials recognized the dominance of the private automobile. Even such reformers and visionaries as Lewis Mumford, Le Corbusier, Charles Beard, and Edith Abbott, who understood the adverse impact of cars on inner-city environments, favored the automobile.[9] Adapting cities to the auto became the principal public policy issue.

Severe administrative and financial obstacles limited urban highway and traffic planning movements in the years from 1900 to 1940. The federal government did create a Bureau of Public Roads in 1906, but it served a limited coordination and research function. Its goals may be judged by the bureau's location—it was part of the Department of Agriculture. Constitutional doctrine had kept the national government away from internal transportation improvements since the 1830s. Certainly Americans never thought of highways for defense mobilization, a key reason for national governments to build highways in continental Europe.

State governments became heavily involved in road building in the early twentieth century. By 1914, every state charged auto registration fees, and most imposed gasoline taxes to finance road improvements.[10] State and county governments built tens of thousands of miles of roads from 1900 to 1940, but less than 10 percent of state highway expenditures were in urban areas.[11] Throughout this period, rural interests were wildly overrepresented in state legislatures. To cite one notorious example, West Greenwich, with a population of 481, had one senator in the Rhode Island legislature, the same as Providence, with a population of 224,326.[12] Under such circumstances, state highway agencies used funds to build secondary (two-lane, farm-to-market) roads, rather than four-lane, intercity roads. State agencies rarely returned the gasoline taxes collected in cities for urban use.[13]

Constitutionally, cities are subjects of state governments.[14] States usually allowed municipalities only the property tax for revenues, rarely permitting them to keep user fees, such as auto registration fees or a gas tax. Most states allowed cities to borrow money only up to 5 percent of their total assessed value. These debt limitations greatly hampered municipal public improvements, such as highways.[15] Finally, central cities could not annex nearby suburbs. Metropolitan areas became politically fragmented throughout the twentieth century.[16] In 1961, Robert Wood estimated that there were more than 1,400 government entities within a two-hour drive of Times Square. This kind of political balkanization made coordinated highway planning just about impossible.[17]

Ironically, the basic design concepts for good urban highways existed before the automobile. Good urban highways depend on three major design elements: (1) limitation of access so that traffic entrances from bordering roads are kept to a minimum; (2) grade separation of traffic so that vehicles cross each other at different levels; and (3) exit and access ramps that allow smooth traffic flows. Cities had built dozens of miles of parkways before the development of the auto. The famous *Chicago Times-Herald* Race of 1894 was conducted on Chicago park roads, and Henry Ford used similar roads in Detroit as a testing ground for his early experimental vehicles.[18] These nineteenth-century highways are the first stops on our tour of urban roads.

The Urban Parkway, 1870–1920

The term *parkway* and the first such roads constructed were concepts developed by the late-nineteenth-century landscape architect Frederick Law Olmsted. Olmsted's primary concern was to provide rural land-scapes and recreation for a group he described as "country boys," who had moved to the city and become wealthy.[19] In his pioneer Central Park (begun in 1857), Olmsted established a series of pleasure drives over which the wealthy might parade their carriages as focal points of the park. He viewed such drives, through an ersatz rural scenery, as per-haps the most important recreational use of the park. New York's Parks Commission, at Olmsted's request, prohibited access for all vehicles ex-cept the private carriages, which only the very wealthy could afford. Cross traffic through the park traveled in an open cut on a separate grade, the first American example of such grade separation. But park drives by themselves could deal only partially with this recreational–transportation need. When Olmsted completed Prospect Park in Brook-lyn after the Civil War, he recommended that the suburban city build a series of parkways, roads with large strips of parkland on either side, to extend parklike amenities to a broad region.[20]

Olmsted's parkway concept served three major functions: recreation, high-speed travel, and amenities for upper-class neighborhoods. Since park agencies banned public vehicles such as omnibuses, wagons, and street railways, parkways provided a pleasure drive for the wealthy owners of private carriages. Olmsted believed that such parkways would maximize the benefits of public parks by bringing recreational land along the margin within reach of a larger population than would have access to a centrally located park. He believed that such roads would also facilitate downtown travel by private carriage, since there would be no slow-moving traffic and few cross streets cut through the roadway at grade. Park strips on either side of the roads would also limit access by local traffic. Traditionally, English common law had provided that governments must provide access to all roads for abutters, but if the new-style parkway had strips of park, not residences or businesses, directly alongside it, access could be limited. Such limitation of access was the key to the parkway concept.[21]

The New York State legislature in 1877 authorized the first of these new roads, Eastern Parkway in Brooklyn. The parkway showed some variations from Olmsted's original concept. There was relatively little isolation from cross traffic, since grade separations were prohibitively expensive. Eastern Parkway followed a straight line, as part of a grid-iron system, instead of winding its way around topographical features, a layout that Olmsted preferred. The parkway incorporated the right, probably borrowed from Commonwealth Avenue in Boston, to control land use for abutters. The Parks Commission could regulate land development adjacent to the parkway by requiring setbacks from the street line, hence widening the parklike effect. The commissioners could prevent the incursion of commercial buildings and require that buildings place ancillary structures, such as outhouses and stables, behind them, rather than on the side facing the parkway. According to Olmsted, these rules, combined with the parkway's aesthetic appeal and traffic restrictions, guaranteed that the area would be a segregated middle- and upper-class enclave.[22] Eastern Parkway was actually more a boulevard than a parkway.

The parkway idea rapidly spread to other cities as a way to foster the development of middle-class neighborhoods and create the segregation by land use and social class that Victorian Americans so avidly sought for their cities. Olmsted's best-known park *system*, the Back Bay complex of parks and parkways in Boston, embodied a new concept: stream preservation. Olmsted planned the Jamaicaway, Fenway, and Arborway segments of that system to create banks for a slow-flowing, marshy stream—the Muddy River. Olmsted believed that building the parkway would eliminate an unhealthy (at least according to nineteenth-century medical ideas) swamp and that putting strips of park along the stream would prevent its use as a dumping ground for industrial wastes, thus assuring the residential character of the neighborhood. Protection of such rural streams soon became an important element of parkway planning. Olmsted went on to a career of developing park and parkway systems all over the United States, with notably successful examples in Buffalo, Chicago, Milwaukee, Louisville, and Washington, D.C. An entire landscape architecture profession, committed to the holis-

TABLE 4.2.
Parks in Major American Cities, 1890–1916

City	Park Area (in Acres)			Park Drives (in Miles)		Largest Park (in Acres) 1912	Miles From City Hall	Miles of Connecting Parkways, 1916
	1890	1900	1916	1890	1916			
New York[a]	1,762	6,909	7,740	47	55	1,756	8	33
Chicago	799	2,151	3,870	87	51	543	8	63
Philadelphia	3,025	4,044	5,500	33	50	3,526	2	0
St. Louis	2,127	2,176	2,479	36	34	1,380	4	0
Boston	1,130	2,618	3,640	6	14	527	5	32
Baltimore	803	1,136	2,278	19	25	674	3	3
Cleveland	93	1,326	1,702	6	43	292	5	12
Buffalo	638	1,026	841	5	17	365	5	15
San Francisco	500	1,192	1,718	17	19	1,013	2	1
Cincinnati	539	530	2,500	12	20	1,000	6	0
Detroit	380	1,055	835	20	29	707	4	12
Washington	539	3,597	3,719	7	25	1,606	2	0
New Orleans	432	552	611	5	6	280	5	0
Minneapolis	1,361	1,553	2,567	17	26	586	3	26
Milwaukee	51	435	721	6	9	150	4	5
Kansas City	0	279	2,120	0	18	1,334	9	21

Sources: U.S. Department of Labor, "Statistics of Cities," *Bulletin of the Department of Labor* 30 (September 1900): 954; U.S. Department of Commerce, Bureau of the Census *General Statistics of Cities: 1916* (Washington, D.C., 1917), p. 53; U.S. Department of the Interior, Census Office, *Report on the Social Statistics of Cities in the United States at the Eleventh Census: 1890* (Washington, D.C., 1895), pp 88–92.
[a]Includes Brooklyn.

tic environmental–residential–recreational–traffic planning embodied in his ideas, sprang up. Horace W. Cleveland helped design major parkway systems in Chicago and Minneapolis in the 1880s. In the 1890s, George Kessler built perhaps the most elaborate parkway system in the United States in Kansas City. Charles W. Eliot, Jr., in Boston developed the first permanent metropolitan governmental authority, the forerunner of an important twentieth-century political form, to create a park/parkway system in Boston's suburbs. By 1900, perhaps 200 miles of urban parkway existed in the United States (see Table 4.2).[23]

Nevertheless, cities stopped building these successful systems early in the twentieth century for complex ideological, political, and aes-

TABLE 4.3.
Articles Dealing Largely with Parkways, *American City Magazine,* **1909–1939**

	Total Articles	General Articles	New York Region	Other Cities
1909–1914	0	0	0	0
1915–1919	1	0	0	1
1920–1924	4	0	2	2
1925–1929	5	2	2	1
1930–1934	11	4	4	3
1935–1939	6	2	3	0

thetic reasons. Chicago added nothing to its parkway system between 1890 and 1925. Conservative urban businessmen had always opposed the parkways. For example, in 1884, a coalition of New York reformers and businessmen, including Abram Hewitt and Theodore Roosevelt, opposed Olmsted's plans for a Bronx park system. They argued that the money would be better spent on schools or docks.[24] After consolidation in 1898, New York added almost no new park acreage until the 1920s. Municipalities could see little reason to expand park and parkway systems when the new pleasure vehicle, the auto, allowed access to nature in the rural countryside. Newer cities, which boomed in the early twentieth century, completely ignored the parkway tradition. Los Angeles, for example, never built any parkways. Detroit, in the boom days of the auto industry, never added to its boulevard and parkway systems, despite the fact that they had proven to be valuable testing grounds for the new vehicle. Some older cities, New Orleans and Cincinnati among them, never built any parkways. Before 1925, few municipalities thought of special auto highways, a function that these parkways admirably served. In the ten years between its first issue in 1909 and 1919, *American City Magazine,* perhaps the most influential and comprehensive periodical for city officials, mentioned parkways only once (see Table 4.3).

There was a revolt against the parkway movement. In the final wave of parkway planning in the late 1890s, planners concentrated on providing special roadways, reserved not just for private carriages but for high-speed trotting teams, the so-called speedways. Boston built one such road, the Charles River Speedway, which even had exit ramps to allow the fast-moving trotters to slow down before rejoining traffic.

These ramps were the third major element in modern highway design, along with the grade separation and limited access pioneered by Olmsted. In 1897, New York opened perhaps the best-known speedway, the Harlem River Speedway in upper Manhattan, built at a cost of $2.25 million. There was substantial public opposition to this speedway. Middle-class New Yorkers had been upset by attempts of park authorities to ban relatively humble bicycles in the 1890s because they might interfere with private-carriage operation. The Harlem River Speedway seemed the final straw. One disenchanted Gothamite wrote his newspaper:

> Thus the trotting fraternity, a small but wealthy and powerful class, secured the great public improvement, pre-empted the river bank and got the exclusive benefit of a vast public expenditure while the crowded masses of the city cried in vain for sufficient parks, schools, baths and lavatories.[25]

The response of the harness drivers, that such projects were acceptable because they "concede a right to a large if neglected class of rich,"[26] could hardly be expected to stir up a wave of popular support for parkways. Courts also became disenchanted with the parkway concept and refused to allow park authorities to ban such vehicles as bicycles. In Massachusetts and some other states (although not in New York), state courts insisted that abutters be allowed access to the road, park strip or no.

Urban Roads and Streets, 1900–1925

The primary reason for turning away from limited-access parkways, however, was the development of the new City Beautiful movement, which dominated American city planning for the first quarter of the twentieth century. This concept, first publicized by Daniel Burnham for the Chicago World's Fair of 1893 and embodied in his famous *Plan of Chicago* (1909), moved against the romantic scheme of Olmsted and his followers. The Chicago plan substituted grand boulevards in the Haussmann tradition for parkways. It focused on downtown improvements, not suburban development. To be sure, Burnham proposed a highway system for suburban Chicago, but this was a minor element of the plan.[27]

Burnham-style boulevards provided incredibly wide (up to 180 feet) streets, apparently using Boston's Commonwealth Avenue, an 1870s design, as a model. Numerous cities built such boulevards in the late nineteenth century, but Burnham popularized them. Examples of such streets included the Grand Concourse in the Bronx (New York City), Boulevard of the Allies in Pittsburgh, and Delaware Avenue in Buffalo. Such roads offered numerous advantages to suburban developers. They had narrow park strips in the center or along the sides. Lot deeds (e.g., along Commonwealth Avenue) included restrictive covenants requiring that houses be set back from the streets, limited house heights, and prohibited commercial uses. Suburban realtors preferred boulevards to parkways as centerpieces for their subdivisions. Boulevards used less land, easily conformed to the rigid gridiron design developers preferred, and created an appropriately natural atmosphere when trees were planted in the medians and setbacks. Prohibitions on trolleys and wagons along such streets protected well-to-do residents from the intrusion of commercial properties or lower-class housing. After World War I, new zoning ordinances in American metropolitan areas reinforced such land-use segregation.[28]

For the largely recreational auto use that predominated in the period, boulevards seemed the ideal solution. Their great width minimized traffic jams and allowed slower-moving traffic to keep to the right. Prohibitions on cumbersome commercial traffic fostered faster travel. Long, straight vistas and parklike design allowed boulevards to serve as pleasure drives in their own right. But by the early 1920s, even these very wide avenues were beginning to jam up. At higher rates of auto ownership, even their great widths were inadequate to handle, not just pleasure riders, but, increasingly, commuters. Abutters used the side lanes to park cars, reducing capacity. Since cross streets were at the same grade as the boulevard, crossing vehicles and those slowing to turn reduced movement further.

Urban governments preferred to respond to the post-1914 traffic jams by regulation rather than highway building. In large part, this was sensible. Debt limits and reliance on the property tax greatly limited the supply of capital available. Even when capital was obtainable, in the rare cases where the need for better intracity highways was perceived, urban voters were not enthusiastic about increasing their property taxes to pay for roads heavily used by suburbanites. City planners in rapidly growing, auto-glutted Detroit proposed a series of semilimited-access roads labeled "superhighways." In a 1924 referendum, Detroit's voters, by a wide margin,[29] refused to approve the new roads. The prominent zoning lawyer Edward Bassett coined the word "freeway" in a 1924 article proposing limited-access highways for both commercial and passenger traffic. Planners widely publicized, but never implemented, Bassett's concept.[30]

Cities did spend massive amounts of money on street improvements, but they concentrated on immediate needs. Automobiles demanded the massive repaving of city streets. Expensive street widenings could eliminate the worst bottlenecks. Removing railroad grade crossings was a top priority for both mobility and safety. Some cities spent money on wide downtown streets, such as the new civic centers in Denver and Cleveland, or on grand boulevards like downtown Philadelphia's $40 million Benjamin Franklin Parkway, but they built them for aesthetic reasons, not mobility. Improving existing street systems absorbed most of the available capital. Nobody anticipated the continued growth of auto use or the rise of auto commutation that would soon overcrowd the newly paved and widened streets.[31]

Urban planners focused on traffic regulation rather than highway construction. In part, this was because regulation provided a cheaper way to deal with traffic; in part also, no urbanite envisioned that massive increases in auto traffic would overwhelm such short-term fixes.

Safety issues demanded such regulations in any case. The early auto drivers, most of them with no prior experience, created carnage on the streets, for states required little, if any, testing for licenses. There was no social, or often legal, taboo against drunk driving. Driving habits were horrible. At one Chicago railroad crossing, fifteen autos hit trains at the seventh freight car or later over one twelve-month period. Police departments, initially without vehicles, could not enforce speed limits. One frustrated Chicago foot policeman went so far as to shoot out the tires of a speeding car that refused to be flagged down. There was no agreement on such simple matters as hand signals for turns and rights-of-way at intersections. Pedestrians rarely thought to look before crossing a street or to cross at intersections. In densely packed inner-city neighborhoods, streets had served as playgrounds and social gathering places from time immemorial. In 1910, more than 6,000 Americans were sacrificed to the automobile; there was 1 motor vehicle fatality for every 68 cars, a rate twelve times greater than thirty years later (see Table 4.4).[32]

Motorists bitterly opposed remedial measures. Traffic police, speed limits, stop signs, and traffic lights were thought to be unwarranted intrusions on personal freedom. State courts sometimes supported these views. William Phelps Eno, who wrote the earliest American treatise on traffic regulation, typified the early motorists. Eno, a wealthy New York auto buff, did not like speed limits and adamantly opposed one-way streets and stoplights. He favored the construction of traffic rotaries, which he argued would not slow the flow of traffic. Even at Fifth Avenue and 42nd Street in New York City, probably the busiest intersection in

TABLE 4.4.
Motor Vehicle Fatalities in the United States

Year	Motor Vehicle Fatalities	Motor Vehicles per Fatality
1910	6,700	68
1920	12,500	650
1930	30,900	745
1940	33,539	819

Source: U.S. Bureau of the Census, Historical Statistics of the United States, Colonial Times to 1970 (Washington, D.C., 1975), p. 719. This source lists fatalities only in five-year averages, so the data are rough approximations.

the country, Eno proposed erecting a signal tower in the center, cutting corners at the curbs and keeping traffic moving by forcing everyone to circle to the right of the tower. Eno's influence lingered long after modern principles of traffic regulation emerged. He endowed the nation's first school of traffic regulation at Yale University and insisted that it teach only the rotary system of traffic control. The large numbers of dangerous traffic rotaries still in use in New England are a dubious tribute to Eno's legacy.[33]

Mainstream traffic engineers also emphasized mobility for autos. According to the historian Paul Barrett, Chicago's traffic engineers measured success by timing the speed at which traffic moved through the Loop.[34] Engineers analyzed traffic lights, rapidly installed after 1917, in terms of their ability to ease traffic flows. By 1923, synchronized light systems were available. Parking bans were a favorite panacea: The engineering firm of MacClellan and Junkersfield estimated that Washington, D.C., could triple its mileage of four-lane roads by imposing a parking ban. New York City inaugurated one-way streets in 1917. Stop signs were first used along urban boulevards to stop cross traffic and increase speeds on these major radial arteries. This gave suburb–CBD traffic the right-of-way over local traffic and pedestrians.[35]

Adopting these techniques facilitated mobility, but planners never envisioned that the newly regulated streets would attract so much additional traffic as to re-create the jams. Trolley companies supported the traffic engineers, since downtown congestion hurt their business. In some cities, the companies actually paid for the first traffic lights. But traffic engineers opposed measures that would have allowed a safer, smoother flow of trolleys. The electric streetcars never got the right-of-way on their own tracks. Cars making left-hand turns always backed up trolleys, but few cities banned left-hand turns in their downtown areas. Traffic engineers frequently opposed safety islands for waiting transit passengers because they slowed down cars. By 1923, the average speed of Chicago's trolleys had declined 25 percent. Although urban rail systems moved more citizens than autos, traffic engineers (with popular support) treated transit vehicles the same as single-passenger vehicles. Beginning in the 1920s, transit lines switched to trolley-buses and buses that had the flexibility to move in traffic.[36]

Other urban policymakers also kowtowed to the auto. Landscape architects began to design scenery that would be attractive at 25 miles an hour, not at a walk. Park planners turned away from the large, natural parks that Olmsted created, reasoning that the automobile opened rural spaces to urbanites. These planners emphasized playgrounds, realizing that automobiles had destroyed the traditional play function of urban streets. Of course, this change reflected the change from passive to active recreation among the urban middle class as well.[37]

The Parkway Revival, 1925–1940

The construction of limited-access parkways did not completely disappear between 1900 and 1920. William Vanderbilt, heir to the railroad fortune, built a limited-access, grade-separated parkway some 20 miles long in 1905–1907 to provide access to his estate on Long Island, the first road in the United States explicitly designed for autos. This motorway was an exotic exception to the norm. It cut through the wealthy estates on Long Island's North Shore and their owners controlled it and for the most part excluded the general public. The road was little more than a private racecourse for a few car-owning millionaires. The Du Pont family built a similar road in Delaware. In Washington, D.C., a limited-access parkway was built through Rock Creek and provided downtown access for newly developed suburbs in the northwest part of the District and in Montgomery County. A few other cities probably constructed short parkways during the period.

The best-known parkway built during this relatively inactive period was the Bronx River Parkway, which ran from the South Bronx to White Plains in suburban Westchester County. This road sparked the revival of the parkway idea in the 1920s. Frederick Law Olmsted had proposed a parkway along the Bronx River when he authored a comprehensive plan for developing the Bronx, newly annexed by New York City, in 1877.[38] He feared the stream might attract industrial plants that would use it for an open sewer as the Central and North Bronx developed. By acquiring strips of park on both sides of the river and building a parkway, New York City might force industry in the new borough to seek sites along the larger, more rapidly flowing East or Harlem rivers, which could flush away heavy pollution. Municipal officials ignored this element of Olmsted's plan. Later, the New York Zoological Gardens (better known as the Bronx Zoo), located on the Bronx River, discovered that its animals were dying after drinking water from the stream. Madison Grant, a wealthy New York philanthropist and Director of the Zoological Society, lobbied the New York State Legislature to have a parkway built to protect the stream.[39] The legislature authorized the parkway in 1907, although construction did not begin until 1913 and completion lagged until 1923, a measure of its low priority. New York City paid three-fourths of the cost, although most of the mileage of the route was outside the city, as an obvious subsidy to the suburbs. The parkway included all three elements of modern highway design: limited access, grade separation of cross streets, and exit ramps. This efficient traffic mover cut travel times to Westchester County in half. As a result, real estate values along its route, especially its upper reaches, increased enormously.[40]

The Bronx River Parkway led directly to the parkway revival of the 1920s and 1930s. It received widespread national publicity among urbanists. The engineer for the parkway, Jay Downer, and its landscape architect, Gilmore Clark, wrote articles for major professional journals publicizing the old concept as a new way to deal with a pressing urban problem, traffic.[41] The new style appealed especially to suburban land developers, at first in New York City, where traffic congestion was greatest. Westchester County, then an undeveloped rural area north of the city, was first to act. The County Park Commissioners hired Clark and Downer to begin construction of the new parkway system for Westchester, including the Saw Mill River Parkway in its western section, the Hutchinson River Parkway in the east, and an extension of the Bronx River. The county built the new roads between 1922 and 1930. The completed system included 250-foot right-of-ways with only 40 or so feet devoted to the roadway. There were few road crossings at grade, and almost all exits were of the cloverleaf type. One major safety problem marred the road: Only short stretches of the parkway had median dividers. The county recouped costs by charging tolls, and the new system proved a spectacular success. Initially it attracted recreational traffic, New Yorkers looking for a ride in the country, then commuters, as Westchester grew. Real estate values in the county tripled, and grew to an even greater extent in the vicinity of the new roads. Travel times were cut one-third, and traffic fatalities on the new roads were fewer than on older highways without traffic segregation and limited access.[42]

Long Island imitated the Westchester systems, building its own parkway system between 1923 and 1928, a system that received enormous public notice, in part because of the spectacular methods of its builder, Robert Moses. Moses developed a new governmental form, the special authority, which collected tolls to pay back its bonds. He took land by a seldom-used New York statute, of dubious constitutionality, that allowed the appropriation of land prior to condemnation. These tactics aroused great hostility from Long Island landholders, but made Moses and Al Smith, the governor who had appointed him, enormously popular in New York City. New Yorkers saw the roads as opening new suburban territory for settlement, thus relieving housing congestion and providing access to Long Island's hitherto inaccessible beach resorts.[43]

The Long Island roads were more carefully designed than their predecessors in a number of ways. Long Island parkways had far fewer accidents than their Westchester counterparts because Moses's engineers eliminated all grade crossings of other roads and provided a generous median strip to reduce head-on collisions. The parkway system provided first-rate recreational amenities, not only because of the wide park area along its right-of-way, but also because Moses opened a number of new parks at the end of the parkway system, most notably Jones Beach, perhaps the best-known public beach in the United States.

As pleasure drives, without regard to recreational facilities, these and other parkways were enormously successful. Moses also wanted to ensure that Long Island would become a middle- and upper-class enclave, that his parkways would enhance land values. As had the older Olmsted-style parkways and the Westchester system, the Long Island roads excluded public transportation, effectively reserving the new suburbs for auto owners. Moses feared that some future generation of politicians might reverse this ban and eliminate the zoning restrictions along the new highways. To counter this, he developed an innovative engineering solution: Ostensibly for aesthetic purposes, the Long Island State Parks Commission engineers used only arch bridges for overpasses. The arches had a height of only 9 feet at the curb, 2 feet lower than the height of buses, thus guaranteeing that the roads would remain permanently for auto use only.[44]

After the proven success of the New York area roads, the parkway became a much ballyhooed urban design alternative in the late 1920s and early 1930s. By 1930, such imitative roads as Lake Shore Drive

(Chicago), the George Washington Parkway (Washington, D.C.), Niagara Parkway (Niagara Falls), Mystic Valley Parkway (Boston), and Palisades Interstate Parkway (New York–New Jersey) were under way. None was as free-flowing as Moses's Long Island parkway system, but all utilized its principles heavily.

Lewis Mumford and Brenton MacKaye's famous 1931 *Atlantic Monthly* article "Townless Highways for the Motorist, A Proposal for the Auto Age, "[45] spread the parkway idea nationwide. Mumford attacked the state and federal highways of the 1920s, such as Route 1, the major artery from Maine to Florida. Because access was not limited, suburban sections of the road had spawned ugly roadside strips of garages, fast-food stands, etc. On rural stretches, frequent billboards screened out the scenery. Motorists bought houses in subdivisions along the new highways, jamming them as soon as they were built. Mumford and MacKaye noted that speeds often returned to those of the preauto age: "Like the fly, the motorist buzzes his wings vigorously; but his feet are stuck to the flypaper of the old-fashioned highway; a spavined horse could often travel as fast as a 120-hp car."[46] They called not only for limited-access parkways on the Bronx River model but for a complete redesign of residential street patterns based on auto use. Here they cited the 1929 New Jersey subdivision of Radburn, which discarded the traditional grid-iron for a pattern of "superblocks" embodying a few through streets with houses on curvilinear cul-de-sacs. This isolated through traffic, even cars taking shortcuts, from residential streets and slowed speed on those streets considerably.[47] MacKaye and Mumford proved good prophets. The limited-access parkway would dominate future highway planning. Radburn-style superblocks have dominated suburban tract planning since World War II.

The parkway movement enjoyed its last flowering in the 1930s. Cities and states began to make major investments in parkways. With the coming of the Great Depression, highway building became a popular form of countercyclical spending. The park components of parkways appealed to the conservationist instincts of the New Deal, an important consideration, since much of the funding came from such New Deal relief programs as the Works Progress Administration (WPA). St. Louis, San Francisco, Los Angeles, Boston, Chicago, Newark, and Baltimore all actively sought WPA funds for parkway construction. Not all cities partook. Despite the recommendation of planners for parkways, Philadelphia's mayor, J. Hampton Moore, refused to accept any WPA money or to engage in deficit spending.[48] WPA funds also financed several partially suburban intercity parkways, most notably the Baltimore–Washington Parkway and the Merritt Parkway in Connecticut. The latter, which connected Hartford to the Westchester County system, was a model of WPA-style planning. To provide jobs for stone masons and other workers, each of the road's thirty-three overpasses used a different design. For aesthetic purposes, the reinforced concrete bridges received stone facing.[49] New Deal–built greenbelt towns disseminated Radburn-style subdivision planning.

Under the aegis of Robert Moses, city planners, while still using the term *parkway,* were beginning to move toward the idea of urban expressways. Moses used his hefty slice of WPA funds to extend the Long Island and Westchester parkway systems downtown. The Henry Hudson Parkway connected Manhattan's West Side to the Westchester system. Across the island, the new East River Drive fed the massive new Triborough Bridge, which connected to the Northern State Parkway on Long Island. Moses built similar connections to Brooklyn. These roads differed considerably from earlier Moses highways. Although commercial traffic was not allowed, park strips dwindled to as little as 5 feet in places. Moses still argued that this provided a recreational resource valuable enough to continue limiting the access of abutters. Elsewhere, planners gutted such existing parks as Boston's Esplanade and Chicago's lakefront park to provide mobility. Ironically, these highways often ran through relatively densely populated areas, where wide park strips would have been more important neighborhood resources than in low-density suburbs.

The parkway planning concept died with World War II. Planners who served in communities, as opposed to theoreticians like Lewis Mumford, had never really seen the concept as anything more than a device to circumvent the common-law requirement of access. Rural-dominated legislatures still preferred unlimited access. Parkways were built only where tolls could be charged. Their planners operated as highway builders rather than park planners and were concerned primarily with mobility. Here, too, Moses led the way. In 1938, he proposed a Port Chester–Pelham freeway to overcome jams caused by truck traffic on the old U.S. Route 1. He later proposed a comprehensive system of freeways for New York to facilitate freight movement, including the famous and never to be built Cross-Manhattan Expressway, which became a major political battleground in the 1960s.[50] Moses took advantage of World War II by advocating the construction of many such urban limited-access, but not parked, roads for civil defense purposes, claiming that they would facilitate evacuation in case of air raids, precisely the justification that the federal government would later use in permitting urban extensions of the interstate system.[51]

Moses was not alone in these ideas. Highway agencies elsewhere in the nation, impressed by parkways as traffic movers, sought ways to take advantage of the mobility that their limited-access characteristics offered, without having to construct expensive abutting parks. The business interests that dominated politics in so many American cities were becoming more interested in the movement of freight than people. Pennsylvania found one solution for intercity roads; it completed a toll road, the Pennsylvania Turnpike, between Harrisburg and Pittsburgh in 1940. Toll roads also offered limited access. The charters of early-nineteenth-century privately owned toll roads, on which the turnpike was based, had allowed limitation of access; otherwise, the right to collect a toll would be valueless. Travelers could merely have entered the road after one toll booth and exited before the next. Toll roads provided another way to create high-speed roads. Such cities as New York and Chicago built them extensively in the 1940s. Toll roads also solved political problems. Roads built with such user fees were not subject to urban debt limits, and so urban governments could build them without control by rural-dominated state highway departments. By 1940, both the New York and California legislatures had removed control of access from the courts by statute laws that allowed the construction of limited-access highways, thus eliminating the common-law requirement of free access, without the often spurious justification of a park strip.[52]

Parkways of the Olmsted or early Moses variety were magnificent examples of comprehensive planning. They met important recreational, environmental, and transportation needs. Moreover, they were politically feasible because they served a local recreational need as well as a citywide transportation function. Nobody who planned them made any pretense that they served other than elite groups in cities. But, in the long run, other groups have also benefited. Some of the new parkway-served suburbs of the 1890s are today's ghettoes. Latin children now play soccer on the park strip of Pelham Parkway in the Bronx where Anglo, middle-class children played football eighty years ago. Certainly the fate of the urban interstates planned in the 1950s and 1960s might have been different had they aimed at providing amenities for neighborhood revitalization as well as roadways for high-speed travel. They were children of the parkways only in their traffic-moving concepts.

Notes

1. Bureau of the Census, *Historical Statistics of the United States, Colonial Times to 1970* (Washington, D.C., 1975), pp. 716, 719.

2. National Automobile Chamber of Commerce, *Automobile Facts and Figures, 1923* (Detroit, 1923), p. 23.

3. Paul Barrett, *The Automobile and Urban Transit: The Formation of Public Policy in Chicago, 1900–1930* (Philadelphia, 1983), p. 130.

4. The discussion of the spread of the auto in this and succeeding paragraphs derives largely from James J. Flink, *America Adopts the Automobile, 1895–1910* (Cambridge, England, 1970) and John Rae, *The American Automobile* (Chicago, 1965).

5. George Hilton and Ross D. Eckert, "The Jitneys," *Journal of Law and Economics* 15 (October 1972): 293–325.

6. Allan Nevins, *Ford, the Man, the Times, the Company*, 3 vols. (New York, 1954), p. 107.

7. U.S. Bureau of the Census, *Historical Statistics*, p. 721.

8. Barrett, *Auto and Urban Transit*, pp. 174–175.

9. Lewis Mumford and Benton MacKaye, "Townless Highways for the Motorist: A Proposal for the Automobile Age," *Harper's Monthly*, August 1931, p. 347; Le Corbusier, *The City of Tomorrow*, tr. Frederick Etchells (Cambridge, England, 1971 reprint of the 1929 translation of the 1924 French edition); Edith Abbott, *The Tenements of Chicago, 1908–1935* (Chicago, 1936), p. 478; Charles Beard, *American City Government* (London, 1913), pp. 243–244, 334–338, 372–376.

10. Flink, *America Adopts the Automobile*, p. 167.

11. U.S. Bureau of the Census, *Historical Statistics*, p. 714.

12. John D. Buenker, *Urban Liberalism and Progressive Reform* (New York, 1973), p. 14.

13. Don S. Kirschner, *City and Country, Rural Responses to Urbanization in the 1920s* (Westport, Conn., 1970), pp. 183–200.

14. Edwin A. Gere, Jr., "Dillon's Rule and the Cooley Doctrine," *Journal of Urban History* 8 (May 1982): 271–298.

15. Seth Low, "The Government of Cities in the United States," *Century Magazine* 42 (September 1891): 730–738. Low, a reform mayor of New York, was a leading supporter of such restriction. Lord James Bryce and John Dillon, probably the leading political philosophers on municipal governance in the late nineteenth century, also supported such measures.

16. Kenneth Jackson, "Metropolitan Government versus Political Autonomy: Politics on the Crabgrass Frontier," in *Cities in American History*, ed. Kenneth Jackson and Stanley Schultz (New York, 1972), pp. 442–462.

17. Robert C. Wood, *1400 Governments: The Political Economy of the New York Metropolitan Region* (Cambridge, Mass., 1961), p. 1.

18. Nevins, *Ford*, 1:223.

19. Olmsted, Vaux and Co., *Observations on the Progress of Improvements in Street Plans with Special Reference to the Parkway Proposed for the City of Brooklyn* (Brooklyn, 1868), pp. 26–27.

20. Ibid., p. 26.

21. Ross Netherton, *Control of Highway Access* (Madison, Wis., 1963), pp. 1–11.

22. Frederick Law Olmsted, *The East Parkway and Boulevards* (Brooklyn, 1877).

23. John Reps, *The Making of Urban America: A History of City Planning in the United States* (Princeton, N.J., 1965), pp. 325–348.

24. "Minutes of a Meeting at Chickering Hall on Behalf of a Bill to Amend Chapter 522, Laws of 1884," typescript at Loeb Library, Harvard University.

25. As quoted in "Editorial," *Rider and Driver* 17 (March 11, 1899): 8.

26. Ibid.

27. Daniel Burnham, *Plan of Chicago* (Chicago, 1909). I can find no reference to autos in the text nor find autos in the *Plan's* lavish illustrations.

28. A. C. Schrader, "Parks and Boulevards," *Journal of the Western Society of Engineers* 5 (June 1900): 157–164; Sylvester Baxter, "Parkways and Boulevards in American Cities," *American Architect* 62 (October 8, 1898): 11. On restrictive covenants, see Andrew Jay King, "Law and Land Use in Chicago: A Prehistory of Modern Zoning," Ph.D. dissertation, University of Wisconsin–Madison, 1978.

29. Mark S. Foster, *From Streetcar to Superhighway, American City Planners and Urban Transportation, 1900–1940* (Philadelphia, 1981), pp. 17–81.

30. Ibid., p. 110.

31. Mel Scott, *American City Planning Since 1890* (Berkeley, 1971), pp. 44–109.

32. Barrett, *Auto and Urban Transit*, pp. 58, 136–138; Flink, *America Adopts the Automobile*, pp. 174–195.

33. William Phelps Eno, *The Science of Highway Traffic Regulation, 1899–1920* (New York, 1920), pp. 26–45; "William Phelps Eno," *National Cyclopedia of American Biography* 47 (New York, 1965): 613–614.

34. Barrett, *Auto and Urban Transit*, pp. 156–169.

35. Miller McClintock, *Street Traffic Control* (New York, 1925), was the most important early traffic engineering treatise. *Proceedings of the National Safety Council* from about 1920 contain many frank discussions by local experts. See also Gordon M. Sessions et al., *Traffic Devices, Historical Aspects Thereof* (Washington, D.C., 1971).

36. Mark Foster, *From Streetcar to Superhighway*; Barrett, *Auto and Urban Transit*, pp. 50, 156–179.

37. Charles W. Eliot II, "The Influence of the Automobile on the Design of Park Roads," *Landscape Architecture* 43, no. 1 (October 1922): 27–39.

38. John J. R. Croes and Frederick Law Olmsted, *Report of the Landscape Architect and the Civil and Topographical Engineer upon the Laying Out of the Twenty-Third and Twenty-Fourth Wards* (New York, 1876).

39. Grant was better known as one of the Progressive era's leading "scientific" racists.

40. John Nolen and Henry V. Hubbard, *Parkways and Land Values* (Cambridge, Mass., 1937), pp. 72–110.

41. Jay Downer, "Parkways and Superhighways," *American Society of Mu-*

nicipal Engineers Proceedings 36 (1930): 49–52; Gilmore Clark, "Our Highway Problem," *American Magazine of Art* 25 (November 1932): 285–290.

42. William A. Proctor, "Problems Involved in the Planning and Development of Through Traffic Highways," M.A. thesis, Stanford University, 1939, pp. 223–231, 305.

43. Robert A. Caro, *The Power Broker, Robert Moses and the Fall of New York* (New York, 1975), pp. 168ff.

44. Caro, *Power Broker*, p. 318.

45. *Harper's Monthly* 73 (August 1931): 347–356.

46. Ibid., p. 347.

47. Ibid., p. 352.

48. Proctor, "Through Traffic Highways," p. 407.

49. L. G. Sumner, "The Bridges on the Merritt Parkway," *Engineering News-Record* 83 (September 23, 1937): 504. Shelley Schaenen, "The Merritt Parkway," B.A. thesis, School of Architecture, Yale University, 1978.

50. *Vital Gaps in New York Metropolitan Arteries* (New York, 1940).

51. Letter from Robert Moses, chairman of the Triborough Bridge and Tunnel Authority, to Mayor Fiorello La Guardia, November 11, 1940, cited in *Vital Gaps in New York Metropolitan Arteries*. In this pamphlet, Moses attacked critics who argued that his highways were copies of Hitler's autobahns by claiming that German engineers had copied the autobahn concept from Moses's Long Island parkways.

52. Proctor, "Through Traffic Highways."

PART II

Water Systems

NO substance is more critical to sustain life, in the city and elsewhere, than an adequate supply of potable water. Although water supply technologies such as the aqueduct have been available for many centuries, urban growth in the nineteenth century necessitated new sources of water and new methods of distribution. Driving the quest for improved water supplies were concerns over fire and health, the need for more copious supplies in the household for personal and sanitary needs, and the requirements of growing industries. But the provision of water was more than a question of the adequacy of potable supplies and the availability of technology. Water had social and political meanings that were often reflected in the pattern of its distribution throughout the city's rich and poor neighborhoods. By the late nineteenth century, American cities were generally providing larger supplies of water to their residents than were European cities, although the Europeans pioneered in methods of filtration. By the beginning of World War I, however, American cities had made large strides in terms of protecting the purity of their water through filtration and chlorination and had moved ahead of Europe in both volume and purity of supplies.

The three essays in this part focus on water supply developments in France and the United States. André Guillerme explores the technical and social questions surrounding water supply during the first half of the nineteenth century in France, as he moves beyond an analysis that focuses on technology alone. He notes that the French often used British technology as a model and lagged behind Great Britain in water provision. He finds the causes, however, in different patterns of urbanization rather than technology adoption. Jean-Pierre Goubert directs attention toward the spread of water and sewerage systems in France from approximately 1850 to 1946. He examines a large sample of municipal systems from a technical, social, and financial viewpoint and finds advances taking place in certain "golden ages," as the importance of the provision of water and sewer services was gradually accepted by the nation's elites. And in what is primarily an economic analysis, Letty Anderson explores the factors responsible for the diffusion of waterworks throughout the network of New England cities and towns in the late nineteenth century. She finds this process a relatively slow one, with the pattern of movement from larger to smaller cities. The rise in the demand for water, she concludes, depended on the availability of information about the relationship between water and health and water and fire.

5

The Genesis of Water Supply, Distribution, and Sewerage Systems in France, 1800–1850

André Guillerme

"Alas! What are you proposing to me?
Water must be like Caesar's wife,
above suspicion."
—Response of an English engineer
to Arago, 1843

THE start of the nineteenth century in France marked the advent there of a true municipal water policy. It was not that water usage changed much—water had always been used for domestic purposes, firefighting, and road cleaning—but the method of distribution and the growing demand for water, both qualitatively and quantitatively, gradually necessitated the abandonment of traditional waterworks and the search for new water sources.

Here, as in other fields of knowledge, the first half of the nineteenth century was full of technical solutions for city problems, but the Second Empire left a majority of them to history—multiplicity of waterworks, wastewater treatment plants, free water for the poor. "The street-fountains have three functions: to serve domestic needs, to water the roads, and to supply pumps in case of fire," wrote Darcy in 1856. "The second use requires that the pumps be placed at the highest locations in the city, so that the cleanings needed for public health can occur throughout the city." But the first function was clearly more important than the second: "The cleaning of the roads," said H. Emmery, "is surely useful. But consult the experts, review all the proceedings of the sanitation commissions, and they will tell you that it is also important to clean driveways, badly aerated courts, convenience areas. . . . They will add that, above all, the underprivileged must be given the ability to clean themselves and their clothes for free. This is how they tell you to attack the problem of urban sanitation in depth; such is the immense service rendered by the public fountains." When Darcy wrote the lines above,[1] the debate was already closed. The Second Empire had just created farmers' societies for water distribution, and a new generation of engineers was launching expensive, but imperial, projects, constructing sewers exclusively.

Aesthetic concerns in regard to water gradually gave way to the shabby fountains Wallace planted in Paris, beginning in 1872. Many projects were proposed from the end of the eighteenth century on, but due to lack of financial and political support, they were simply filed away. Paris received water from the Ourcq at the end of the First Empire, but had to wait until the Restoration to witness the creation of the first water supply, distribution, and sewerage networks.[2]

The Forecast

During the first half of the nineteenth century, forecasts for domestic water use established by hydrologists emphasized increasing demand. If Deparcieux had proposed 20 liters per day for each Parisian around 1760, many would have considered this number a great exaggeration. In 1802, Bruyere, the inspector of bridges and highways, who had previously declared that the average individual demand in Paris would never exceed 5 liters per day, changed his estimate to 7 liters per day;[3] in 1817, Prony proposed to raise it to 10 liters.[4] In Great Britain, Manche estimated this minimum quantity at 20 liters, 30 for those who considered clean linen to be of vital importance. With Emmery and Rambuteau in Paris and d'Aubuisson in Toulouse, eyes turned toward England, which stopped at 20 liters, the generally accepted number until 1850. But then other demands, neglected by the First Empire, became indispensable in the forecasts. Rambuteau estimated 75 liters as the daily need per head of livestock or dairy animal, 3 liters a day per square meter for gardening, 20 liters needed to clean a cart, and 1.5 liters a day per person to clean the pavement.[5] These are qualitatively different numbers and cannot be added together, since water for household tasks does not have the same destination as that reserved for irrigation. Nevertheless, high-ranking civil servants confused these dual needs. For them, a uniform distribution of water was needed. Here one does not find the idea of a dual system like that found in sewerage. At least the separation of waterworks from other water systems relieved officials of the system of the ancien régime of porters, who had been cursed as a public service burden.

Funding

If municipal councils were favorable toward these liberal and some-times utopian ideas, they withdrew their enthusiasm after seeing the price tags for the networks: 120,000 francs for Chartres, 6 million francs for Lyon. Occasionally a large sale—for example, the sale of a forest in Gray but more often a wealthy benefactor—provided the resources. At Toulouse, the municipal magistrate Lalanne conferred 50,000 pounds, to be given on the death of his widow (1817); at Reims, 100,000 francs was offered; in Soissons in 1880, Madame Finfe gave 270,000 francs to the town to build a church, a magnificent fountain, and two schools.[6] But of course these funds remained insufficient for the needs of mayors, chief administrators from commissions of elected officials, technicians, doctors, and architects entrusted to find proper solutions.[7] Some municipalities, among them Chartres, Rouen, Toulouse, and Marseille, resorted to loans; others, such as Amiens and Dijon, accumulated part of the budget each year and spent it after four or five years when the commissions produced the estimates; still others—for example, Reims, Amiens, Grenoble, Chaumont, and Béziers—counted on the influence of their prefect in the Cabinet of the Interior to engage larger expenditures.

Unlike the situation in Great Britain, the creation of private com-panies in France failed for all practical purposes until 1853, which marks the founding of the Compagnie Générale des Eaux (General Water Company). The first water project dated from 1783 when the Perier brothers estimated that there could be annual profits of more than 3 million francs for water distribution in Paris. The brothers, supported by Beaumarchais but violently criticized by Mirabeau in 1784, based their calculations on the participation of 20,000 houses. By 1786, only 40 participants had been obtained. The company declared bankruptcy and relinquished its privileges to the city. The second project took shape at the end of 1829 under an edict from Charles X, authorizing the city of Paris to create a company that would purchase the works and hold the concession (for ninety-nine years) to distribute water within the city limits. But again, insufficient participation caused the plan to be abandoned even before the July Revolution.[8]

Lines or Networks

In all the cities, two methods of distribution appeared at the start of the Restoration. The first, inherited from the ancien régime, consisted of "bringing water directly to neighborhood centers via independent paths. The most elevated point above ground is chosen as the center, and the conduit is terminated by a magnificent fountain or a water castle." The second method was more of a network: "The principal pipe follows the median line of the serviced sector, with branches leading down the more populated or more important roads";[9] at the extremities public fountains were to be placed. The two methods were diametrically opposed: The linear one described a monumental fountain that goes looking for a source, whereas the network was obliged to start from a point of entry and predict the flow for each of its "terminals." The technology necessary for water mains was simple and had been proven dependable since Mariotte; network technology was much more complex with respect to pipe dimensions and remained only approximate until about 1839 when Mary, professor at the Ecole Centrale of Paris in 1839, and, more important, Darcy, around 1855, were able to establish the abacus of the calculus for diameters.[10]

Aside from technological issues, there were financial and ideological considerations. The linear system used a minimum of small-diameter piping, while the network multiplied the lengths or the pathways, requiring large volumes of water; this necessitated very resistant conduits at prohibitive cost, the debit being proportional to the price of the material used. The cost of a simple conduit quadruples if you double its weight,[11] and the costs of transportation, moving equipment, and groundwork double. As an example, in 1825 Dijon planned a linear system with six huge fountains, estimated at 60,000 francs; fourteen years later, it opted for a network that cost 1.25 million francs, twenty times as much. In Toulouse, the first project, a linear system, was not to exceed 50,000 francs in 1817; it cost more than a million francs in 1827. The stakes of water distribution thus outweighed economic dimensions and acquired gigantic proportions. If London estimated 90 liters per person daily, Grenoble aimed for 60 liters, Toulouse exceeded 80 liters, and Dijon surpassed all domestic records (240 liters), although soon overtaken by Besançon (246 liters). At the start of the Second Empire, the hydraulisticians watched, dumbfounded, as New York City estimated 1 cubic meter per day per citizen.

This emulation, found in the more industrialized cities where engineering was in the hands of alumni from top schools, increased the per capita needs tenfold in a few decades; these were to be doubled again when considering the urbanization that occurred frcm 1815 to 1850. Of course, more modest, less industrialized towns, such as Chartres and

Auxerre, limited extravagance to 20 liters a day. They were also limited to linear systems and artesian wells.

Water and Social Inequality

The distribution of water reveals urban segregation; at possibly no other time was the inequality between the upper and lower classes in cities so strongly felt between those areas in which water was within reach and those in which it was not, between the artisans and workers who washed in the rivers and the more well-to-do who lived "on the hill." An even more flagrant inequality dealt with the bourgeois distribution of wealth. All the projects tended to separate the city into two parts and reestablish the socioeconomic inequality of the citizens by unevenly distributing running water: two-fifths for downtown and three-fifths for uptown in Amiens;[12] one part for downtown, two parts for the plateau at Toulouse (despite affirmations to the contrary by d'Aubuisson), although reports on the surface area served indicated the reverse;[13] thirty streets "downtown" and ten "uptown" at Chartres, according to the Pellerin project (1838);[14] six fountains for the lower town versus twelve for the upper town at Béziers (1831). In London, a model town, the distribution of water was composed of a "lower service" supplied by certain reservoirs for downtown and an "upper service" connected to pumps supplying a few floors,[15] a system developed in Lyon soon after 1850.[16]

The topographical distribution of water reveals political trends. In Toulouse, d'Aubuisson fought to prevent the cleaning of "upper class" roads with "lower class" water; in Paris, Emmery "looked upon the closing of a fountain as a calamity, as if it increased the mortality of the underprivileged class,"[17] and forced himself to oversupply the poorer neighborhoods; Darcy, very critical of d'Aubuisson, affirmed that "public fountains must be gradually brought closer to the most populated class. Water must force the people to adopt new habits, and the municipal council of Dijon has not spared any expense in carrying this out, in any way."[18] In contrast, in Amiens and Rouen, where the conduits installed in the eighteenth century were extended and modernized, care was taken to avoid creating too many fountains in the "lower" towns. Subsidies were lowered, and crowded conditions made the work long and dangerous.[19]

Obviously, by cleaning the "upper" city with large amounts of water, the "lower" city accumulated all the waste. The creation of sewers and

receptacles for runoff water was not always possible, and many municipalities hesitated to provide the needed funds. In Paris, the prefect of
police, Gisquet, "disgusted by the putrid lakes on the lower parts of the
roads,"[20] unblocked the funds necessary for the construction of 14,000
meters of sewers in 1832 and 1833. To do this it was necessary to raise
road levels to avoid counterslopes, divert wastewater to underground
galleries, replace convex pavement with V-shaped pavement, and add
sidewalks to separate gutters from residents.[21] Toulouse was less fortunate: d'Aubuisson contented himself by choosing a few gathering points
and the edge of the plateau for driving out the water, and used the ancient system of discharge: "We have cleaned the entire center of town
and more than three-quarters of the entire area: we have disposed of
the water into all the sewers; which, if cleaned continuously, will no
longer discharge their foul odor in the warm and dry seasons, which
is a real scourge for the neighborhoods which have an outlet/vent."[22]
In Dijon, Darcy sent all the wastewater into the Suzon and raised its
banks, transforming it into compartments, the prevailing style of the
seventeenth century. The cost of the operation represented merely 5.5
percent of the total cost of a water system.[23] Not until the filling up of
cesspools and the large undertakings of the Second Empire were the
precepts of Emmery applied: raising the ground level and establishing
covered galleries to be used as wastewater-flushing canals. Public health
was still concerned only with the wealthy and remained a dream to the
less-well-off, despite the recommendation of doctors who had studied
the two epidemics of cholera, notably that of 1832.

Innovation in Water Quality

River water was free and abundant, but slowly, during the eighteenth
century, some of the elite began to find it distasteful to consume water
laden with suspended particles, often drawn from unknown sources.
The market for clear water was ready to be captured, and although it
was not yet well organized, it was to provide substantial profits. In 1820,
Genieys estimated that "one waterworks service could deliver ten times
as much water to a home and for the same price as a porter."[24] By the
second half of the eighteenth century, private organizations had set up
along the Seine in Paris, where the water was filtered through gravel
and sand. One of these plants had been installed in Port-à-l'Anglais
(Vitry) in 1771, another on the outskirts of Ile-Saint-Louis. The output
was reasonably weak (barely 12 cubic meters a day);[25] even so, the water
of the Seine did lose a third of its particles. Other processes were being

developed at the beginning of the nineteenth century, notably under the direction of the navy, which wanted to keep potable water on board its ships. Pulverized coal, animal or vegetable, became a substitute for nature, and the Academy of Science encouraged new methods, notably that of Loowitz.[26] Bertholet proposed to line cisterns with charcoal. James Smith, Cuchet, and Montfort showed that, once passed through a thick layer of charcoal, even the dirtiest waters of Paris become potable; the testimony of members of the institute assured publicity.

In 1806, the first plant was installed in Quai des Célestins in Paris and drew from the Seine 2.22 liters a minute. The filters were composed of a layer of gravel (3 centimeters), a thick layer of coal laced with sand, and another a layer of gravel. The water rose in lead vases lined with sponges that were changed every two or three hours.[27] Two other establishments using the same process were set up in la Boule Rouge and Gros Cailloux.

In Rouen, the first purification plant was established at Toulouse[28] in 1823 with the founding of a private company charged with reselling water through a system of porters. Installed in the outskirts of Saint-Sever, it drew water from the Seine. Dubuc's chemical tests, demanded by the enterprise, revealed that, thanks to the activated charcoal, the water lost one-third of its suspended particles (one-half compared to the municipal fountains).[29]

But charcoal filters were inconvenient; they had to be replaced often. And Genieys estimated in 1835 that for a demand of 150 cubic meters a day, the minimum surface required would be 7,000 square meters.[30]

If the anticipated benefits were large enough to lure private enterprise,[31] they turned into losses when taken over by municipal services. In most instances, one had to be satisfied with drawing water in the proximity of drains and improving natural filters, such as alluvial gravel banks. So in Toulouse, the Fountain Commission, created in 1817, rejected the insufficient ascending purifying system and dug a long basin, 1,080 square meters, as a receptacle for the filtered water of the Garonne.

The other big problem was the clogging of filters. If in a private system the sponges had to be changed every two or three hours, for general distribution one had to clean the gravel and replace the sand every week in the two basins (4,800 and 4,000 square meters) of Marseille, every two months in Cherbourg, and each year at Chartres. Such a spread suggested, as d'Aubuisson claimed in the case of Toulouse, that it was useless to clean the filters. Darcy repeated the experience in Dijon and, using results obtained from British plants, proposed that the superficial layer of sand be regenerated by 2 centimeters.[32]

After trying other filters, engineers came to prefer natural filtration because it involved less manpower;[33] its only major inconvenience was that it was slow if one did not have riverside property close to the populated areas. What about drawing directly from the river? This solution, possible in the eighteenth century with a system of porters, was no longer feasible with a system of conduits because suspended particles quickly sullied and clogged the pipes. Above all, spring floods carried quantities of matter; around 1853, Guerard and Terme showed that it took nine to ten days for the water to clear after the floods. If a new system were put in place in Lyon—a cement-filtering gallery with a gravel apron [34]—it would remain a prototype. It was preferable, for a few million additional francs, to fetch the water at its source, not far from the city if possible, and have it come downhill solely by gravity, or by pressurized conduits, of which much was known. Spring water was devoid of suspended particles; it was described by Darcy and Pasteur as pure and virginal. River water, considered best since the Renaissance, was abandoned because, even filtered, it could carry microbes and cause epidemics. Certain municipalities looked for new springs; others turned to chlorine, alum, or ozone to disinfect the river water. Public health was saved.

The Elevation of Water

The fundamental problem was to find a means of bringing water to the highest point in the city so that distribution could work via gravity. Once the hope of finding a suitable artesian well was lost, three choices remained: Do nothing or ameliorate the present system; use canals to bring the water to an elevated point very close to the city; or use energy to raise the water.

As a solution inherited from history, the aqueduct offered the advantage of being able to supply a reservoir easily with large quantities of water. The engineers and the emperor were lured by monumental, and thus expensive, works. The canal of Ourcq made Paris the equal of Rome and Girard. Whoever the promoter was, he knew how to flatter Bonaparte: On January 1, 1816, the total expenses of the canal were 14,353,118 francs, and they ended up at more than 24 million francs. A folly, exclaimed Genieys,[35] and with calculations in hand, he showed that with equivalent funds, the installation of steam engines/pumps would have been less costly.[36] Under the Restoration, the aqueduct was still to be found in Nîmes; in Saint-Etienne (a project for a canal 100 kilometers long); in Lyons, in the form of a proposal in 1823 (the chief road and bridge engineer was ready to divert the Sereine or the Ain up to the gates of the city); in Marseille, where Seguin, in 1836, proposed a metal aqueduct for carrying the waters of the Roquefavour; and in 1856 in Sens, where Belgrand decided to reuse the old Gallo-Roman aqueduct.

Noisome and sometimes explosive, the steam engine was rejected by people who lived by the rivers, who did not hesitate to answer public utility questionnaires, both mandatory and nonmandatory. Many mayors, despite the pleas of architects and engineers, decided to "wait for steam engines, which still left much to be desired, to become perfected, and to become widespread enough in France to assure their superiority over other systems. Each day, there were reports that new improvements had been made by mechanical wizards."[37] Furthermore, the technology was still very expensive: Genieys estimated 1,500 francs per steam-horsepower, but he stayed well below actual costs in other places—3,000 francs in Amiens, 3,200 francs in Chartres, 3,300 francs in Gray; and, before 1835, improvements did not increase effective output.[38] The steam engine also consumed an enormous quantity of coal, a rare commodity in France and one that was expensive to develop. Fuel and upkeep cost as much as the interest on the loan needed for a whole system.[39] Finally, the machine was difficult to regulate; the appropriate pressure depended on the dexterity of the engineer who controlled the fire by limiting or feeding it. Sometimes the water did not flow; sometimes the engineer became apprehensive, and it became necessary, at Béziers and Chaumont, to appoint a supervising machinist.

While waiting for the "mechanical revolution," water was pumped by men and horses. Four horses worked continuously in Toulouse to activate pumps;[40] four men (eight after 1826), paid by a charity fund, turned the Archimedean screw in Amiens.[41] These systems were about 50 percent efficient.

Others thought of the possibilities of hydraulic power. An old system in use at the pumps of la Samaritaine in Paris had the advantage of having a regular output. Pellerin proposed to the municipal leaders of Chartres that they revamp the mill of Saints-Pères using a Fourneyron turbine that guaranteed an output of .6 horsepower, which was better than the steam engines' .3 horsepower. Ulrich, in Strasbourg in 1824, hoped to use the mill in Ill to bring water to the town square in front of the cathedral.[42]

Water Reservoirs

The reservoir is the cornerstone of a network. Its location and capacity occupied three years of the engineers' time in Toulouse and Marseille, and even more time in Chartres and Tours. To take everything into account, the reservoir must be on firm and nonporous ground and must be near the drawing source and the customers, because it is at the head of the supply route. It must protect the pumps and the central artery. It requires vast terrain so that it can accommodate the pipes, fuel supply, and disposed wastes, and the property of the keeper and the caretaker. In Toulouse, the extremity of the bridge on the Garonne was chosen; in Chartres, it was a platform near the ramparts. Both placed their reservoirs 35 meters above ground, the height required to send the water throughout the town. The water tower was an urban "monument," and the Buildings Council supervised construction;[43] to emphasize its function, a monumental fountain was placed on one facade.

If the water reservoir is inherited from the ancien régime, the underground reservoir seems to be a copy of the British model. Grenoble built one in 1823, and Dijon acquired two in 1835: "that of Porte Guillaume is extramural, in the center of a circular park. It is covered with a meter of ground and a pavillion by which a beautiful fountain is situated. That of Montmusard is more vast, and thus of less architectural quality. Together there was a capacity of 5,490 cubic meters, in sum 20 liters a person and six days of reserve."[44] The major concern was a lengthy breakdown of the machinery, which would prevent the town from getting supplied. Darcy affirmed that reservoirs with great capacity were most subject to breakdowns, while elevated reservoirs could store only a few hundred cubic meters.

The System of Pipes

In France at the end of the eighteenth century, cast iron began to replace lead in urban water conduits; England and Germany had been using cast iron since the Renaissance.[45] At the beginning of the changeover, one must first recognize the developments in metallurgy that were constantly lowering the cost of iron. It was competitive with lead by the eighteenth century, half as expensive in 1825, and four times less expensive in 1850.[46] The development and upkeep of the road network, especially during the Restoration, reduced the price of transportation, which fell, between 1790 and 1830, from a quarter to an eighth of the manufacturer's initial cost.[47] Meanwhile the price of lead, which was often reused or recycled, remained steady during this period.[48]

A thorough search of a technical library of the time would show that cast iron symbolized a modern material, whereas lead was more traditional. Two factions were sharply divided on which material to use for pipes; there was more agreement on other technological "components": pumps, steam engines, public fountains, etc. There was Boudsot, engineer of the Ecole Centrale, d'Aubuisson and Galle, engineers of the Mines, and Vicat, Girard, and Emmery, engineers of the *ponts et chaussés*, on the one hand; Janvier, Biston, and Roret, mechanic–metallurgists, and Chaussey, architect–futurist, on the other. The first group of men, powerful because of their education, always looked to the example of Great Britain, and saw in iron the material of the future. The self-educated men, dependent on their experience, chose lead, the historical material of fountains. As the networks spread, cast iron replaced lead because power flowed more and more to formally educated engineers.

In fact, between 1815 and 1830, the latter imposed their norms on all piping. Until then, builders sold their pipes by weight, giving the thickness as one-twelfth the interior diameter[49] taking into account the difficulties of manipulation, these pipes barely extended 1.2 meters long. Genieys, in charge of the Parisian system, modified the thickness after having tested its mechanical resistance. He estimated that for larger diameters, under a hydraulic pressure of 10 atmospheres, a thickness of 7 percent of the diameter sufficed.[50] In reality, the thickness of the same pipe varied according to the fabrication process; because the center of the mold had a tendency to wander, the thickness was greater at the bottom than at the top of the pipe. Air bubbles and slag could also locally reduce the strength of a conduit.

But in the 1820s a new system of casting—vertical casting—gave a more uniform thickness, which could be sensibly reduced. Genieys recommended 3 percent of the diameter. These propositions, adopted by the builders, reduced the weight of a pipe by half and extended the length to between 2 and 2.5 meters. The second innovation came from d'Aubuisson of Voisins, who affirmed, using the calculations of Jardin of Edinburgh, that a thickness of 1 millimeter would suffice if it were not for certain small imperfections, and thus made the builders charge by length and not by weight, as was done in Britain.[51] In a short time, this system was adopted by project engineers;[52] they reduced the price and the weight fourfold. In twenty years (1810–1830), the cost of a cast-iron pipe could be divided by twelve.[53]

The second advantage of cast iron was its malleability. Whereas lead had to be welded with much care,[54] cast-iron pipes were assembled end-to-end with a bridle and, gradually, starting in 1820, by jointing, which permitted easy and rapid disassembly and increased impermeability.[55] These materials did not require such highly trained personnel to be at the work site.

Finally, the mechanical resistance of cast iron was three times higher than that of lead. It was therefore chosen for the large diameters of the principal conduits recommended by engineers with experience in waterworks and sewers. Resistance to time and wear was also important. Enthusiastic people estimated that iron lasted two thousand years, while lead, like clay, lasted 300 years, and wood less than 100 years.[56] The only inconveniences some people saw in cast iron were its frequent breakage and internal roughness, which increased loss of flow. But from 1830, boring technicians reduced the roughness, just as the first bitumen pipes went on sale.[57] The future lay in cast iron.

A serious problem did appear in the pipes of Grenoble in 1834: The interior of the pipes became covered with pear-shaped, blackish nodules. The engineers called on scientists to find an explanation for the phenomenon, which threatened their prestige and their faith in iron.[58] Despite much creative effort, no solution was ever found, though the growth of nodules did stabilize after a few years, to the relief of the technicians.[59] But the estimated life span of 2,000 years was severely reduced around 1835, because of corrosion. Gaudin showed in 1851 that wear was on the order of .4 millimeters of thickness per year; so, for a pipe 18 centimeters in diameter, a life span of 44 years.[60] For the smaller diameters (less than 5 centimeters), lead was used more often than iron.[61] A more supple material, lead could easily be bent or crushed, and its technology was improving: Since 1787, it had been forcibly stretched to reduce its thickness; since 1819, it had been stretched on a draw-plate; and in 1825, hydraulic pressure was first used to produce pipes continuously.[62] In fact, linear costs did decrease, but less than those for cast iron, and manufacturing remained confined to large cities.

Wood and clay slowly faded away. Wood was cheap,[63] but it had the inconvenience of not being long-lived, it swelled with humidity, it provoked heavy losses, and might have burst under the high pressures of steam engines. Engineers found clay pipes too soft, heavy, and short (they were less than a meter long) and so relegated them to rural areas. Cement made a timid showing—it was used for the first time by Fleuret in Meurthe-et-Moselle in 1805—but as long as it was not reinforced and bituminized, it remained relatively unused.[64]

Castings—conduits, faucets, public fountains, etc.—generally constituted a quarter of the costs of the distribution system in 1830;[65] it had been 90 percent fifty years earlier.

From the Water Carrier to the Concessionaire

Individual or collective distribution? Beyond technical evolution and economic reasons, the connection to water was primarily social, and the grip of water was present daily in urban communities. The sociability might be difficult to perceive in abundantly supplied agglomerations, but in the others, it was valued in the form of water carriers, "vile men and braying women who desolate the inhabitants around the public fountains," who were vituperated by J. F. Blondel.[66]

The *éviers*, as they were called in Chartres, received a modest salary, but played a role in the fight against fires.[67] Above all, they were the ears and the voice of public opinion that was unreachable by the press. They informed, amplifying or reducing by whim, the noise of the city. They were in the front row when Etienne Marcel was assassinated (1858), and in 1789 they headed the protests at Versailles.[68] In fact, they were feared and protected at the same time. In 1845, when Chartres decided, after a check on the artesian well of the "upper" city, to raise the water of the Eure, some municipal councillors protested because of the disappearance of the *éviers* and forced a compromise: Part of the water would be sold at the fountains, which would be auctioned, and a water meter would be installed.[69] In 1778, Périer saw an opportunity to hire fountain porters to open all the faucets of its Parisian network two or three times weekly.[70] In Amiens, in contrast, the municipality was obliged to have "an ordinance against evildoers at the public fountains and related buildings." In 1755, "many libertines, vagabonds, and people with evil intentions bothered and insulted the entrepreneurs, controllers, and public workers, notably at the fountains, of which the system, having been desired for a long time, was useful and necessary; they obstruct timely execution by degrading and destroying the works, by throwing garbage in the trenches, mutilating with rocks and sticks the water reservoirs, as well as fountains and other buildings, . . . clogging the pipes."[71] The object of this vandalism was, of course, the new distribution system —"each concessionnaire would have his own small basin from which his pipe would lead"[72]—which removed a part of the sociability aspect by eliminating the water porters.

"A question of political economy is herein raised which, it seems to me, cannot be ignored," wrote an anonymous reader to the *Journal du*

TABLE 5.1.
Water Supply Systems in France, Selected Cities

City	Year System Opened	Population (thousands)	Cost (thousands)	Usage (m³ per day)	Usage (liter per person per day)	Cost (Francs per person)	Length of System (in meters)	Number of fountains
Paris	1808	714	24,326	4,000	7	34,09	13,000	124
Agde	1810	7		210	30			
Toulouse	1817	65	1,084	4,000	60	16,68	12,835	111
Grenoble	1826	18,6		2,060	112		3,800	
Lyon	1830	134	6,000	20,000	85	25,59	75,000	121
Chaumont	1831	6,5	156	92	14	25,06	2,900	15
Béziers	1831	13	97	138	12	24,61	3,200	18
Cherbourg	1836	16	226	500	28	14,12	2,426	
Gray	1836	6,5	238	150	23	36,61	2,560	3
Chartres	1839[a]	15	247	600	40	16,47	6,500	42
Dijon	1839	25	1,250	6,100	240	49,46	14,360	149
Bruxelles	1844	250	6,600	20,000	90	26,40		
Strasbourg	1845[a]	72	820	1,840	255	11,40	20,250	289
Besançon	1848	35	1,600	8,600	246	45,71	12,320	
Bordeaux	1851	132	4,200	22,000	170	31,84		
Nantes	1854	100	950	6,000	60	9,50		

[a]Estimated date.

Génie Civil on the eve of the July Revolution in 1830. Concerning the distribution of water by porters, he wrote:

> This industry provides a readily available resource for capable individuals who, out of a job or without hope of finding steady employment, would rather earn their living than be reduced to begging. And since we are concerned about finding ways to wipe out begging, we must take care not to eliminate—especially if there is no overall, general or especially very clear advantage—those jobs which might sustain any person with a good pair of arms and the desire to work. If we encourage only those enterprises which make the rich richer, we are going to end up—and the example of England is there to show us—concentrating the fortune of this country in just a few hands, while we make of the masses of the population a class reduced to living off a scant poor tax; thus the poor will become, in reality, veritable Helots, condemned to eternal humiliation by discouragement and impotence.

How many *éviers* were there? It is difficult to know the number because, as noted by Genieys, they were neither taxed nor registered in any employment agency. Nevertheless, by tabulating the overall water distributed daily and by assuming an average of a dozen trips daily, the number of water carriers in Paris might have been 2,000 in 1815 [73] and about 30 at Chartres, 2 per every 1,000 people.[74]

Within towns, the price of water distributed by porters was fixed, independent of the distance from the source, but it varied between towns, from one to four units: 1.25 francs per cubic meter in Chartres, 2.5 francs in Mulhouse, 4.4 francs in Paris.[75] Sometimes it varied according to volume, and traditionally the tariff of someone with a cart was less than that of someone without. As filtering enterprises and distribution systems grew, they offered diverging tariffs: In Rouen in 1823 the treatment plant on the outskirts of Saint-Sever sold water for .1 franc for 2 buckets or 1.5 francs for 32 buckets or 4 francs for 4 barrels,[76] numbers similar to those in Paris. The price essentially covered services rendered in the distribution of water to the home by porters; some of the money went to fund the filtration plant. This water reached only 1 to 2 percent of the urban population. The rest supplied itself with groundwater or water taken directly from rivers.

With the ancien régime, the cities that had "linear" distribution networks had granted a few concessions to individuals. It was a way of rewarding an ancient mayor, a monastery, a cathedral, or a rich bourgeois who had contributed something to the city;[77] in fact, the concession was free. Certain agglomerations, such as Amiens or Reimes, had attempted to build their own system according to their means. But, as explained by the priest Féry or the economist Riquier, it was very difficult to measure the volume used with a minimum of precision.[78] For the rest, the Revolution abolished the privilege of the free concession, which was partially reestablished by Napoleon.

In fact, between 1780 and 1815, decisive progress was made in gauging water. Research was done by d'Alembert and Bossut on hydrodynamics, and it was succeeded by that of Dubuat on the resistance of fluids and the speed of water in conduits. During the Revolution, Prony and Girard studied mechanical applications of hydraulics in the empire.[79] By the Restoration, they could measure and figure water expenses as much by the level of the reservoirs as by the output at the extremities.

If the distribution via public fountain was free, the individual concession needed to be paid for. D'Aubuisson affirmed that concessions must be bought and not given away (as was done in Paris before the Revolution). On the contrary, the privileged who paid the concession helped fund the municipality, and hence paid personnel. In Toulouse, there were four kinds of concessions: by day, by night, monthly, and longer than six months. The flow was continuous (at the faucet), and the sales brought 7,400 francs, while the cost for maintenance and upkeep cost the town 4,873 francs. "Before the water service, the inhabitants of Toulouse paid 150,000 francs to have water; today they pay a twentieth that."[80]

In Dijon, fifteen years later, a system of concessions had evolved. Thanks to the use of faucets and the precision of meters, two kinds were available: by given quantity or by valuation.[81] Individual installation, with everything included, came to 100 francs. In Rouen, where the system had slowly evolved since the Middle Ages, free concessions were abolished in 1794. At the end of the Empire, the municipality elaborated on the regulations (in 1812 and 1817), extending some concessions to perpetual status by the act of a notary, averaging 1,000 francs and 30 francs in annual rent. The new, higher prices drove people from the system: 60 francs in 1840, 78 in 1850, 70 in 1868.[82] "The success of the enterprise," declared Genieys in 1830, "is essentially founded on this new mode of taming water at all its stages. As long as we will need someone to supply us with water, if only to raise it a few feet, we would not expand its availability, and we would prefer to buy it from actual porters. But it is so simple to place reservoirs in the most elevated parts of the home: it is by this example that we shall succeed."[83]

A fundamental principle of the system was to cover, with the help of a conduit, a maximum surface. Emmery, Genieys, Darcy, d'Aubuisson

—all the engineers, unlike other professional groups—used the average distance to the nearest public fountain: one fountain for 150 meters of pavement in Dijon, or "50 meters to travel to get to one";[84] "the population of Toulouse is less than 200 meters or three to four minutes from a public fountain";[85] and Genieys stated, a few years earlier, that "all these pipes form a network which embrace all the roads of the area under consideration."[86]

"The British Mirror"

The outstanding technical innovations that marked the development of water distribution in France, and that acted as so many assets in the achievements of the second half of the nineteenth century,[87] must not hide the fundamental role played by British technology, which was so often used as a model by French builders and thinkers.

At the same time that Paris was trying in vain, and for the second time, to create a distribution company, a "report on water distribution in London, printed by order of the Parliament" indicated that eight private companies took care of the needs of the city, supplying close to 140,000 cubic meters per day, with one-third pumped from the Thames, half from the Lea, and the rest from wells in Chandwill; 177,000 dwellings were supplied, using seventeen steam engines. All available technical options were present and in competition with one another. For example, the three companies on the right bank of the Thames, which supplied the suburbs, did not possess a reservoir and thus sold water more cheaply to the working classes. But as soon as the steam engines stopped working, the load on the network became overwhelming, increasing the use of the hydraulic ram at start-up time. On the right bank, the New River Company—the largest company, distributing 13 million gallons per day—owned four reservoirs and divided its services into high and low; the East London Company (6 million gallons pumped from the Lea) was not responsible for the high service, but individuals could put up their own pipes at whatever height they wanted; the West Middlesex Company (2.25 million gallons from the Thames) stored its water in reservoirs with inverted vaults so as to improve decantation, the same as the Lambeth Company, which owned a basin with a capacity of 14,400 gallons. London could thus claim to be better watered than Paris, but the quality of water was not as good (this was the purpose of the report by the Chamber of Districts). The water was not filtered, and its clarity depended entirely on the weather and the time of year. It was full of insects, filth, and heavy chemical and bacteriological pollution, and so the inhabitants of London could use it only for washing, while continuing to draw their drinking water from wells.[88]

Even more than London, the capital of Scotland was admired by French technicians. To provide for its water needs, Edinburgh had succeeded in reconciling both science and money. In 1810, two commissions were created, one made up of the most influential and well-informed citizens, the other of the best engineers and professors of chemistry, mineralogy, and physics. While Playfair and Jardine set about looking for the best sources of water, Hope carried out analyses on the water, and Telford made the plans for the work to be undertaken. The final report was turned in to the municipality in 1813; once the necessary funds had been raised, the works were begun (at the end of 1819) and were completed in November 1822. The springs chosen flowed from an uninhabited valley covered over with an immense bed of sand and gravel, 40 feet deep and more than a mile long. The water was collected in a huge vaulted reservoir built of cut stone and then taken by an aqueduct (buried underground in two places) to the gates of the city, where it was distributed into two branches through underground tunnels and metal water mains, built to withstand pressures of 1,000 feet of elevation. The total costs were calculated at £145,000 for 1.2 million gallons of water a day (40 liters per person per day). Among the 130,000 inhabitants of Edinburgh, Leith, and Cuthbert who were supplied in 1829, 40,000 persons in the old part of the city and the southern sections obtained their water free of charge at public fountains; charitable institutions were also provided with water free of charge. Those who paid for the service were not obliged to pay more than 7 percent of the total amount they paid for rent, in keeping with standard practice throughout the kingdom. The project undertaken by Jardine, who reduced the thickness of the pipes considerably, and supervised by Telford and Rennie, who built the first dam-reservoir next to the wells, with an earth wall some 60 feet thick and a holding capacity of 100,000 cubic meters was highly praised by Stevenson in 1825.[89] Most certainly, the Scottish model was studied by d'Aubuisson de Voison in Toulouse, and by Masclet in Paris; it was proposed by Mary in his courses at the Ecole Centrale in Paris between 1834 and 1840.[90]

Water distribution in France thus lagged somewhat behind that in Great Britain. But one should not, as we have seen, seek the causes in technology—the situation was quite the opposite.[91] Instead, the causes would seem linked to ethics and kind of urbanization.

This was an era when the French were very hesitant about the idea of paying for water, "this substance which, along with the air, is practically the only gift which Nature has exempted from tyranny!" claimed Mirabeau four years before the French Revolution. "The privilege of a water company is forbidden by the very nature of its product," he added.[92] Forty years later, the academician Girard, the grand master of the canal of the Ourcq, wrote the same thing in more modest terms.[93] It was not until the upper classes came to power in 1830 that "paying" projects were multiplied and carried out according to a rule whereby the most wealthy would pay for the free water of the poorer citizens.

If we compare the success of the British to that of the French, we must question the urban phenomenon and, to conclude, limit the comparison to London and Paris in 1830:

• London had only one free public fountain; Paris had over 150.

• London had no water carriers left; Paris still had 2,000. "The water purchased from carriers is but a rather small fraction of total consumption, because the wells with which most dwellings are provided supply the household needs for washing and eventual fires; the fire regulations stipulate their use. In London, on the other hand, the wells or pumps set up within the houses supply drinking water, from which it results that a considerable quantity of the water necessary for washing is furnished by the water companies."[94]

• London used coal as its chief fuel, which meant that more frequent washing was necessary; Paris still used wood.

• London had 200,000 dwellings, most of them of modern construction in brick and cement mortar. Paris had only 30,000 dwellings, most of them quite old, with lime mortar,[95] and "the owners refuse to have running water because of the harmful infiltrations to the foundations and other buildings."[96]

• In London, there were approximately 6 persons per dwelling; in Paris, 27.

• In London, "much of the water is used by industry, including what is needed to run machines." In Paris, industries preferred to drill artesian wells or take their water directly from the Seine.

London served as the model. But the technique of water distribution essentially followed (although with some difficulty), the science of hydraulogy. French engineers had dominated the area of hydraulic applications since the mid-eighteenth century. Scientific terminology and concepts in France remained continental: The concept of "networks" emerged from a combination of the scientific thought of the Enlightenment and the milieu of the elite engineering schools.

Notes

1. Darcy, Les Fontaines publiques de la ville de Dijon. Exposition et application des principes à suivre et des formules à employer dans les questions de
distribuiton d'eau (Paris, 1856), p. 337; Emmery, "Egouts et bornes-fontaines,"
Annals de Ponts et Chaussées (hereinafter APC) 2(1834): 235.
2. Principes d'hydraulique (Paris, 1786), p. x.
3. Rapport du 9 Floréal an X pour éclairer l'administration sur les moyens
à employer pour fournir l'eau nécessaire à la consommation de Paris, p. 7.
4. "Mémoire sur le rapport de la measure appelée pouce fontainier," Histoire de l'Académie des Sciences (Paris, 1817), p. 422: "We recognized that in
Paris, a family of ten consumed on the average 70 liters of water daily." According to Genieys, Essai sur les moyens de conduire, d'élever et de distribuer
les eaux (Paris, 1829), p. 53, the actual consumption in Paris was five liters per
person per day.
5. The préfecture of Paris, arrested January 1, 1846.
6. Pécheur, Les Rues de Soissons (Soissons, 1897), p. 47.
7. The commission to build the fountains in Toulouse included d'Aubuisson, an exceptional military engineer who served as the chief engineer of
bridges and highways and also as the municipal architect.
8. Mirabeau, Sur les actions de la Compagnie des eaux de Paris (London,
1785). Mallet, "Notice historique sur le projet d'une distribution générale d'eau
à domicile dans Paris," Journal du Génie Civil 8 (1830): 272–278.
9. Genieys, Essai, p. 155.
10. Darcy, Recherches expérimentales relatives au mouvement de l'eau
(Paris, 1857), vol. 1, chap. 6.
11. As long as the price (P) is calculated according to weight.
12. Archives départementales, Somme (hereinafter AD, Somme), OaAo, no.
4: 16/40 and 24/40 cubic meters.
13. "I answer . . . that vulgar people shall always blame us for distributing
the public fountains unequally. The truth of the matter is that the problem is
not to have equal distribution, but to maximize the potential cleaning area."
D'Aubuisson, "Histoire de l'établissement des fontaines de Toulouse," APC 2
(1835): 289.
14. Pellerin, Distribution des eaux de la rivière d'Eure dans la ville de
Chartres (Paris, 1838), p. 8.
15. Genieys, Essai, p. 440; from 9 feet under street level to 5 feet above for
the ground floor.
16. Darcy, Les Fontaines publiques, p. 552: 15,000 cubic meters per day
for the ground-floor service, 5,000 cubic meters per day for the upper. For a
discussion of the Weltanschauung, see Guillerme, "Des nappes souterraines aux
couches sociales," Milleux, nos. 3/4(1980): 105–108.
17. Darcy, Les Fontaines publiques, p. 338.
18. Ibid., p. 337.
19. AD, Somme, Oa10, no. 13, June 11, 1825. In urban areas, the work usually

took place during the summer (with fair weather) when the citizens were working in the fields.

20. Mille, "Mémoirs sur le service des vidanges publiques de la ville de Paris," APC 1(1854): 136.

21. Emmery, "Egouts et bornes-fontaines," APC (1834): 243, 248. A convex pavement costs half as much as a V-shaped one.

22. D'Aubuisson, *Fontaines de Toulouse*, p. 289.

23. Darcy, *Les Fontaines publiques*, p. 432; 3 percent for Toulouse, 8 percent for Paris.

24. Genieys, *Essai*, p. 154.

25. Jaubert, *Dictionnaire raisonné universel des arts et métiers*, 2nd éd. (Paris, 1773) 3:543–545.

26. "Mémoire sur l'emploi du charbon dans l'épuration de l'eau," *Mémoire de l'Académie de Saint-Petersbourg*, 1800; see also Crellis, in *Annales de chimie*, 1802; Th. de Saussure, ibid., 37; and, Bussy and Payen in *Journal de Pharmacie*, 8.

27. Genieys, "Clarification et épuration des eaux" APC 1(1835): 64–65; du Commun du Locle, *Mémoire sur l'épuration et la clarification des eaux de l'Ourcq et sur leur emploi dans Paris en concurrence avec les eaux de Seine* (Paris, 1836). The purification project in Chartres in 1778 projected a "reservoir containing 34,770 cubic feet (1,000 cubic meters) to receive the water of the Eure after its passage through a filtration canal of ten fathoms and its cleaning through sand." Bethouard, *Histoire de Chartres, 1789–1900* (Chartres, 1904), p. 16.

28. D'Aubuisson, *Fontaines de Toulouse*, p. 263 (around 1780).

29. Dubuc, "Notices et observations sur les différents degrés de pureté de l'eau ordinaire servant aux usages de la view, dans les arts, etc.," *Précis analytique de l'Académie de Rouen* 28(1826): 50: .1 g/l of residue; .15 for the Seine; .2 for the fountains.

30. Genieys, *Essai*, p. 76.

31. In 1817 all the factories combined treated 1,500 cubic meters a day or 2 million francs annually; of the 6.2 million francs collected from the distribution system, 4.3 million went to the porters who took water at the fountains. Genieys, *Du projet d'une distribution générale d'eau dans Paris, considéré sous le rapport financier* (Paris, 1830), p. 18.

32. Darcy, *Les Fontaines publiques*, p. 561.

33. Gaultier de Claubry, "L'emploi du charbon pour le filtrage en grand des eaux destinées aux usages domestiques," *Annales d'Hygiène publique* 26 (1839): 386; Darcy, *Les Fontaines publiques*, pp. 553–555. The Souchon procedure was used in Paris and in Nantes about 1848–1850.

34. Ibid., p. 572.

35. Genieys, *Essai*, p. 130.

36. Ibid., p. 133. With steam, it would have cost 16.5 million francs to elevate 140,000 cubic meters to 25 meters.

37. AD, Somme Oa10, letter from the mayor to the prefect on May 6, 1824;

clogging also in Toulouse in 1823, d'Aubuisson, *Fontaines de Toulouse*, p. 123.

38. In 1855, steam engines from Nantes that generated 60 horsepower cost 90,000 francs apiece. Darcy, *Les Fontaines publiques*, p. 553.

39. Chartres, projet Lefort (1838): 11,220 francs a year for coal and up-keep (without manpower); the reimbursable interest is 11,250 francs. Projet Hubert (1839): 14,500 francs a year (interest 11,500 francs), Bethouart, *Histoire de Chartres*, I: 372. For Toulouse in 1819, 60,000 francs a year for coal was predicted. See d'Aubuisson, *Fontaines de Toulouse*, p. 162.

40. D'Aubuisson, *Fontaines de Toulouse*, p. 112. It took all of d'Aubuisson's prestige to persuade the municipal council to use the pumps.

41. Archives municipales, Amiens (hereinafter AM, Amiens), resolution of the municipal council of May 8, 1830.

42. For Chartres, see Pellerin, *Distribution des eaux de la rivière*, p. 6; for Strasbourg, see Lornier, *Mémoire sur l'alimentation en eau de Strasbourg* (1856), p. 42. See also Claude, *Strasbourg, projet urbain, 1850–1914. Assainissement et politiques urbaines* (thesis EHESS, Paris, 1985), pp. 92–95.

43. In Gray, a ceiling above the steam boilers had to be torn down; see Boudsot, *Compte rendu de l'établissement des fontaines de la ville de Gray, Haute-Saône* (Gray, 1838), p. 2; in Amiens, the criticism is more severe: "Mr. Chaussey feared giving the project monumental characteristics. The main building gives a mediocre impression, appearing more like a type of pavillion than a water castle. The accompanying niche is small and paltry." AD, Somme Oa10, August 18, 1823.

44. Darcy, *Les Fontaines publiques*, p. 245.

45. Buffet and Evard, *L'eau potable à travers les âges* (Liège, 1950), p. 145.

46. The conduits of Auxerre from Creusot cost 5,124 pounds in 1790 or 1 franc a kilogram: Demay, "Les procès-verbaux de l'administration municipale en 1790," *Bulletin de la société historique et scientifique de l'Yonne* (1954), p. 474. They cost .39 francs a kilogram in Paris in 1834: Emmery, "Egouts et bornes-fontaines," *APC* 1 (1834): 286; in 1840 the cost was .29 francs a kilogram in Dijon. The castings came from the mills of Merley (Meuse) and of Busy (Haute-Marne).

47. The cost was 1,000 pounds in Auxerre: see Demay, "Les procès-verbaux," p. 483; in Toulouse, it was 80 francs a meter: see d'Aubuisson, *Fontaines de Toulouse*, p. 297.

48. The city of Toulouse paid 6.50 francs a meter for castings in 1828, but in 1839 the Hubert project in Chartres cost between 14 and 33 francs a meter. Lead cost 1.2 francs a kilogram in Toulouse, Chartres, and Amiens (in the last city, the lead came from Paris), .58 in Dijon in 1840, but the average price in France according to Heron de Vellefosse, *Recherches statistiques sur les métaux en France* (Paris, 1825), was .52 francs for a kilogram of pig iron and .65 francs for a refined kilogram: see Babbage, *Traité de l'économie des machines et des manufactures* (Paris, 1833), p. 220.

49. D'Aubuisson, *Fontaines de Toulouse*, p. 298: 1 line per inch of diameter.

50. Genieys, *Essai*, p. 179.

51. D'Aubuisson, *Fontaines de Toulouse*, p. 300.

52. In 1824, in Amiens, the formula used was $e = D/10$ (AD, Somme, Oa 10, no. 9), but in Chartres in 1839 the formula of d'Aubuisson was used.

53. The thickness is divided by 8 and the price of delivered castings by 1.5.

54. The tin used was usually expensive: 3 francs per kilogram or eight times the cost of the casting. See AD, Somme Oa10, no. 9 (1824), no. 13 (1825), no. 14 (1826).

55. "We understand bridle to mean a double crown of iron or copper, being bolted, taped and then screwed one to the other." Anon., *L'art du plombier et du fontainier* (Paris, 1773), p. 112. The joints were made from a lead washer, three turns of bituminous cord and a last lead washer. This method also provided for movement of the pipes, either when they were installed, or later as the ground settled. The joint also reduced the swelling of the pipe. In Amiens, in 1823, the architect recommended that cast iron be used in order to permit the eventual reuse of the pipe. See, AD, Somme, Oa10, no. 6.

56. Dupin, *Géometrie et mécanique des arts et métiers et des beax arts* (Paris, 1825–27), 2:143.

57. Darcy, *Les Fontaines publiques*, p. 412.

58. Vicat, Crozet, Gueymard, Correze, Chaper, and Breton, "Rapport sur la situation des conduites d'eau des fontaines de Grenoble," *APC* 1 (1834): 355–363. The other known cities are Aubenas, Saint-Etienne, Cherbourg, Saint-Chamont, Annonay, and Paris.

59. The conduits in a magnetic south–north direction would be in phase with the terrestrial magnetism, the lead of the bridle, and the iron of the pipe to form a battery; ibid., p. 362; introduction of organic matter would produce conferva; see Fournet, "Observations sur la production des tubercules ferrugineuses dans les tuyaux des fontaines de la ville de Grenoble," *APC* 2 (1834):175; for Payen, "Mémoire sur les oxidations locales et tuberculeuses du fer," ibid. 2 (1835): 114–117. They are provoked by alkaline solutions. For Gras, see "Note sur les obstructions," ibid., 118–125. Their origin is sulfur of the mastic. See Gueymard and Vicat, "Compte rendu des expériences faites à Grenoble en 1834, 1835 et 1836 sur les enduits propres à prévenir le développement des tubercules ferrugineux dans les tuyaux de fonte," ibid., 249–264. The minimum thickness had to be 2.5mm. See also Gaudin, "Note sur la conduite d'eau de la ville et du port de Cherbourg et particulièrement sur les moyens d'enlever les tubercules formés dans son intérieur," ibid. 2 (1851): 341–342.

60. Vicat *et al.*, "Rapport sur la situation des conduites Grenoble," p. 360; Gaudin, "Note sur la conduite d'eau Cherbourg," p. 352.

61. Janvier and Biston, *Manuel*, pp. 187–188; and Fremont, *Origine et évolution du tuyau* (Paris, 1920), p. 72. Around 1850 vitrifiable lead pipe appeared on the market; see Darcy, *Les Fontaines publiques*, p. 208.

62. In 1828 a "physical tube" made of a table of casting, rolled around a mandril and welded, was used; Janvier and Biston, *Manuel*, p. 190, claim that these are not solid.

63. In 1820, the cost was 6.77 francs per meter versus 15.97 francs for the

clay, 16.09 for the cement, 17.19 francs for cast iron, and 39.67 francs for lead; see Genieys, *Essai*, p. 210.

64. Its maximum diameter does not exceed 8 centimeters; ibid., p. 190. Fleuret, *L'art de composer les pierres factices aussi dures que le caillou* (Pont-à-Mouson, 1807).

65. For Gray, 85,800/240,525 francs; for Amiens, 42,000/124,000 francs; and for Toulouse, 372,000/1,070,000 francs.

66. Patte, ed., *Traité de la décoration* (Paris, 1776): 2: 402.

67. Archives municipiale, Chartres C1c (1612) and C2h (1706); for Auxerre, see Leboeuf, *Mémoire concernant l'histoire civile et ecclésiastique d'Auxerre* (Auxerre, 1856), p. 283.

68. Lefevre, *Annuaire statistique, administratif, commercial et historique d'Eure-et-Loire pour 1845* (Chartres, 1845), p. 323.

69. Bethouart, *Histoire de Chartres: 1789–1900*, p. 383.

70. Decambre, "Notice sur la vie de Périer," *Histoire de l'Académie des Sciences*, 1818, p. 59.

71. AM, Amiens, DD 247 (July 5, 1775), and Riquier, *Traité d'économie pratique* (Reims, 1780), p. 154. Thomas, "Les Fontaines," p. 34.

72. Féry, *Mémoire sur l'éstablissement des fontaines publiques dans la ville d'Amiens* (Amiens, 1749), p. 39.

73. It is estimated that they delivered 30 units or 600 cubic meters in about twelve trips a day and made 1.2 francs. Genieys, *Essai*, p. 154.

74. The numbers are relatively constant because there were about 200 porters in Paris serving Philippe le Bel and about 100,000 people. See Geraud, *Paris sous Philippe le Bel d'après les documents originaux* (Paris, 1837).

75. Bethouard, *Histoire de Chartres*, p. 67; for Mulhouse, see Penot, "Note sur l'analyse de quelques eaux de Mulhouse," *Bulletin de la Société Industrielle de Mulhausen* 2 (1829): 460; for Paris, see Genieys, *Essai*, p. 154.

76. *Journal du Rouen*, June 11, 1823.

77. In Paris, the members of the royal family had had this privilege since the beginning of the eighteenth century.

78. Féry, *Dissertation sur le projet qu'on formé de donner des eaux à la ville de Reims* (Reims, 1747), p. 17; Riquier, *Traité d'économie pratique ou moyens de diriger par économie différentes constructions* (Amiens, 1780), p. 152.

79. Dubuat, *Principles d'hydraulique*, 3 vols, 3rd ed. (Paris, 1816); Prony, *Recherches physico-mathématiques sur la théorie des eau courantes* (Paris, 1804); Girard, *Essai sur le mouvement des eaux courantes* (Paris, 1804); Lahiteau, *Traité d'hydraulique expérimentale* (Paris, 1824). For a general discussion, see A. Guillerme, *The Age of Water: Cities, Water and Technics* (College Station, Texas, 1988), pp. 212–217.

80. D'Aubuisson, *Fontaines de Toulouse*, p. 313.

81. Darcy, *Recherches expérimentales*, p. 22.

82. Cerne, *Les fontaines de Rouen* (Rouen, 1898), p. 34.

83. Genieys, *Essai*, p. 48.

84. Darcy, *Recherches expérimentales*, p. 37.

85. D'Aubuisson, *Fontaines de Toulouse*, p. 84.

86. Genieys, *Essai*, p. 49.

87. Darcy found the law of soil permeability, which bears his name and which is the fundamental law of soil mechanics, by studying the manner in which Dijon was furnished with water; see Darcy, *Recherches expérimentales*, p. 22; the Artesian well technology was used for finding petroleum in the late nineteenth and twentieth centuries; and, dam technology was critical in regard to electricity generation.

88. Mallet, "Rapport sur la distribution des eaux dans Londres imprime par ordre de la chambre des communes," *Journal du Génie Civil* 2 (1829): 271–294. The Thames was polluted by the "manufacture of gas, by steam boats, by the sewers which empty into it, and by detritus which came from agricultural production. Some suction pipes are located in front of sewer outlets" (p. 276). "This water has killed off the fish in some places, thus even when filtered it must still be considered suspect in regard to its drinkability" (p. 277).

89. Masclet, "Système hydrauli-économique de la ville d'Edinburgh," ibid., pp. 441–462. The letter by Stevenson mentioned on p. 450 is dated October 11, 1825.

90. *Notice et table déstinées a faciliter les calculs des divers elements d'une distribution d'eau* (Paris, 1839).

91. It should be remembered that the mechanics prize given in 1789 by the Royal Academy of Science in Paris was awarded for "the best method of distribution . . . for a given volume of water in the different sections of a city, with consideration for the unevenness of the ground." See also Ducrest, *Traité d'hydrauferie* (Paris, 1809).

92. Mirabeau, *Sur les actions de la compagnie des eaux de Paris*, p. 2; and Mallet, "Notice historique sur le projet d'une distribution générale d'eau à domicile dans Paris," *Journal du Génie Civil* 8 (1830): 275.

93. *Des eaux de Paris, depuis l'établissement des pompes a feu de Chaillot et du Gros-Caillou* (Paris, 1829), p. 5.

94. Julliot, "Sur un projet de distribution d'eau à Paris," *Journal du Génie Civil* 9 (1830): 148.

95. A. Guillerme, "From Lime to Cement: The Industrial Revolution in French Civil Engineering (1770–1850)," *History and Technology* 2 (1986): 25–85.

96. Mallet, "Notice historique," p. 276. Cf. Masclet, "Approvisionnement d'eau de la ville et des usines de Greenock, sur la Clyde, en Encosse," *Journal du Génie Civil* 7 (1830): 473–486.

6 The Development of Water and Sewerage Systems in France, 1850–1950

Jean-Pierre Goubert

WATERWORKS history can be conveniently analyzed by a study of the evolution of water systems in Paris and the large and medium-size towns in France during the nineteenth century. The development of waterworks can be ascribed to certain "golden ages." The First Empire and the Restoration witnessed both the planning and construction of several projects; the Second Empire saw funding for many towns in addition to Paris; the Edwardian era saw smaller and more conservative towns (e.g., Limoges), as well as more progressive towns (Toulouse), inaugurate or complete their systems; finally, the period between the two world wars resulted in considerable technical evolution and the expansion or replacement of older works.

The first important factor involved in the development of waterworks is the direct correlation between periods of military strength, political power, and economic growth and the boom of urban sanitation equipment. This is not surprising; if a country wishes to be strong, it needs to keep its citizens healthy for economic (production), military (respect for conquest), and financial reasons, and for social (keeping the peace) and cultural order (providing emotional security). Thus, only strong and powerful states set up modern hygienic systems. The state movement for improving urban sanitation usually began in the larger cities, the most suitable locations during the nineteenth century for sustaining a strong populace.

A second important factor concerns the social geography of water systems and the distribution of public or communal water services. Army barracks, hospitals, and schools had first priority because they collected more pollution and because they were the backbone of public order and education. A second priority involved construction in important and older urban areas, which were believed to be more polluted because of their age. In all locations, the upper class demanded water-works to reaffirm their power and accentuate their social position. By 1840, however, only a few privileged beneficiaries had home distribution, and waste systems controlled by municipal authorities were quite limited in their scope of operation. In Paris, for instance, the water distribution system was "an imitation" of a much older system.[1]

A third reason for the rise of waterworks in urban areas was because the "geometry" of water and sewerage systems favored the center and neglected the periphery. Boroughs with train stations constructed in the second half of the nineteenth century were an exception to this rule because steam engines required ample water supplies.

Another major factor in the development of waterworks was the monumental character of the systems constructed after 1860. Visible in the form of above-ground and ground-level aqueducts, ostentatious when seen in splendid fountains (such as in Paris, Montpellier, Rennes, Toulouse, and Angers), noticeable in water castles, and later in reservoirs and decantation and purifying plants, these systems, when in operation, solemnly inaugurated and symbolized the industrial production of water in the nineteenth century.

Municipal officials and professionals involved in waterworks construction, among them doctors, architects, and engineers, modeled their work on Greco-Roman culture, and they often followed Roman models,[2] whether in France, other occidental European countries,[3] or North America.[4] The engineer G. Bechmann, for example, at the beginning of the nineteenth century, when he was studying ancient water systems,[5] expressed admiration for the Roman works. And Teissier-Rolland wrote in 1843: "In a very well constructed aqueduct, large-scale maintenance and repair should be nonexistent for 2,000 years; the example of all the Roman aqueducts is here to prove it."[6]

Municipalities in the First and Second Empires and, to a lesser degree, the Third Republic, were seduced by these Roman aqueducts, and in drawing on them for their own systems, reproduced the strongly built models of Mediterranean antiquity. But if they were conscious of this heritage, their engineers (such as Belgrand in 1875) knew how to recognize the technical distance between themselves and Rome. Assuredly they made use of the Roman heritage by adopting the large Roman models, but they also knew that they could surpass the Romans.

Their vision and achievements are considered here from a technical viewpoint, a social viewpoint, and a financial viewpoint. From the technical point of view, the engineers demonstrated their prowess by undertaking totally new works and questioning older systems. From the social point of view, they aimed at serving each household with purified water and sewerage evacuation systems. Thus Bechmann defined, for the first time in a modern way, the distribution of water as "a collection of works conceived and combined to bring to all of a community water suitable for every use."[7] Lastly, from the financial point of view, first the communes, and after 1902 the state, allotted considerable sums for waterworks projects. The Ourcq Canal (a project begun in 1802) cost Paris 85 million francs.[8] The Durance Canal cost 36 million francs in 1851.[9] In 1847, the city of Lyon contracted to borrow 7 million francs[10] to realize a grandiose project conceived a few years earlier.[11] In the same period, Bordeaux spent 4.2 million francs to divert the Taillan. In cities such as Brussels, London, New York, Philadelphia, and Amsterdam, financial costs for waterworks reached similar levels.

Despite the focus on urban centers, supplying towns and villages was also essential in France by the end of the nineteenth century. To market towns, villages, hamlets, and isolated individuals, the gigantic water projects constructed for the large French cities represented an image of utopia, much like Jules Verne's *Hygeia*. As clearly expressed in the *Britannica Review*, which appeared in Brussels at the beginning of the nineteenth century, "a water mania, which we can only compare to the recent railroad mania,"[12] was born. This mania on the part of intellectuals, technicians, and politicians for water and sewerage systems can be explained by their conviction that these technologies would bring a higher standard of living and better public health.

From the first half of the nineteenth century, large and small French villages were dedicated to constructing water and sewerage systems. Small towns such as Morbihan or Nivernais often competed, on a modest scale, with their chief regional centers. This competition resulted in a distribution of the knowledge and techniques necessary for the completion of modern sanitary projects. In most French villages, their development was supported in the early years of the nineteenth century (around 1830) by administrative bodies, the medical community, and scholarly experts. The use of outside "experts," those who did not reside in the area, was a common practice since the time of Lumiers. Evidence of the rural proliferation of water systems would ideally require an exhaustive "urban prosopography," but for this essay a limited sample, based on the archives of the Superior Public Council of France for 1884 to 1902, is utilized.

Analysis of a Sample, 1885–1902

The sample was gathered from the files of 582 communes in 25 departments.[13] It was chosen in a way that did not favor rural or urban areas, wealthy or poor districts, educated or uneducated areas, or the hydraulically favored or unfavored cities. This sample is supposed to represent the "true France" of 1900. Three-quarters of the studied communes (76.2 percent) had fewer than 2,000 people, and one-quarter (23.6 percent) had from 2,000 to 30,000 inhabitants.

France in the years from 1885 to 1902 was attempting to evaluate and improve its water and sewerage systems as illustrated through projects submitted to the Superior Council on Public Hygiene. In these years, small communes (those with fewer than 1,000 people) affirmed their involvement in the modernization of public health. Although small communes made up 58.9 percent of the total number of communes, they contained only 14.6 percent of the people, versus the communes of greater than 5,000 people, which accounted for 49.4 percent of the population. Despite a great effort on the part of rural communes, modern hydraulic and sanitation equipment (waterworks and sewerage systems) remained a typically urban phenomenon.

From the previously constructed sample, the distribution of waterworks in rural communes confirms our expectations. Waterworks were altogether absent from eight of twenty-five departments, but they were licensed in the small and medium-size cities of a dozen departments. In neither of the two cases cited does this presence or absence seem related to one or more of the traditional determining factors of social historians, namely geographic, socioeconomic, or cultural factors.

It is true that the low number of registered (licensed) observations does not permit us to test this correlation statistically. For each department, our sample does not provide a bell-shaped curve for given dates, but it does give us an overview of the whole period from 1885 to 1902.

In cases where the archives of the Superior Council of Public Health contain contradictory data about certain communes, it is not surprising to find that their water-distribution equipment was limited and often "archaic." In contrast, documents in the archives show that water supply systems were in greater supply than we would have expected. In agricultural areas such as Doubs, for instance, the problem of watering livestock led to the development of plans for an efficient water provision system before the 1880s. This system provided high-quality water in ample quantities, using the criteria of the time. Second, where nature was bountiful, many wells were dug or adapted, springs harnessed, and fountains built or enlarged. The fountains provided an average of nearly 20 percent of the water distribution, the wells more than 25 percent, springs close to 10 percent, and running water greater than 10 percent. In the country, domestic and agricultural needs were combined to pool the uses of one or more sources. In towns, the density of the population, the importance of removing human and animal wastes (horses were a problem), the needs of industry, and familiarization with progressive sanitary standards created a strong demand for water before the 1885–1902 period.

What appears to be new around the 1890s is the realization by certain communes that their current systems were outdated and not adaptable to their new lifestyles. There was a general absence of scientific criteria for determining water quality during a time when purely formal criteria was in vogue. At the end of the nineteenth century, clear and running water, originating deep in the earth and gathered from fountains, springs, and wells, was favored. In contrast, the use of groundwater or stagnant water was almost unheard of. The absence of water-distribution technology and the lack of private distribution systems reflected this traditional vision of what constituted a good water supply.

In the time of Pasteur, technical means of providing water, such as reservoirs, water galleries, pumps, cisterns, and water-carrying barges, occurred in extremely small numbers, with an estimated 1 percent of communes possessing any water-distribution technology. Street fountains were present in only about 2.5 percent of the communes studied. Even if we assume that these communes were among the best equipped, there is still a net difference between the towns, which had equipment, and the countryside, which ignored all modern hydraulic equipment. Though communes possessed springs or fountains and the majority had washhouses (56.8 percent), and a small minority had watering places (5 percent), in general they were more populous communes and more closely associated with the market town than with the small village.[14]

From the same sample, it is also clear that systems of wastewater removal were not prevalent around 1900. Sewers, often in embryonic stages, existed in only 26.5 percent of the communes and less frequently in important cities, many of which contained more than 5,000 people. Also, the flushing of wastes and wastewater was practiced in a "primitive way" not only in the countryside but also in the medium-size towns, and even in the cities.

In town, the method was "everything-in-the-street." Wastes were often thrown from windows into the street or dumped in gutters, which eventually led to a pit or cesspool. The contents of these were disposed in a pit, on a dunghill, in a neighborhood waterway, or again in a gutter, where one existed.

In the countryside, wastewater disposal still took advantage of ground and subsoil absorption. Swamps, streams, brooks, groundwater, well water, dunghills, courts, and gardens made natural receptacles; cesspools, basins, and cisterns were rare.

Flushing, transporting, and removing sludge remained the prerogative of large market towns and small cities.[15] Water-treatment plants were the most rare. Only 6 of 253 towns and cities owned treatment plants, and these communes contained an average of 5,075 people. In the cities, as in the countryside,[16] wastes and wastewater were used as fertilizer; 8 percent of the communes (21 of 253) "recycled" their night soil.

In contrast, there were ample numbers of public toilets around 1900. From the sample, 70 percent of communes possessed at least a few public toilets; they were found principally in public buildings and among inhabitants who were more affluent and more concerned with public health. The presence of public toilets in each building and, more significantly, a toilet in each house, was not the norm for most communes, although around 1900, 21 percent of them claimed to have a toilet in each house. Size of the community seemed to be the determining factor in regard to the possession of public and domestic waste facilities. Only those in cities paid attention to this aspect of sanitation. As a rule, cities with more than 4,000 to 5,000 people had public toilets, while the communes with fewer than 1,500 people were not provided for.[17]

The story for sewers is similar. The average size of communes with sewers was 10,000 inhabitants, whereas those without averaged 1,400 inhabitants. Quality of the system, however, seems to be relatively independent of size of the commune. Equal quantities of porous and nonporous gutters were in the villages, as well as in the countryside. Similarly, wastewater evacuation from public washhouses did not depend on size of population. In towns, as in the countryside, the watercourses, ponds, pools, and the sea were usually used as receptacles.[18]

A major finding of this analysis is that the diffusion of technical innovation was assured by the hierarchical network of communes. The one reservation is that the quality of the "product" furnished by this equipment remained largely independent of all socioeconomic determinants.

Forms sent by the central administration in the inquiries of 1892 and 1912, and by the Public Health Council of France on a regular basis beginning in 1884, informed local public health officials of the "defects" of supply, distribution, and evacuation of water in their communes. Also, the projects officials submitted to this council showed a negative balance sheet for the equipment necessary to carry out their projects. The principal defects of water supply systems were primarily insufficient quantity (49 percent of the communes) and quality (31.5 percent), and, less importantly, remoteness (13.5 percent) and poor condition of equipment.

Among the factors contributing to insufficient quality, 57 percent of those responding blamed the wells. In reality, this meant the absence of a correct supply and evacuation system. After 1892, when the form was modified by the Superior Council of Public Health, public health officials sought a relationship between the poor quality of water supplied and an etiologically defined mortality, caused by typhoid, cholera, and dysentery. In total, between 1892 and 1901, 168 of 378 communes (44.4 percent) were touched by these three diseases, primarily typhoid, although it was abnormally absent in our sample in Haute-Marne[19] and in Mayenne.

Once this negative balance sheet is drawn, the works projects can be regarded as an aspect of modernization that touched all of France during the nineteenth century. I shall limit myself to retracing the chronology and the geography around the period from 1880 to 1900 for the waterworks projects and, to a lesser degree, for the sewerage projects.

From December 1884 to the end of 1897, 791 projects received favorable reviews by the Consulting Committee of Public Health of France.[20] From 1885 to 1897, the number of favorable reviews delivered by the committee rose regularly for France (except during 1896 and 1897). There were 46 in 1885, 60 or so from 1886 to 1889, 80 to 90 in 1892, around 140 from 1893 to 1895, and more than 100 in 1896 and 1897. In our sample of twenty-five departments, from which some 600 projects were gathered, the opinion of the committee was favorable in 67.7 percent of the cases, guardedly favorable in 9.6 percent, unfavorable in 9.4 percent, and absent in 14.3 percent of the cases. In the twelve years under study, the total number of projects presented before the committee rose to about 1,200. The rhythm of construction at the end of the nineteenth century was slow: about 100 projects per year were spread among 87 departments, which included both the mother country and Corsica. Thus, the average was little more than one project per year per department.

Once again, the geography of development does not seem to correlate with the diverse parameters that usually encapsulate a country's degree

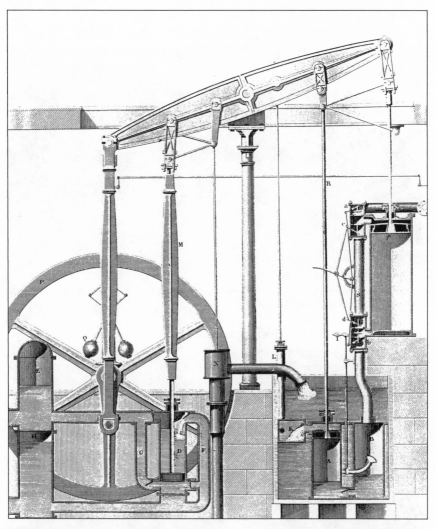

Steam machine installed at Gros-Cailloux, Paris, 1838. This machine was patterned on Watt's system. Drawing from Genieys, *Essai sur les moyens de conduire, d'élever et de distribuer l'eau* (Paris, 1829), vol. 2, plate 6.

A water-storage reservoir of a type commonly used in French towns at the turn of the century. Courtesy Jean-Pierre Goubert.

A fountain in Nantes, La Place Royale, about 1890. Courtesy Jean-Pierre Goubert.

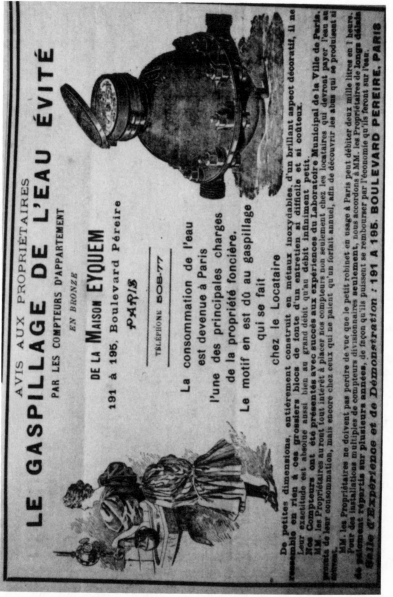

An advertisement in a Parisian newspaper (circa 1900) for an apartment water meter to reduce water waste by tenants. Courtesy Gabriel Dupuy.

of modernization. Neither educated France, wealthy France (agricultur-ally and industrially based), nor physically healthy France (established from recruitment registration) equipped itself faster than less educated, poorer, and less healthy segments. At the turn of the century and during the period from 1830 to 1860, it was no longer only urban France that became modernized.

Let us review this unusual geography point by point, while refining it, from 1884 to 1890 and from 1890 to 1897. The first time span is the period when administrative machinery was inaugurated. The debut was rapid, taking only one year. After 1886, projects came from all areas. Meanwhile, from 1884 to 1890, the annual departmental average was still low: .55 projects versus .86 from 1890 to 1897. But during the 1884–1890 period, as in the following 1890–1897 period, the spatial distribution was the same. Overall, eastern France equipped itself fastest, excluding the Mediterranean Midi and Gard. Industrial France of the north, the Parisian region, rural western, central, and southwestern France were the slowest in undertaking projects.[21] If we outline a geographic balance sheet over fifteen or so years on a national level, a minority of the departments seem "privileged," having developed three times faster than the national average. These are, in alphabetical order: Ain, Aude, Côte d'Or, Doubs, Drôme, Eure-et-Loir, Gard, Isère, Haute-Marne, Meurthe-et-Moselle, Haute-Saône, Savoie, and Vosges. All of these departments, except Eure-et-Loir, are situated in the eastern half of the country, and they testify to the new geography of France at that time—to the north–south opposition at the end of the eighteenth century and the beginning of the nineteenth century, and to the east–west configuration of the twentieth century (as studied before the war by many geographers). Although developing waterworks was not peculiar to this set of departments (out of a total of eighty-seven departments, sixty-five were equipping themselves during 1884–1890 and 1890–1897), development is more apparent in the twelve cited departments either because they had an initial (pre-1890) advantage or because, as in Aude, Doubs, and Eure-et-Loir, they had "jumped on the bandwagon" before 1900.

So the choice was to modernize. The proof is that more than 80 percent of the forms in our sample that were sent in by the communes were correctly filled out. The chemical and bacteriological analyses were also successfully carried out in the same proportion. In addition, the average market volume was almost never less than the scientific norm of the 1880s, namely 100 liters of water per person per day. This was often exceeded because, for twenty-five departments, the projected market volume was around 370 liters. Finally, cast iron supply pipe was used in two-thirds of the cases studied. It was adopted particularly in those cases where the water was of high quality and met expert standards (spring water in 68 percent of the cases). The subsidies granted by the state fed an increasing demand and permitted normalization of the waterworks.

The mode of distribution was already selected; it was public in three-quarters of the cases. This shows that tapwater for all homes (private distribution) was still a distant goal. The most urgent priority seemed to be the establishment of public fountains sufficiently supplied with clean water and in close proximity to private homes, in order to serve the majority of the population.[22] When this was achieved, the first level of development was attained; water—clean water—was brought closer to man.

This modernization, contrary to the common stereotype, was not identified only with urban areas. Heretofore in our analyzed sample, the average size of the 582 communes is only slightly more than 1,300 people. Also, 208 of the communes studied had fewer than 1,000 inhabitants and constituted 61.3 percent of the total number of communes, even if they had only 14.6 percent of the total population of the sample.[23] The same figures appear when we analyze the data of Henri Monod; they concern the waterworks projects submitted to the Consulting Committee of Public Health of France between 1890 and 1897.[24] Of the 633 communes equipped, 382 (62.9 percent) had at least 1,000 people. Yet cities of more than 10,000 people (25 total) comprised 74.2 percent of the total population of the sample and formed 4.1 percent of the communes. This double deficit (in number of communes and number of inhabitants) changes the stereotype of people in towns versus people in the country. If the construction of waterworks was made more urgent by the concentration of people in the cities, there was also a demand for them in rural areas, at least in areas where the people had confidence in their leaders.

Therefore, the modern equipment, identified by its installation, remained typically urban until around 1900. In the departments of the North, from a prefectural study of 1907,[25] 6.2 percent of the communes had total or partial waterworks, and only 3.2 percent had total distribution. The cities monopolized 75 percent of the distribution and comprised most of the fully equipped communes. In the other communes, principally rural, the normal source of water was wells and fountains, which were guaranteed abundant and "very pure" by mayors. Many individuals owned wells, and often the communes equipped themselves to pump water from the community well or wells.

Research in underseries F8 of the National Archives[26] confirms an impression of two kinds of waterworks, one mainly rural, with town wells being used until 1940 or 1950, and the other essentially urban. Thus, as we leave the nineteenth century, modern equipment was spreading over the land, starting in 1912–1913[27] with medium-size towns (20,000 people) of the French mainland (Aisne, Basses and Hautes-Alpes, Bouches-du-Rhône). This modernization affected waterworks as well as sewers.

In this sample, which dates from 1912, some regional differences are clearly visible, with each department representing one region. If an average of these eleven departments was established among that of the remaining departments for the distribution of water (77.4 percent) and for the distribution of sewers (63.3 percent), the departments of Cantal, Gers, Gironde, Ille-et-Vilaine and Nord are clearly below average, some for waterworks, others for sewers. These percentages are high because the sample contains only 73 communes (of a total of 122) that contained fewer than 10,000 people, and 10 communes containing fewer than 2,000 people. Even so, the Nord department, well supplied with water by nature and man (i.e., wells) has a low percentage. Only 47 percent of the communes of the Nord that answered the questionnaire owned waterworks, and 71 percent owned sewers; Aisne claims, respectively, 70 percent water and 80 percent sewage for communes of comparable size.

In 1912, as in 1890, the preference was still for spring water (57.4 percent of the cases), followed by stream water (12.1 percent), which was principally captured in plains along the large waterways (Bouches-du-Rhône, Gers, and Gironde). Meanwhile, even in urban communes, the use of wells remained great. Of the five departments that answered the form, the use of wellwater was present in nearly one-fourth of them (21.5 percent). The marked preference was for spring water, which was drawn and purified before distribution. Using river water made this procedure a necessity, at least in one-fourth of the most important communes of one department (Bouches-du-Rhône) and in half the communes of another (Gironde).

The oversight and analysis of distributed water varied greatly by department: Hautes-Alpes and Basses-Pyrénées put their faith in nature; Basses-Alpes, with reason, did not. Gironde and Bouches-du-Rhône analyzed distributed water insufficiently, even considering their rate of water purification. The other departments' analysis of distributed water was sufficient and sensibly above the norm of 42.4 percent.

If there was a geography to waterworks in 1912, as there was in 1890, it does not hold for sewers in 1912. Of the eleven departments studied, sewers existed in two-thirds or three-fourths of the communes, except in Gironde (54 percent). In the majority of the small and medium-size towns, an embryonic purification system became an issue. All the sewers of Paris stood out here because of their size and, of course, certain very large cities, such as Bordeaux (with 106 kilometers of sewers) and Marseille (230 kilometers), also had sewers. In effect, the majority of the towns, small and medium, owned between a few dozen and a few hundred meters of sewers. The main function of these sewers was to transport rainwater and household wastewater (64.8 and 57.5 percent of the cases); more rarely, they transported wastes from toilets (30.3 percent) and industrial wastewater (34.7 percent). Industrial wastewater was, more often than not, dumped directly into the nearest river.

A considerable gap existed just before World War I between rural communes, small and medium-size towns, and regional capitals in the development of public health systems. The differences became overwhelming between the two world wars, as shown by research in the underseries F10 of the National Archives,[28] and in a statistical analysis of the series 2-0 of the department of Aube.

In the department of Ain, which initiated waterworks modernization in 1890, the boroughs and the countryside equipped themselves from 1910 to 1930. So Versonnex (161 people) was not satisfied with "a water pump placed in the proximity of stables, stalls, and dungheaps . . . which constitute a veritable public health danger."[29] The community expected to exploit two sources, divert 3 liters per second, and therefore supply 1,600 liters per person. In accordance with hydraulic engineers, the town council accepted an appropriation of only 90 liters per minute, giving 800 liters per person. This was not a considerable amount because these figures do not take into account farm animals, mostly horses and cows, although they could each consume 20 to 50 liters of water per day. The project was drawn up in 1911, endorsed and subsidized in 1914, taken up again in 1922 after an interruption due to the war, resubsidized in 1923, revised in 1926, and executed from 1927 to 1930. The Versonnex case is not an isolated one; many communes had projects from before 1914 that were not executed until between 1920 and 1930. The priorities of World War I overcame sanitation priorities of the "back country," and waterworks close to military forts and major railways were developed first.

Thus, on September 25, 1908, the minister of war asked his colleague from the agricultural department[30] to create a "hydraulic hookup" with the train station of Montreuil-sur-Ille, "with the objective of assuring the mobilization intensive circulation of military trains on the Rennes–Saint-Malo line."

Nevertheless, from 1920 to about 1940, ameliorating and extending the distribution of potable water, and not creating it, was the primary concern. Research reveals that innovations for transporting water were more numerous in the "backward" departments of the center and west, particularly in the market towns. Menetreols-sous-Vatan (Indre) asked for a subsidy in 1939 to build a water trough with a supply reservoir of 600 cubic meters, which would be supplied by two springs, because "the market town . . . is actually supplied water by a few wells whose levels are insufficient in times of drought. To supply themselves, the farmers resort to bringing barrels of water to satisfy their domestic and agricultural needs, and for this they make long trips."[31] The project, accepted on October 9, 1940, by the Superior Council of Public Health, was to receive the amount of the agreed-upon subsidy only in 1946, *after* the completion of the construction.

As a general rule, small and medium-size towns improved and expanded the waterworks already begun. For example, Château-la-Vallière, like other market towns, and Capelle Blanc Saint-Martin in 1930,[32] created a water supply and a distribution system in 1928. Yet, in both the most anciently equipped departments and in those familiar with rapid development (e.g., resorts with hot springs and those on the sea), the work undertaken between 1900 and 1940 often consisted of projects to enlarge and extend the networks of water supply and distribution for daily use. At Gahard (Ille-et-Vilaine), for instance, the project of the commune concerned the extension and distribution of potable water destined for the public school and the communal bathhouse.[33] In 1907, the serviced population rose to 300 people (of a total of 1705). The project was finished in 1895, and distribution was accomplished with the help of five public fountains and two special concessions. For 307 people and about 150 head of livestock, the consumption was 5 cubic meters per day per person or animal. To meet these conditions, the plan entailed replacing a section of the main conduit, building a hookup with the boys' school (.2 cubic meters), which had 105 pupils, lengthening the principal conduit to the public washhouse (4 cubic meters), and arranging for its evacuation. The daily water allotment jumped from 16 liters per serviced person to 24 liters. This included the estimate that one-third of the students were already part of the community. The promise of this project was judged "of an uncontestable interest."

Construction in the heart of France was a slow affair. Instead of having "positivism" penetrate the spirit, it rested on the undeniable success of wiser men—technicians and entrepreneurs—and on the symbolic tradition of water. Progress was made by thousands of obscure projects, in inverse proportion to the grandiose and ostentatious hopes of the large cities. A drinking trough, a washhouse, a pump, a diversion, a ramified steel conduit, a few well-situated public fountains, a connection to the network of a town[34]—these are the works undertaken on a modest but real scale that, resulting in a transformation of spirits, transformed customs in their wake.

From this perspective, the department of Aube, though not necessarily the "national average," presents a convincing example. As a part of eastern France, which, between 1885 and 1897, equipped itself faster than the rest of the country, Aube showed the tenacity of the men who undertook small parts of the gigantic public health effort that typified nineteenth-century France. For example, studies of potable water, which effectively did not exist until around 1880, began to grow thereafter, and the hopes that sparked the development of water supply projects that started around the 1860s were fulfilled during the 1930s. The great public wells and washhouses were built between 1830 and 1850 from a collection of small "revolutions" that followed and built on each other. In this essentially rural department, dominated by a network of small and medium-size towns, two revolutions overlapped: that of the well and washhouse (1830–1900), and that of the waterworks (1880–1940). Holdouts were conquered, and necessary funds were allocated for modernizing the infrastructure of public health, even though the chronological difference between the initial project and the beginning of construction averaged twenty-two years.[35]

In fact, the connection of running water and sewers to households was far from complete at the beginning of World War II, despite the boom of public baths and showers, descended from the Russian, Turkish, and Oriental baths that were in vogue during the first half of the nineteenth century.[36] The majority of the buildings (83.1 percent) in rural communes in 1946 had neither running water nor sewer connections. In contrast, urban communes were endowed 46 to 76 percent of the time.[37] In 1946, inhabitants of rural communes still fetched their water in the courtyard (35.3 percent) or from a pump or fountain (22.2 percent). The percentages (between 12 and 2.6) are clearly inferior for towns.

Similarly, sanitary bathrooms in homes were rare in rural communes in 1946: 1.6 percent of rural homes possessed a bathtub, compared to 7 to 16.6 percent in cities; 2.8 percent of rural homes had a washstand, compared to between 7 and 9.8 percent in the various categories of cities.[38] Still in 1946, 29 percent of rural homes had no toilets, compared to 7.3 to .1 percent in cities. In the countryside and in towns of up to 100,000 people, toilets were not often connected with sewers; it was much more commonly done in Paris (97.7 percent) and in towns of 50,000 to 100,000 people in the Paris suburbs (76.6 percent).[39] Finally, the use of toilets in 1946 was not reserved to one (nuclear) family in the cities; between 29.4 and 47.3 percent of toilets were used "in common with strangers."

The census of 1946 and the results of investigation in 1941–1942 showed that comfort was still not widespread in the country. Many buildings (1 of 25) in towns of more than 30,000 people were razed during this period. According to the 1946 census, running water existed in only 37 percent of French buildings, and connections to sewers, typically "Parisian," were made in only 12 buildings of 100 for the entire country. Indoor plumbing was not much more advanced: In 1946, 5 of 100 homes had a bathtub. Finally, according to the 1941–1942 inquiry on constructed property in rural communes, 28 percent of the houses had no comforts; 14 percent had water and electricity; and 1.2 percent water, electricity, and heating. The average distance from a home to a point of water was 29 meters (national average).

Despite these mediocre results, waterworks systems underwent a revolution in the 1930s, at least in cities. In fact, the arrangement of water in proximity to the house no longer satisfied the people of the towns; they wanted water in the home, at the kitchen tap, and in the toilets. In Lozère during the 1930s, the number of subscribers having a water meter was on the rise. In Langogne, from 1930 to 1934, it increased from 510 to 720. In Mende, from 1930 to 1934, it went from 239 to 333; but at Malvieu-Ville it remained limited to "hotels and other establishments which use much water."[40]

Running Water, an Industrial Product

Water, whether as a gift from God or Nature—in any event, a free gift —became an industrial and commercial product during the nineteenth century. Society was compelled to recognize the trade value already granted water by a flourishing capitalism, instead of being left to discover the value of water by its own free will. Schools, barracks, the press, and even hospitals extolled the virtues of a source of pure, plentiful running water.

Once a commodity reserved for the upper classes, water was later allotted to technical and hygienic specialists.[41] From then on, it became an interesting economic window for big private entrepreneurs in England, as well as in France and the German Empire. In France, the Paris water company (1778–1815) patterned after that of London, was an early leader.[42] Its assets were concession, the relative purity of the product, and subscription.

But the first big private society on the French national scale dated from 1853. In that year, the Compagnie Générale des Eaux was created as a limited society and was allowed to exist by imperial decree. The report made by the board of directors on October 26, 1853, dealt with "general reflections upon the forming of the company"; "in the new era which is opening up, be assured of it, gentlemen, millions will be devoted to water supply, just as in the two preceding eras millions were used for railways, roads, sailing in the navy. . . . We are opening a mine whose wealth has not yet been gone through and whose best veins we can choose and exploit, as it is the rightful privilege of the first owner." [43] Indeed, even before its creation, the company was acclaimed by French and European business circles. One of the Rothschilds, the Laffittes, the Pereire brothers, a secretary of state (Persigny), and Napoleon III's half-brother (Morny) were among the first and foremost shareholders.[44]

From then on, water carrying was doomed to disappear; it stopped in Paris around 1890. The numbers of household-filtering fountains, wells, and washhouses decreased, at least in towns. The number of subscribers increased little by little, and they became widespread in French towns by the end of the century. Planning and building supply networks, which was finally achieved in the 1855–1860 period,[45] became necessary despite their high cost. In the same way, filtering and raising water were gradually mastered; at least, cast-iron pipes were used, thanks to a more efficient method of melting, yielding a better and less expensive product. Water supply networks appeared early in London; they were started soon after 1800 in the large towns of both France and New England (United States), and later came to Belgium. In many towns and districts they could be seen at fire hydrants, public fountains, or the collective tap of a building. But, for a variety of reasons, the "revolution of running water" did not occur in one day. People objected to the price or preferred water with a "better" taste; landlords and tenants fought over water supply and citizens objected to towns meddling in private sanitary affairs; and, water meters were only belatedly adopted. All these factors caused the "revolution of running water" to stretch over a few decades, if not longer.

The dream appeared two centuries ago; it was made possible in the

1850s; and it came true in rich parts of the New and Old Worlds. It meant being supplied with plentiful running water not only near one's lodging but also "at home." It meant at last mastering the cycle of water. From then on, the utopian ideas of the British hygienist F. O. Ward were adopted. A convinced organilist, he founded his sanitary system on the pattern provided by "the great Harvey, the famous enlightener about (blood) circulation in the individual body." So Ward aimed at being the "imitator of [Harvey's] discovery, a strictly analogous one, in the social body." Ward's system was essentially "based on the continuous circulation of water which arrived in a town pure, and on the ceaseless movement of waste that must be cleared out."[46]

Our century still bears the mark, even more so than does the previous century, of the omnipotent "fluids": water, electricity, and the various means of long-distance communication.

Distributed principally in the towns, but not in any clear pattern by the end of the period under consideration, hydraulic equipment testified to the change of attitude toward water in the nineteenth century, and toward social and sanitary uses due to the breakthroughs of intellectuals and technology. A dream, which seemed crazy at the end of the ancient regime, progressively came to fruition in the framework of a conquering civilization, at once urban and industrial: to have at its disposal, not far from home, or even at home, pure and abundant water. The twentieth century continues in this respect what the nineteenth century inaugurated, which, in spite of social division and political and military separation, benefited from an economic boom without precedent and finally inspired the country to concern itself with the health of all the people.

Notes

1. C. F. L. Panckoucke, ed., Dictionnaire des sciences médicales par une société de médecins et de chirurgiens (Paris, 1814), 10: 467–471.

2. P. Lavedan, Histoire de l'urbanisme (Paris, 1956), 3:45.

3. Liliane Viré, La Distribution publique d'eau à Bruxelles, 1830–1870, vol. 1 (1973); and Revue Brittanique, vol. 1 (1850).

4. Nelson Manfred Blake, Water for the Cities—A History of the Urban Water Supply Problem in the United States (Syracuse, N.Y., 1956).

5. G. Bechmann, Salubrité urbaine. Distribution d'eau, assainissement (Paris, 1908).

6. Teissier-Rolland, Etudes pour procurer de l'eau à la ville de Nîmes, 3 vols. (1843–1844), vol. 1.

7. Bechmann, Salubrité urbaine, p. 23.

8. Teissier-Rolland, L'eau à la Nîmes, 3:678.

9. From Rapport sur la situation des travaux du canal de Marseille au 31 decembre 1851, p. 122.

10. From Rapport sur le projet de distribution d'eau à Lyon (1847).

11. From Rapport sur le projet de derivation et de distribution d'eau de source (Lyon, 1843).

12. Revue britannique (1850) 1:284.

13. The list of departments is: Ain, Aisne, Alpes-Hautes, Alpes-Maritimes, Ariège, Aude, Bouches-du-Rhône, Calvados, Cantal, Doubs, Eure, Finistère, Garonne-Haute, Gironde, Ille-et-Vilaine, Isère, Marne-Haute, Maine-et-Loire, Mayenne, Nord, Puy-de-Dôme, Pyrénées-Basses, Rhône, Seine-et-Oise, Vendée.

14. In the sample (352 communes), the average population of towns with washhouses is 2,342 people; that of those without is 1,761 inhabitants.

15. Nineteen of 253 communes use the method of evacuation; they have on the average 7,000 people.

16. Of main interest are the rural communes with between 500 and 1,500 people and a minority of the urban ones (2,500 and 25,000 people).

17. From the 39 communes that supplied the information, the average size for those with public toilets is 2,832 people; those without average only 985 people.

18. In the vicinity of 79 percent of the sample.

19. Jeans-Louis Maigrot proclaims his ideas of the nineteenth century in thesis of the third cycle, La Morbidité en Haute-Marne au 19th siècle (1780–1914) (Dijon, 1976).

20. From Henri Monod: "From 1890 to 1897, 845 projects examined by the Committee . . . produced 856 reports, of which 35 had unfavorable conclusions," L'Alimentation publique en eau potable (Melum, 1901), p. 7.

21. This apparent slowness hides a disparity between regions: Northern France in this era is already well equipped, even if only partially, and even in the towns; southern and western France are generally much less equipped.

22. In the sample the serviced population is, at the end of the century, around 79 percent, representing a significant advance when compared to the percentage at the beginning of the century (in comparison to the total population and in rounded numbers about 424,000 and 537,000 people).

23. On the other hand, the 32 communes count more than 5,000 inhabitants including 49.4 percent of the considered population.

24. Monod, L'eau potable, pp. 103–125.

25. Archives départmentales, Nord, M 285.

26. Archives Nationale, F8, 215, 216, 217, 218, 221, 223 (search of 1892); Departments of Aisne, Basses-Alpes, Hautes-Alpes, Bouches-du Rhône, Calvados, Cantal, Gers, Gironde, Ille-et-Vilaine, Nord, and Basses-Pyrénées.

27. From the results of the inquiry of January 9, 1913, from a sample of 11 departments including 122 communes with a population of 2,488,424.

28. From Archives Nationale, F10, 2225, 2228, 2254; Minister of Agriculture, service of agricultural hydrology.

29. Archives Nationale, F10, 2225, February 26, 1914.

30. Ibid., F10, 2254, Ille-et-Vilaine.

31. Ibid., May 4, 1939 (in "champagne berrichrome").

32. Ibid., Indre-et-Loire.

33. Ibid., December 4, 1912.

34. In 1930 Saint-Brac connected with the waterworks of Dinard. Archives Nationale, F10, 2254.

35. From Archives départmentales, Aube, "under-series" 2-0, 1800–1944, enlargement of waterworks in 33 communes.

36. My search did not come across any statistics (national or regional) on the expansion of public baths and showers between 1900 and 1940. The "Turkish baths" of Paris had had their heyday. For more information, see C. Lambert, *Traité sur l'hygiène et la médecine des bains russes et orientaux* (Paris, 1836).

37. Rural communes equipped in 1946, 12.9 percent; urban communes with 2,000 to 50,000 people, 46.1 percent; provincial towns with 50,000 to 100,000 people, 57.2 percent; cities with more than 100,000, 65.7 percent; Paris, 76.5 percent. From *Etudes et conjonctures*, October-November 1957, p. 1223.

38. Of those polled, 49 percent and 58 percent did not declare whether or not they had washbasins or baths. Their silence poses a problem of interpretation.

39. The percentage of people who did not declare if they had a toilet is also large. Interestingly, it decreases with the largeness of agglomerations (43.3 percent to 0.9 percent).

40. A. D. Lozère, 2-0-60, *Enquête sur les compteurs d'eau* (1935).

41. See Chapter 5 in this volume.

42. See Jean Bouchary, *L'eau à la villes de Paris au 18th siècle* (Paris, 1946).

43. For information about the Compagnie Générale des Eaux, see *Centennaire* (of the Compagnie Générale des Eaux) (Paris, 1953).

44. Archives Nationale, F12, 6785 (1853).

45. See Jean-Pierre Goubert, *La Conquête de l'eau: L'Avènement de la santé à l'âge industriel* (Paris, 1986).

46. See F. O. Ward, *Circulation ou stagnation?* (Brussels, 1852), p. 8.

7

Fire and Disease:
The Development of Water Supply Systems
in New England, 1870–1900

Letty Anderson

For most of the nineteenth century, the pace of urbanization seemed to outstrip the development of means to deal with the resulting urban problems. Two such consequences of unprecedented levels of urban population and population density were epidemic diseases and urban fires. In the early decades of the century, the response pattern of cities to these particular shocks indicates that (1) there was a substantial time lag between the appearance and perception of the problems qua problems; (2) there was another substantial interval until it was realized that the solution lay in the installation of citywide water supply systems; and (3) there was yet a third delay before the central supply was implemented, during which a city tried to find a plan that would minimize the uncertainty of the results. The problem was how to overcome disease and combat fire; the solution was the installation of a central water supply system that could provide pure water and water in sufficient volume to deal with major fires. The time lags in the decision process shrank as more cities found solutions.[1]

Investments in water supplies were minimal in 1800; they grew very slowly and relatively steadily until the 1870s and 1880s, when there was an apparent boom in water supply investments. A breakdown of this investment pattern by cities indicated that the earliest cities to install central water supply systems were large and had had severe problems of epidemics and fires. The cities that composed the investment boom of the 1870s and 1880s were smaller and probably were not faced with the severe problems to which the earlier cities had responded directly.

Parts of this article appeared earlier as "Hard Choices: Supplying Water to New England Towns," in the *Journal of Interdisciplinary History*, 15, no. 2 (Autumn 1984). It is reprinted here with permission.

TABLE 7.1.
Water Utilities, United States and New England

	United States		New England		
Year	Number of Water Utilities	Percentage Publicly Owned	Number of Water Utilities	Size of Town[a]	Percentage of Urban Population Supplied with Water
1800	16	6.3	—	—	—
1810	26	8.7	10	8,271	41.6
1820	30	16.6	10	—	—
1830	46	20.5	10	—	—
1840	54	27.8	11	7,702	—
1850	83	39.7	12	11,375	47.6
1860	136	41.9	24	8,877	43.3
1870	243	47.7	42	8,439	54.9
1880	598	49.0	100	9,591	74.4
1890	1,878	42.9	211	4,451	81.6
1897	3,196	53.2	275	2,449	—

Sources: M. N. Baker, ed., The Manual of American Water Works (New York, 1897); idem, "Water Works," in Municipal Monopolies, ed. E. W. Bemis (New York, 1899); U.S. Bureau of the Census, Population, each census 1800–1900.
[a] Average size of town at date of waterworks completion, for the nine years preceding the year indicated (e.g., the figure for 1897 is for the years 1890–97).

Decade figures for numbers of waterworks in the United States, to the year 1897, are presented in Table 7.1. At the beginning of the century, when 16 water supply systems existed, there were 33 cities with populations larger than 2,500. By 1897 there were 3,196 water utilities, but only 1,797 cities with a population larger than 2,500.

As can be seen in the table, in New England the average size of a town as of the date of completion of its water supply system fell by almost half in the 1880s and 1890s. Given that Philadelphia's population was 41,220 when its water supply system was completed in 1801, it seems clear that the late-century investment activity took place among small towns that did not have the same major population-associated problems as towns that had been moved to invest in the earlier decades.

There are several striking characteristics of the pattern of water supply investments in the nineteenth century. The first, already mentioned, is that the largest cities were apparently pioneers in water supply construction. Two of the first cities to build water supply systems were New York and Philadelphia. This pattern of the appearance of water supply systems in cities with large populations continued until 1860. At that date, the sixteen largest cities in the United States, all of which had populations over 50,000, had invested in water supply systems, as had several hundred smaller cities.[2]

The second striking characteristic is that the years from 1870 to 1895, a period in which the number of cities investing in water supplies expanded very rapidly, were characterized by a predominance in the market of private companies. These companies received municipal franchises that were tantamount to monopolies; the standard contractual agreements gave the corporation the right to acquire property through eminent domain and to use the streets to lay pipes, and usually granted the corporation the exclusive right to supply the city with water.[3]

A third fact to note is that there is little regional variation in the pattern of investments. Investments in all regions accelerated in the 1870s, with an even larger increase in the 1880s. The next few years, 1890 to 1897, show a split, with a decline in investments in four regions and an increase in four others.

This study discusses the development of a widespread demand for water systems. It arose from the efforts of public health organizations and the fire insurance industry, as well as from local sources. Problems of inadequate municipal financial structures were met by easing state restrictions on local borrowing and by the alternative of private corporate ownership of water companies. On the supply side, there were few constraints after 1880, by which time domestic manufacture of water pipes and pumping equipment was sufficient to meet demand. In part, the rapid spread of water systems reflected the efforts of two major pumping manufacturers to promote their equipment.

The diffusion of water systems after 1870 cannot be seen as supply constrained or demand constrained. Because of advances in technology that had taken place by the early 1870s, and the nature of marketing waterworks by equipment manufacturers and later by other entrepreneurs, availability of waterworks expertise or equipment was not a limiting factor in the rate of adoption. The ability of cities to finance waterworks systems might be thought of as a constraint on demand; however, in times of municipal financial crisis, this problem was alleviated by the readiness of private companies to obtain the necessary capital.

A more serious constraint was in the technology, which was site specific. Even though the basic techniques of water supply construction were well developed by 1870, a town that wanted to invest in a system had to have one designed specifically for the town site. This characteristic added considerably to what might be thought of as the cost of information.[4]

Viewed as a whole, the process by which American municipalities invested in water systems resembles the diffusion of any new technology. The pattern of city investments in water supply systems is really an indication of the rate of adoption of this new technology. Because of the geographical location of adoption points in the market, the adoption pattern has a spatial as well as a time dimension. Hypotheses about the determinants of the diffusion of water systems were tested with data from 275 New England towns. The spread of municipal water supply investments is seen to have occurred over at least a thirty-year period. The existence of a stable population threshold at which investment became feasible was rejected as a determinant of the rate of diffusion, as was a possible decline over time in the per capita cost of investment. In fact, populations of investing towns fell, and per capita costs rose.

Water supply technology appears to have diffused via a "trickle down" process throughout the urban size hierarchy, from large cities that were regional centers to remote small towns. The technology is not directly transferable from one site to another because of variations in local topography and water availability. Its successful implementation required the services of a civil engineer. This fact led to a final hypothesis, that the probability of a given town's receiving relevant technical information was the major variable in predicting the date of the town's investment in a water supply system. The data upheld this hypothesis and is the central conclusion of this chapter.

Some Demand Factors

Until the isolation of bacteria in the 1880s and the advent of the germ theory of disease, the connection between pure water and public health was a matter of conjecture. Although some medical practitioners believed diseases were contagious, the miasmatic theory of disease prevailed. Ironically, the miasmatic theory, and the connection between urban filth and miasma, led city officials to promote water supply systems as a check on epidemic diseases. The water systems begun in 1798 in Philadelphia and New York were undertaken in order that each city would be able to clean the streets effectively in time for the anticipated summer epidemic of yellow fever, almost an annual event in the late 1790s. Once these cities had obtained pure water for at least some of their populations, a second influence on water supply investments elsewhere became the demonstration effect. New York to some extent and Philadelphia (which had a very large, municipally owned system) to a considerable extent were successful in reducing morbidity and mortality during the cholera epidemic of 1832. Consistently, in the cholera epidemics of 1832, 1848–1849, 1866, and 1873, places with water supply systems had lower death rates than places without. In the 1873 epidemic, water supply systems in large cities were sufficiently widespread that in many states the urban mortality rate was lower than the rural rate. The demonstration effect was strengthened by the work of Dr. John Snow, who in 1854 traced the source of cholera in the Soho district of London to a single polluted well, the famous Broad Street Pump.[5]

The suspicion that water was the transmission medium for cholera and other epidemic diseases could not be proved until the era of bacteriology. The extent of pollution of local wells was not quantifiable until bacteria counts were taken, beginning in the 1880s. After this time, pollution levels provided strong incentives for investments in water supplies.

An important factor in disseminating what was known about water and disease, and later in promoting widespread construction of urban water supply systems, was the American public health movement. From its formation in the late 1830s, it gained ground slowly until the Civil War and quickly afterward, emerging as a national organization, the American Public Health Association, in 1872. From the beginning, the public health organizations were concerned with the quality of drinking water, and the increase in their organizational strength was paralleled by increasing pressure for municipal water supply systems.[6]

Fire was probably as important as disease in influencing the demand for municipal water supplies. Although cities with water supplies demonstrated that piped water improved the efficiency of firefighting, it was in the 1850s, with the appearance of steam fire engines, that firefighting technology changed. Steam fire engines required continuous maintenance and a great deal of water. As a result, volunteer fire departments were replaced with paid municipal organizations, and there was increasing demand for water.

As with health concerns, formal organizations also promoted the adoption of water supply systems for purposes of minimizing fire losses. The value of water supply in combating fires was recognized by the American fire insurance industry as early as 1800, and was incorporated into local and then nationwide rate structures, providing a financial incentive for municipalities to do something about the water supply question. The nationwide rate sheet published by the National Board of Fire Underwriters in 1872 incorporated a rate differential for towns with and without water supply systems. By the 1880s, a town that obtained a water supply system could be certain of a reduction in its fire insurance rates. These rate differentials were monitored by the American Water Works Association; in 1888, the association reported that, as a general rule, towns that introduced waterworks could expect a reduction of 20 to 50 percent in fire insurance premiums on buildings.[7]

The growth of demand for water also reflected commercial considerations. Municipal boosterism and the desire to attract industry counted in municipal decision making. Local businessmen, including real estate developers, were a significant influence on city councils when the water supply decision was pending. Among the lobbyists in support of New York's Great Croton Aqueduct project, for example, were the city's bakers, brewers, tanners, sugar refiners, bar owners, and hotelkeepers. In these industries, water quality was as important as water quantity. The level of industrial use of water is difficult to estimate because most water systems were unmetered before 1900. A 1907 survey of water use in Holyoke, Massachusetts, gives the expected result that the larger users were paper mills, cloth mills, and wire mills, followed by machine shops and steam boiler shops. Although industrial needs for water were an important component of overall demand, municipal treatment of manufacturers varied from town to town, from subsidization to exploitation. Wilmington, Delaware, reportedly supplied water to industrial users without charge; municipal officials in Rochester, New York, estimated that the city made a 100 percent profit on water supplied to manufacturers.[8]

From an economic standpoint, a major determinant of the level of demand is ability to pay. Effective demand for municipal water supply systems—demand that resulted in their being built—also depended on whether, and how, they could be paid for. A town desiring a water supply system had the choice of building one at public expense or granting a franchise to a private company. The attraction of a franchise was that it relieved the town of a financial burden; the attraction of public ownership was that it usually resulted in much more satisfactory prices and better service.

In Table 7.1, it is notable that the 1880s, during which the number of waterworks trebled, is also the only decade in which the percentage of publicly owned systems fell. Restrictions on municipal indebtedness following the Panic of 1873, combined with rapidly increasing demand for municipal water supplies, created in these years a unique opportunity for private capital, although publicly funded construction of water-works did not become as rare as it might have. Municipal water bonds, because of their relative stability and good payment record, were not as affected by debt restrictions as were other kinds of municipal borrowing. But from 1873 to about 1890, public sentiment was against borrowing for any reason, and many municipalities were restricted to funding out of current revenues. In these places, there was no alternative to private capital. By the mid-1870s, private capital for building waterworks, under what came to be known as the "franchise system," was readily available.[9]

The franchise system resulted from the development in the late 1860s of standard waterworks pumping engines, most notably those of the Holly and Worthington companies. These equipment manufacturers actively sought franchises for municipal systems in the early 1870s, and they succeeded rapidly; by 1878, 105 towns had installed Worthington engines or systems, and 70 towns had Holly systems. These earliest franchises were apparently highly profitable, and their success led to the entry into the market of imitators, regularly organized construction companies that had their own agents and their own hydraulic engineers. As the franchise system spread, the terms of the franchises granted changed. Before 1875, city franchise agreements often were made extremely attractive in order to ensure the interest of private companies; some of the early city franchises granted charters in perpetuity, exemption from taxes and from obligations for street paving and maintenance, and freedom from price regulation, as well as the standard agreements giving the corporation the right to acquire property through eminent domain, the right to use the streets to lay pipes, and the exclusive right to supply the city with water. After 1875, franchises more commonly were of limited term or gave the city an option to purchase at any time.[10]

The most common provision contained in waterworks franchises, and the most important, was the agreement between the company and the city regarding the number of fire hydrants and their annual rental. This clause guaranteed the survival of a private company in its first years of operation; rental from the city always covered the company's interest obligations incurred on the cost of construction, although annual hydrant rental rates varied from place to place. In 1888, these reportedly ranged from $15 to $125 per hydrant; in 1905, the reported range was from $21.50 to $79.25. The variation in rental rates probably reflects differences in total construction cost of the systems as well as the terms of the franchises. Guaranteed income for the companies from hydrant rentals did not always mean guaranteed fire protection for the cities, however. In the latter years of the nineteenth century, many lawsuits, often initiated by insurance companies, were launched against private water companies for recovery of damages for fire losses resulting from insufficiency or failure of the water supply. As of 1909, the liability of water companies for fire damage had been upheld in three states and rejected in nineteen others.[11]

A city typically had two major problems with a private water company: rates and service. Rates charged by private water companies in the 1890s were between 40 and 43 percent higher than rates charged by municipal works. According to some sources, this difference was attributable to the fact that municipally owned plants operated at a loss and were subsidized by taxes; according to others, private companies were overcapitalized, the value of stocks and bonds being well in excess of total construction cost. The city government was most directly affected by fire hydrant rental rates, and disagreement over hydrant rentals was the major source of friction between cities and private companies.

Problems with service arose because of the apparently conflicting goals of short-run profit maximization on the part of the water company and adequate supply, for household consumption and especially for fire protection, on the part of the municipality. Typically, private companies concentrated on the most profitable areas of the city, ignoring the outlying districts and the poorer sections.[12]

The move toward municipal ownership was a direct outgrowth of dissatisfaction with private companies. By the 1890s, when cities were better able to finance the construction or purchase of water supply systems, the percentage of municipally owned utilities rose once again. The large change in the percentage of municipally owned waterworks between 1890 and 1897 reflects in part municipal purchases of existing private companies. By 1899, there had been 225 changes of ownership of water utilities, 205 of which represented municipal purchase of private water companies. Typically, the larger systems were publicly owned: In 1890, 66.2 percent of the urban population supplied with water was supplied by a publicly owned system, even though only 43 percent of all plants were owned by municipalities. The two forms of ownership have continued to coexist. As of 1958, approximately 30 percent of the water utilities in the United States were operated under private ownership.[13]

Technology and Supply

A water supply system is a water transportation system. If it is to function properly, adequate water to supply the needs of a town or city must be collected, stored, and transported to consumers. This is done by means of pipes or conduits and pumping engines, sometimes with the addition of reservoirs or standpipes.

Cities may use as a water source springs, wells, rivers, lakes, man-made reservoirs, or combinations of these. Smaller towns in the last century characteristically relied on springs and deep wells, which provided pure water, although it was of limited quantity. Springs had the additional advantage that pumps often were not required to transport the water, although the saving on pumping machinery was partially offset by the cost of longer aqueducts. Larger cities relied on lakes (e.g., in Chicago, Milwaukee, Cleveland) and rivers (in Philadelphia and New York), both of which had the advantage of supplying large quantities of water but the disadvantages of poor water quality and the need for pumps. Natural reservoirs combine the advantages of relative purity and a large supply, many reservoirs holding 150–200 days' supply. The expense of building a dam, however, greatly adds to the initial cost of the system. A reservoir system may work by gravity or may require pumps, depending on elevation. Filtration was nonexistent until the last quarter of the nineteenth century and was still uncommon in 1900, and so the source had to be as pure as possible, since the water was not purified before it was consumed.

The system of pipes may include a large main or aqueduct that connects the source to a central distribution point and a system of mains and smaller-distribution pipes that form a network through the city. A good deal of the expense in installing water pipes is attributable to the labor involved in digging and filling trenches, laying pipe, and sealing joints. It is important from a cost standpoint that once the pipes are laid, they do not have to be taken up and repaired or replaced.

Pumping systems may include a pumping station that has a relatively low-level water intake and that pumps water into a high-level reservoir or water tank, from which the water is distributed by gravity, or else pumps it directly into the city water mains. A system that works completely by gravity, or at least has a gravity flow from a water tower or standpipe, was thought preferable to a direct pumping system (in which water is pumped directly into the mains) because in gravity or partial gravity systems, the flow of water to put out fires does not depend on the pumps being in working order. Another advantage of gravity is uniform water pressure. Few nineteenth-century pumps worked smoothly enough to provide even water pressure where direct pumping was used.[14]

The set of requirements for a technologically successful nineteenth-century water supply system included the following: the source had to be far enough removed from the city that pollution was unlikely, but close enough that transportation costs did not become prohibitive; the tank or reservoir had to be used in areas where the amount of supply from the original source was stochastic (e.g., dependent on rainfall); pipes had to be of sufficient strength to transport water under pressure and take the pressures of distance, pumps, and elevation changes without rupturing or leaking at the joints; pumps had to be powerful enough to lift water into the storage space and push it through the distribution system, and had to perform reliably. The extent of pumping and storage, and lengths of pipeline required, varied with source and local topography. The initial cost of the system consisted of construction of storage systems, laying pipe, and purchasing pumps. Operating expenses included maintenance and repair of these components plus labor and fuel for the operation of pumps.

Major technical improvements occurred in both pipe and pump construction. In the 1850s, vertical casting techniques for iron pipes, along with the development of a tar preparation that prevented ferric hydroxide build-up inside them, solved the problems of strength and durability. Although other pipes were used, this type prevailed. Pumping equipment improved steadily, and by the end of the nineteenth century, some standardization existed for steam pumps in use. By 1890, the vertical triple expansion engine was virtually standard for a system that pumped 10 to 30 million gallons of water a day. Large flywheels permitted expansive use of the steam and provided steady water delivery. For systems pumping 2 to 10 million gallons a day, the most widely used engine was the horizontal, cross-compound, crank, and flywheel design.[15]

If any facet of waterworks pump manufacture tended to speed the rate of adoption of water supply systems, it was probably the change in technology and market organization in the pump manufacturing industry after 1860, when two major companies emerged: the Worthington Pump Manufacturing Company and the Holly Manufacturing Company. Worthington, a designer and manufacturer, had invented the duplex steam feed pump in 1859. In this pump, one steam engine activated the valves of a second, allowing an even flow of water, an unusual attribute of a direct-action pump. The Worthington pump thus had the advantages of a direct pump, but was also smooth and quiet. It was adopted widely after 1863. During the 1860s, the Holly Manufacturing Company devised a marketing technique for its rotary water pump; this involved the development of an entire system in which the pump could be used to force water directly into the mains, eliminating the need for a reservoir or water tower, thus greatly reducing initial construction cost. By 1878, the Holly system reportedly was in use in more than 70 towns; by 1894, this figure exceeded 2,000. One of the reasons for Holly's success was its marketing strategy. The company demonstrated its system by taking prospective buyers to towns where it was operating successfully. In addition, Holly offered an entire package—source recommendation, engineering surveys, plans, and construction—as well as the pumps.[16]

A survey of 169 towns mentioned in *Engineering News* between March 1881 and December 1882 indicates that the two companies had a significant share of the waterworks market. Of towns mentioned, 21 percent used Worthington equipment exclusively, and 24 percent had only Holly engines. In addition, 33 percent of the towns used Worthington equipment exclusively or in conjunction with another pump; the figure is 25 percent for Holly. Of the 46 towns that used neither Holly nor Worthington pumping engines, and that reported the type of pump used, twenty-two other types of engines were listed, including water power and windmill pumps. The third most frequently listed engine was the Knowles pump, found in only 8 towns. Since both the Holly and Worthington pumps were of the horizontal cross-compound type, they must have been used extensively in systems requiring lower pumping capacities.[17]

Water supply technology is site specific; even though the technology is generally known, each location is different in terms of elevation, source characteristics, capacity requirements, rainfall variation, and storage requirements. A municipality considering a water supply system had to have access to an engineer trained to apply the general technology to a specific situation. Civil engineers concerned with water supply were responsible for many of the improvements in system components during the last half of the nineteenth century and for the rapid adoption of European advances. They were in many ways the purveyors and implementers of the new technology, and the numbers of water supply systems constructed depended on the availability of their services. The rapid increase in the number of formally educated professional engineers after 1850 allowed a corresponding rise in the number of water supply systems constructed.

In the first half of the century, most practicing engineers were Europeans trained in Europe or Americans with no formal training who had come up through the apprenticeship ranks. Two of the more famous of the latter type were E. S. Chesbrough and William McAlpine. Chesbrough began as a chainman on surveys for the B & O railroad, advanced through railroad engineering, and later worked with John Jervis, a famous canal and water supply engineer, on the Boston Cochituate Aqueduct project in the 1840s. By 1846, Chesbrough had become an engineer in the Boston Waterworks System, becoming the city's first chief engineer in 1851. He resigned from this post in 1855 to head the Chicago Sewage Commission. During his later career, he acted as a consultant in Indianapolis, New Haven, Milwaukee, and Memphis regarding sewerage systems, and in Pittsburgh, Jacksonville (Illinois), Detroit, and Toronto for water supply systems. McAlpine began his career apprenticed to Jervis from 1827 to 1836, at which date he succeeded him as chief engineer of the eastern division of the Erie Canal. Several years later, McAlpine became chief engineer of the government drydock in Brooklyn, New York. This job, technically difficult, established McAlpine's reputation. McAlpine later applied his expertise to water supply systems for Chicago, Brooklyn, and Buffalo in the 1850s, Montreal in 1869, Philadelphia and San Francisco in the 1870s, and New York and Toronto in the 1880s. A number of smaller cities in the East—Albany and Oswego in New York, and Fall River, Worchester, and New Bedford in Massachusetts,—also employed McAlpine as a consultant during this period.[18] The careers of Chesbrough and McAlpine, exhibiting a pattern of frequent job changes and numerous consulting activities, were typical. Competent, imaginative engineers, formally trained or not, were relatively rare, at least until the end of the Civil War, and widespread consulting activity on their part was not unusual. It was in this manner that the diffusion of their talents was accomplished, even though their numbers were relatively small.

Most engineers confined their consulting activity to one region, but not to the larger towns. Smaller towns appear frequently in the lists of places for which a single consulting engineer drew plans for systems. In addition, some engineering services were provided by private companies.

After 1850, the engineering profession grew rapidly, and a professional organization emerged. During this time also, formal university education began and flourished. By the 1880s, waterworks engineering was a well-defined field of specialization. The census for 1880 reports 8,261 civil engineers, compared to 512 in 1850. The American Water Works Association, founded in 1880, had 192 active members in 1887; the New England Water Works Association had 181 active members in 1888. In this decade, the exchange of information was in full swing, with articles about waterworks systems appearing weekly in Engineering News and quarterly in association publications.[19]

Despite the increasing numbers of engineers, evidence that engineering services were still in short supply in the late years of the nineteenth century can be seen in the average salaries of engineers, the lack of concern by the American Society of Civil Engineers (ASCE) about members' salaries, and the activities of practicing engineers. Salaries for water supply engineers, as for most others in the civil engineering profession, were high. It is reported that the superintendent of the Hartford waterworks received a salary of $10,000 per year in 1868 and that his earnings were not unusual. The consulting fee of $10,000 paid to George Baldwin for the Quebec City Waterworks was reported to have been comparable to scale for other chief engineers on municipal projects. Further evidence that salaries for engineers generally were high is in the lack of concern for salary structures among members of the ASCE. Apparently, the organization was one of very few professional organizations that did not concern itself with salaries during this period, nor was it concerned with placement of members; there was simply no need. Not surprisingly, the activities of practicing engineers after 1865 follow a pattern similar to that of Chesbrough and McAlpine: They worked in many cities, moving from project to project.[20]

By 1880, the supply of civil engineers, and of waterworks engineers in particular, seems to have been large enough and sufficiently dispersed that obtaining their services in water supply projects was not difficult, although it may have been expensive. Even small towns could obtain the services of well-known engineers such as Jervis and McAlpine if they so desired; if not, they could opt for the less expensive lesser lights of the profession, with reasonable certainty of satisfactory results.

By 1870, water supply technology was available, and demand was growing. Over the next twenty-seven years, 2,953 systems were constructed in the United States; this is twelve times the number that existed in 1870. The rapidity of the increase is typical of the middle stages of a diffusion process.

The observed time path of the diffusion of a new technology maps the percentage of the potential adopting population who actually do adopt against time. This percentage increases slowly in the early stages of diffusion, then rapidly, then slowly again as the market for the innovation approaches saturation. This characteristic S curve of adoption has been found in studies of the diffusion of machine tools, for example, and hybrid corn.[21]

In order to identify and estimate the S curve, the initial and final levels of adoptions must be known, as must the rate of increase throughout the entire adoption process. Unfortunately, information about

American water supply systems is incomplete. The number of waterworks in the United States today is probably unknown. The last reliable
estimate was published in 1899 by M. N. Baker and was based on data
published in *The Manual of American Water Works* of 1897. After the
close of the nineteenth century, statistics on waterworks were compiled
by the Census Bureau for cities with populations greater than 30,000,
but smaller towns were not accounted for. Known data points for aggregate numbers of waterworks in the United States consist of Baker's
estimate of 3,196 in 1896 and an educated guess of 7,500 to 8,000 works
in 1918.[22] Given the absence of aggregate data, an S curve cannot be
estimated, since the equilibrium or final level of adoptions is unknown,
as is the time path of adoptions between 1896 and 1918; however, this
does not preclude testing other hypotheses about the determinants of
the rate of diffusion in the portion of the process that is observable. To
perform these tests, the database was restricted to New England because
its experience is comparable to that of other regions and to the United
States as a whole, and because it is a manageable sample.

Factors shown to be of primary importance in determining the
speed of diffusion of a production technology can be grouped into
information–uncertainty and cost–profit categories. The former category includes as important the availability of technically sophisticated
people who can implement the new technology, the existence of organizations that help disseminate information about the new process or
its advantages, the reliability of the technology, and the likelihood of a
potential adopter receiving the information necessary to make a decision. Cost–profit considerations include the long-run profit advantage
of the new process over the old, as well as the cost of adopting the new
technology relative to, say, the assets of the firm or the cost of capital.[23]

Since water supply systems in New England were constructed first
by larger municipalities and then by smaller ones, their diffusion is
characterized by a "filtering" process in which the innovation trickles
down through the hierarchy of city sizes. Several reasons for hierarchical filtering can be posited, among them a market-search process, in
which an expanding industry exploits the profit potential in the larger
markets first, then moves to the smaller ones (a cost–profit argument),
or a simple probability mechanism in which the probability of adoption depends on the chance that an entrepreneur in a given town will
receive necessary technical or marketing information (an information–
uncertainty argument). This probability declines with size of town.

As a test of some cost–profit considerations, I considered initially
two usual determinants of the rate of adoption: threshold populations

TABLE 7.2.
Size Distribution of New England Towns

Population	Number of Towns in Each Size Group				
	1850	1860	1870	1880	1890
1,000–2,500	657	655	610	573	484
2,500–5,000	178	189	176	181	186
5,000–10,000	47	64	59	63	82
10,000–25,000	17	21	28	46	44
25,000–50,000	2	7	10	10	16
50,000–100,000	0	0	2	4	7
100,000–250,000	1	1	0	1	1
250,000–500,000	0	0	1	1	1
Total number of towns with more than 1,000 population	902	937	886	879	821

Sources: For 1880, 1890: U.S. Bureau of the Census, Eleventh Census, 1890, Population, p. 1; for 1850 to 1870; U.S. Bureau of the Census, Ninth Census, 1870, Population.

and declining costs. The threshold population argument is that towns attaining a certain size will invest in a water system, whereas towns below that size will not. Benefits, particularly social benefits such as fire control and disease prevention, increase with population. The population threshold hypothesis, however, assumes that the population threshold is stable. As Table 7.2 shows, this was not the case. The average population of a town investing in water supply systems remained relatively stable at around 8,500 from 1850 to 1880, then fell almost by half in each of the two succeeding decades. Thus, if a population threshold for investment operated at all, something caused it to fall after 1880. One possibility is that costs fell after 1880, allowing smaller towns (or private companies) to earn higher private and social returns to investments in water supplies. In fact, per capita cost measured in current dollars fell from an average of $30.89 in the 1860s to $24.95 in the 1880s, but rose to $27.72 in the 1890s. Per capita cost measured in constant (1914) dollars rose steadily from $8.86 in 1860–1869 to $30.05 in 1890–1897. These figures may reflect only a change in the type of investment a town was likely to make, however, since characteristics of waterworks constructed after 1880 may not have been the same as those constructed earlier. For instance, fire insurance standards for reliability of the water supply and for water pressure were somewhat more exacting after 1880.

Evidence also shows that towns building waterworks later invested in more sophisticated systems and built plants with excess capacity in order to allow for future growth. This expected growth was in fact often overestimated, especially in New England before 1825.[24]

The absence of a stable population threshold and evidence that investments did not take place in response to declining per capita costs indicate that cost–profit variables were not important in determining the rate of diffusion of water supply technology. The spread of water supply systems took place downward through the size hierarchy, and this process of hierarchical diffusion has a spatial aspect as well. Suppliers of equipment and other private entrepreneurs may have exploited the market in large towns and then moved on to smaller ones, or the explanation may lie in the fact that smaller, more remote towns were simply less likely to receive good information about waterworks technology. The strength of a demonstration effect, in which the presence of a working system nearby is sufficient to sway a city council, is another consideration. In any event, the location of a town with respect to towns that had waterworks was thought to be significant in determining the probability of the town obtaining its own system. This is an implicit gravity model of information flow, in which the probability of receipt depends on the size of the town and on distance from towns that have the necessary information. At the same time, it is consistent with the hypothesis of a "market searching" process by private waterworks companies, since they often demonstrated successful systems to prospective clients.[25] A pattern develops of early adoptions of waterworks in larger cities that are relatively spread apart, followed by adoption in smaller towns that are relatively close to them. These are concurrent in the later decades. Location and size were important in determining the date of adoption by an individual town, upholding the importance of receipt of adequate information.

Conclusion

The process through which American towns and cities obtained water supply systems in the nineteenth century was a relatively slow one, with the appearance of water systems in progressively smaller cities becoming more frequent as more information became available. The rise in demand for water depended on the availability of information about water and health and about water and fire. By the 1870s, this information was disseminated by health organizations and the fire insurance industry. Additional information, in the form of technical plans for a specific locality, was required to construct the water supply system. This information was available only through a water supply engineer. Engineering services became more easily attainable after 1870, concomitant with an increase in technical training and the incorporation of engineering services into the marketing organizations of private waterworks syndicates and equipment manufacturers. The private sector was particularly crucial from 1875 to 1890 when states restricted municipal bond issues. Tests of hypotheses regarding the spread of water supply systems in New England suggest that the market was perhaps exploited by private firms in order of descending size of town. The tests are also consistent with a hypothesis that private companies did not matter as much as size and location of town, which determined the probability of receiving the information necessary to precipitate water supply construction.[26]

Notes

1. For example, Boston officials debated the water question for twenty-one years before deciding in 1846 to construct the Cochituate reservoir system. In New York, various water proposals were heard from 1803 to 1834, the year in which the great Croton project was begun. (It was completed in 1842.) Philadelphia's first water supply system, completed in 1801, was generally regarded as a disappointment. It was not improved significantly until 1822. In comparison, towns that built works after 1880 typically reported a time lapse of about a year from initial consideration to completion. See Nelson Blake, *Water for the Cities* (Syracuse, N.Y., 1956), pp. 5, 18, 172–173; Fern L. Nesson, *Great Waters: A History of Boston's Water Supply* (Hanover, N.H., 1983), p. 76; Moses N. Baker, ed., *The Manual of American Water Works* (New York, 1897).

2. Blake, *Water*, pp. 5, 18, 272–273.

3. Edward C. Kirkland, *Industry Comes of Age: Business, Labor, and Public Policy, 1860–1897* (Chicago, 1967), p. 241.

4. H. R. Vallentine, *Water in the Service of Man* (Baltimore, 1967), pp. 193–203.

5. Wilson G. Smillie, *Public Health* (New York, 1976), p. 138; Charles E. Rosenberg, *The Cholera Years* (Chicago, 1962), p. 194; J. J. Cosgrove, *History of Sanitation* (Pittsburgh, 1900), pp. 91–107.

6. Smillie, *Public Health*, pp. 308, 318; Elisha Harris, "Report on the Public Health Service in the Principal Cities, and the Progress of Sanitary Works in the United States," *Public Health: Reports and Papers of the American Public Health Association, 1874–75* 2 (1876): 1–7; Henry I. Bowditch, *Public Hygiene in America* (Boston, 1877), p. 62.

7. Bayrd Still, *Milwaukee: The History of a City* (Madison, Wis., 1948), pp.

250–253; *Engineering News*, March 19, 1881; F. L. Fuller, "The Indirect Income of Water Supplies," *Proceedings of the American Water Works Association*, 1888–1889, p. 101.

8. Eugene V. Moehring, "Public Works and the Patterns of Urban Real Estate Growth in Manhattan, 1835–1894," Ph.D. dissertation, City University of New York, 1976, pp. 53–58; James L. Tighe, "Water Consumption, Waste and Meter Rates," *Proceedings of the American Water Works Association*, 1907, p. 76; G. W. Rafter, "The Cost of Meters at Rochester, New York," *Proceedings of the American Water Works Association*, 1907, p. 65.

9. Paul Studenski and Herman E. Kroos, *Financial History of the United States*, 2nd ed. (New York, 1963); p. 196; Fred Emerson Clark, "The Purposes of the Indebtedness of American Cities, 1880–1912," Ph.D. dissertation, University of Illinois, 1916, pp. 25, 27–29; M. N. Baker, "Public and Private Ownership of Water Works," *Outlook* 59 (1898): 78.

10. *Engineering News*, August 1, 1878, and February 16, 1887; Emery Troxel, *Economics of Public Utilities* (New York, 1947), p. 51; Eliot Jones and T. C. Bigham, *Principles of Public Utilities* (New York, 1932), p. 110; Kirkland, *Industry*, p. 241. For the distribution of Holly and Worthington pumps in U.S. cities, see Letty Anderson, "The Diffusion of Technology in the Nineteenth Century American City: Municipal Water Supply Investments," Ph.D. dissertation, Northwestern University, Chicago, 1980, pp. 23–28.

11. D. F. Wilcox, *Municipal Franchises* (New York, 1910), p. 401; William Reinecke, "Water Works Securities," *Proceedings of the American Water Works Association*, 1891, p. 67; J. Nelson Tubbs, "Yearly Rental Value of Fire Hydrants," *Proceedings of the American Water Works Association*, 1888–1889, p. 157; S. J. Rosamund, "What Would Be a Fair Basis for Fixing Water Rates for Fire and Domestic Service?" *Proceedings of the American Water Works Association*, 1905, p. 134; Chester H. McFarland, "Liability of Water Companies for Fire Losses," *Proceedings of the American Water Works Association*, 1909, pp. 609–618. The three states that upheld the principle of liability were Kentucky, North Carolina, and Florida.

12. Baker, "Public and Private Ownership," p. 78; Allen Ripley Foote, "Cost of Service to Users and Taxpayers," address before the National Conference of Mayors and Councilmen, 1897; John W. Alvord, "Equitable Hydrant Rentals and Better Methods for Apportioning Fire Protection Cost," *Journal of the American Water Works Association* 1 (1914): 98; Blake, *Water*, p. 77.

13. M. N. Baker, "Water Works," in *Municipal Monopolies*, ed. Edwin W. Bemis (New York, 1899), p. 26; U.S. Public Health Service, *Municipal Water Facilities. Inventory as of January 1, 1958*, Publication 775 (1960), vols. 1–9; Anderson, "Diffusion of Technology," pp. 103–125.

14. Baker, *Manual of American Water Works*, pp. 1–84; Louis P. Cain, "An Economic History of Urban Location and Sanitation," *Research in Economic History* 2 (1977): 337–389; *Engineering News*, June 21, 1890; W. K. Burton, *The Water Supply of Towns* (London, 1907), pp. 43–44, 232.

15. John W. Alvord, "Recent Progress and Tendencies in Municipal Water

Supply Practice," *Journal of the American Water Works Association* 4 (1917): 288; Burton, *Water*, pp. 241–242; Vallentine, *Water*, p. 111; Octave Chanute, "History and Progress of American Water Works," *Scientific American*, August 14, 1880; H. G. H. Tarr, "Fifty Years' Reminiscence in Water Works," *Proceedings of the American Water Works Association*, 1912, p. 52; *Engineering News*, March 5, 1881, and April 2, 1881; Charles Singer et al., eds., *A History of Technology* (Oxford, England, 1958), pp. v, 550, 559; Harold E. Babbitt and James J. Doland, *Water Supply Engineering* (New York, 1949), p. 560; Ernest W. Whitlock, "Concrete Pressure Pipe in Today's Water Industry," *Journal of the American Water Works Association* 52 (1906): 1244–1250.

A technical note is in order here. Development of steam and other engines proceeded in a number of different directions so that, as in the case of pipe material, water supply engineers after the Civil War had a wide choice of engines to use. After 1845, the compounding principle was adopted for steam engines. This involved addition of a small, high-pressure cylinder that exhausted into a low-pressure cylinder, allowing the steam to expand partially in each, and increasing the power of the engine. Later engines were compounded using three cylinders and further improved by changes in the placement of the pistons. In 1850, an alternative to the vertical engine was invented by James Nasmyth; this engine had a cylinder at the top and a crankshaft below and was called an *inverted vertical*. Engines with cylinders side by side in a horizontal position with a central flywheel between the cranks, called *horizontal cross-compound engines*, appeared after 1845. *Triple expansion engines*, using three cylinders instead of two, and employing a flywheel, were a further refinement in compounding. Finally, engines without flywheels appeared. The flywheel had the advantage of distributing more evenly the steam power produced by the pistons, but it was heavy and cumbersome. Direct-acting pumps had many advantages over crank and flywheel engines: They were smaller and lighter, cheaper to construct, and required lighter foundations and less maintenance. One disadvantage was that, generally, the flow of water from such a pump was not as smooth as it was from a crank and flywheel engine. See Singer, *History of Technology* 5:133, 137. See also L. T. C. Rolt, *Victorian Engineering* (London, 1970), pp. 64–65.

16. Dumas Malone, ed., *Dictionary of American Biography* (New York, 1936), pp. xx, 539; *The National Cyclopaedia of American Biography* (New York, 1926), p. 108; idem, 27:207. See also *Engineering News* for the following dates: June 24, 1876, August 1, 1878, November 12, 1881, November 19, 1881, December 17, 1881, February 4, 1882, December 23, 1882.

17. *Engineering News*, weekly issues from March 5, 1881, to December 23, 1882.

18. Louis P. Cain, "Raising and Watering a City: Ellis Sylvester Chesbrough and Chicago's First Sanitation System," *Technology and Culture* 13 (1972): 353–371; Daniel H. Calhoun, *The American Civil Engineer: Origins and Conflict* (Cambridge, Mass., 1960), p. 53; Raymond H. Merritt, *Engineering in American Society, 1850–1873* (Lexington, Ky., 1969), pp. 30, 31; *Dictionary of American*

Biography 11:549; *National Cyclopaedia*, 9:35. See also *Engineering News* for the following dates: March 26, 1881, March 30, 1881, April 9, 1881, May 21, 1881, June 4, 1881, June 11, 1881, July 2, 1881, July 9, 1881, July 23, 1881, August 13, 1881, October 15, 1881, November 26, 1881. On apprenticeship patterns, see Nesson, *Great Waters*, pp. 3–4; and Anderson, "Diffusion of Technology," pp. 23–34.

19. *Proceedings of the American Water Works Association*, 1888–1889; *Journal of the New England Water Works Association* 4 (1889–1890): 5; Merritt, *Engineering*, pp. 30–31; Alba M. Edwards, *Population: Comparative Occupation Statistics* (Washington, D.C., 1943), p. 87.

20. Merritt, *Engineering*, pp. 113–117.

21. Zvi Griliches, "Hybrid Corn: An Exploration of the Economics of Technological Change," *Econometrica* 25 (1957): 501–522; Edwin Mansfield, "Technical Change and the Rate of Imitation," *Econometrica* 29 (1961): 741–766.

22. M. Baker, *The Manual of American Water Works*; idem, "Water Works," p. 16; John M. Goodell, "The Magnitude of the American Municipal Water Supply Business," *Journal of the American Water Works Association* 7 (1920): 16; Henry C. Hodgkins, "Franchises of Public Utilities as They Were and as They Are," *Journal of the American Water Works Association* 2 (1915): 739–758.

23. See Griliches, "Hybrid Corn"; Mansfield, "Technical Change"; Paul David, "The Mechanization of Reaping in the American Midwest," in *Industrialization in Two Systems*, ed. H. Rosovsky (New York, 1966), pp. 3–39; Brian J. Berry. "Hierarchical Diffusion: The Basis of Developmental Filtering and Spread in a System of Growth Centers," in *Growth Centers and Regional Economic Development*, ed. N. Hansen (New York, 1972), pp. 108–138; John C. Hudson, "Diffusion in a Central Place System," *Geographical Analysis* 1 (1969): 45–58. For the long-run advantage of water supply systems, see Letty Anderson, "The Rate of Return to Water Supply Investments, 1850–1900," *Explorations in Economic History*, forthcoming.

24. Paul M. Berthouex, "Some Historical Statistics Related to Future Standards," *Journal of the Environmental Engineering Division, Proceedings of the American Society of Civil Engineers*, 1974, pp. 423–426. Anderson, "Diffusion," p. 148.

25. For example, *Engineering News* (June 24, 1876) reported that the Holly company had invited officials from Fort Wayne, Indiana, to Indianapolis (a distance of about 100 miles) to observe the Indianapolis Holly system and for a test of its engine. Information/uncertainty arguments can also generate the S curve.

26. For a more complete description of the econometric work associated with testing various hypotheses, see Letty Anderson, "Hard Choices: Supplying Water to New England Towns," *Journal of Interdisciplinary History* 15, no. 2 (Autumn 1984): 211–234.

PART III

Waste Disposal

AS human habitats and as consumers and producers of goods and services, cities generate vast quantities of wastes. Historically these have been primarily of a liquid, solid, or gaseous nature: sewage, garbage and ashes, and smoke and fumes. If an excess of nuisance is to be avoided, and if the public health is to be protected, these wastes must be removed from the city environment. As cities grew larger and denser and more polluted in the nineteenth century, and as concern over the effects of this pollution on the public health increased, a variety of approaches to disposing of urban wastes were attempted. These involved not only the use of technology, but also required new methods of administration and new organizational structures. What forms of technology would most efficiently and economically cleanse the city, however, was often a matter of dispute among urban professionals and elites. Eventually, the development of environmentally related technologies, as well as a growing understanding of disease etiology, produced striking public health improvements in both American and European cities.

In his essay on the construction of urban sewerage systems in the United States from 1850 to 1930, Joel A. Tarr examines the environmental, health, and economic conditions that produced these networks and their effects on various aspects of city life. Georges Knaebel focuses on sewer building within the context of one city, Bielefeld, located in the Federal Republic of Germany. For him, the most revealing aspect of the construction of sewer networks was the manner in which these underground "machines" reflected the power relationships and "moral outlook" of the larger urban society. And Martin V. Melosi uses his study of a garbage-incineration technology, the "British Destructor," to explore the process of the transfer and diffusion of technology between Europe and America within the context of the urban public sector.

157

PART III

8

Sewerage and the Development of the Networked City in the United States, 1850–1930

Joel A. Tarr

ONE of the most important developments in American cities during the nineteenth and early twentieth centuries was their increasing dependence on interconnected technological systems for a variety of urban services. Water, fuel, messages, freight, and people, as well as human and industrial wastes, were carried in and out of the city through a network of wires, pipes, and tracks. During these decades, the sought-after ideal for many urban theorists and planners was the completely networked and automated city, in which the messy decentralized and labor-intensive technologies of the preindustrial city would be replaced by sanitary, centralized, and capital-intensive systems.[1]

As these centralized technological systems developed in emerging industrial cities, they transformed urban life and society in a number of ways. These effects included population decentralization and centralization, spatial changes in regard to housing and industry, institutional and organizational developments, health and sanitary improvements (but also health deterioration), the formation of new engineering professions, public value changes, and alterations in the qualilty of life. While the late-nineteenth-century ideal of the fully automated city has never been realized, the late-twentieth-century city is more technology dependent than at any point in history.[2]

This essay discusses the evolution and effects of one of the most significant of these urban networked technologies—the water carriage, or sewerage, system. Cities began constructing these systems in the middle of the nineteenth century, and by World War II, all large American cities were almost completely networked. Rather than focus on this chronological development, the study develops several themes highlighting the effects of the technology on aspects of urban life and society in the

159

years from 1850 to 1930. These themes include a discussion of the development and adoption of sewerage systems; the various positive and negative externalities produced by the technology; and the professional, governmental, and institutional changes the technology stimulated.

Water Supply and Waste Collection, 1800–1880

Until well into the second half of the nineteenth century, most American urbanites depended for their water supplies on local surface sources such as ponds and streams, on rainwater cisterns, or on wells and pumps drawing on groundwater. Under these supply conditions, water consumption per capita probably averaged between 3 and 5 gallons a day. Householders disposed of wastewater from household functions such as cleaning, cooking, or washing by simply throwing it on the ground or placing it in a street gutter, a dry well, or a leaching cesspool.[3] Human wastes were normally deposited in cesspools and privy vaults located close by residences or in basements. Some of these were just shallow holes in the ground: others were more elaborately constructed receptacles lined with brick or stone.[4] Usually cesspools were permeable so that the ground could absorb the liquids: privy vaults were supposedly impervious, requiring frequent emptying. When the ground around the cesspools became saturated with wastes or vaults were filled to overflowing, they were often covered over with dirt and new ones dug.[5]

Most large cities instituted periodic cleaning by private scavengers under city contract or by city employees, but services were inefficient and irregular regardless of who supplied the service. The cleaning technology utilized for most of the nineteenth century was rudimentary—dippers, buckets, and wooden casks. The process was labor intensive and inefficient, creating both aesthetic nuisances and health problems, primarily through pollution of groundwater and wells.[6] Scavengers collected the wastes in "night soil carts" and disposed of them in nearby bodies of water or in dumps on the land, on farms for fertilizer, or sold them to reprocessing plants to be made into fertilizer.[7] Although larger cities such as New York and Boston possessed both private and public underground sewers early in the nineteenth century, they were intended for stormwater drainage rather than human waste removal. These sewers were often constructed of stone or brick in circular or elliptical shapes, and were usually large enough so that a man could enter them for cleaning. In some cities, ordinances forbade the placing of human wastes in sewers. The majority of nineteenth-century municipalities, however, had no underground drains. Street gutters of wood or stone, either on the side or in the middle of the roadway, provided for surface storm-water and occasionally for human wastes. Private householders often constructed drains to the street gutter to remove wastewater from cellars.[8]

Demographic and technological factors combined in the middle of the nineteenth century to stress the cesspool–privy vault system, causing its eventual collapse and replacement. The two most important factors involved were urban population growth and the development of new urban water supply systems with the consequent adoption of household water fixtures.

The nineteenth century was a period of rapid urbanization in the United States. By 1860, about 20 percent of the population lived in places over 8,000 population, and by 1880 the percentage had risen to 28.[9] As cities attracted population, densities in their central cores rose considerably. Poor public transportation limited the distance that people could move from jobs and urban institutions. As the city became more congested with both people and structures, the existing waste-collection system became increasingly inadequate. Overflowing privies and cesspools filled alleys and yards with stagnant water and fecal wastes, often polluting wells and other water sources. In addition, the paving of streets reduced their ability to absorb rain and increased the possibility of flooding and the gathering of pools of stagnant water.[10]

Urban adoption of the technology of piped-in water increased the demands on the cesspool–privy vault system. The technology of a piped-in water supply dated back at least to the ancient Roman water systems, with more modern examples available from seventeenth-century London and several colonial towns. The movement in nineteenth-century cities away from reliance on a localized and labor-intensive water supply system to a more capital-intensive system that utilized distant sources occurred primarily for four reasons in addition to population increase: (1) local sources were often contaminated, presented taste and odor problems, and were suspect as a cause of disease; (2) more copious supplies were required for firefighting; (3) water was required for street flushing at times of concern over epidemics; and (4) developing industries required a relatively pure and constant water supply.[11]

Philadelphia was the first city to build a waterworks (1802), followed by other large municipalities such as New York, Boston, Detroit, and Cincinnati. By 1860, the nation's 16 largest cities had waterworks, and there were 136 systems in the country; by 1880, this number had increased to 598. The availability of a source of constant water in the household caused a rapid expansion in usage. Chicago, for example, went from 33 gallons per capita per day in 1856 to 144 in 1882; Cleveland increased from 8 gallons per capita per day in 1857 to 55 in 1872; and Detroit from 55 gallons per capita per day in 1856 to 149 in 1882.[12] These figures reflect unmetered usage and include industrial and other nonhousehold uses, but they are still indicative of greatly increased water consumption over a relatively short span of time.

While hundreds of cities and towns installed waterworks in the first three-quarters of the nineteenth century, few cities simultaneously constructed sewers to remove the water because of high capital costs and poor forecasts about future water use. In most cities with waterworks, wastewater was initially diverted into cesspools or existing stormwater sewers or street gutters; householders often connected their cesspools with the sewers via overflow pipes. The introduction of such a large volume of contaminated water into a system designed to accommodate much smaller amounts caused serious problems of flooding and disposal.[13] This situation was exacerbated by the adoption by affluent urbanites of the water closet, or flush toilet.

The widespread urban installation of the water closet was largely an unanticipated consequence of the development of systems of piped-in water. The water closet actually dated back centuries, but was only patented in the United States in 1833.[14] In cities with waterworks, affluent families were quick to install closets in order to take advantage of their convenient in-house location and comparative cleanliness. In Boston in 1863 (population about 178,000), for instance, out of approximately 87,000 water fixtures, there were over 14,000 water closets. In Buffalo, in 1874 (population about 118,000), there were 5,191 dwellings supplied with water and 3,310 with water closets. By 1880, although the data are imprecise, it can be estimated that approximately one-quarter of urban households had water closets (usually of the pan or hopper type), while the remainder still depended on privy vaults.[15]

The water closet had many advantages, but it also produced nuisances and sanitary hazards. Water closets were usually connected with cesspools, which often became surcharged by the greatly increased flow of waste-bearing water. Occasionally they were linked by overflow pipes with surface gutters or sewers intended as stormwater drains, contaminating them with fecal matter. The soil near homes became saturated, cellars were "flooded with stagnant and offensive fluids," and frequent emptying was required.[16]

The spreading of fecally polluted water also created both real and perceived health dangers. During most of the nineteenth century, physicians generally divided into two groups: contagionists and anticontagionists. Contagionists maintained that epidemic disease was transmitted by contact with a diseased person or carrier, whereas anti-contagionists held that vitiated or impure air was the cause. The vitiated air could arise from any number of conditions, including miasmas from putrefying substances such as feces, exhalations from swamps and stagnant pools, or human and animal crowding. By the latter half of the nineteenth century, the majority of physicians were anticontagionists who believed that filthy conditions accelerated the spread of contagious disease, thus creating demands for cleaning the urban environment.[17] Public health officials warned of the health dangers created by connecting overflowing cesspools with water closets. As late as 1894, Benjamin Lee, secretary of the Pennsylvania State Board of Health, complained that householders persisted in connecting water closets to "leaching" cesspools, thereby distributing "fecal pollution over immense areas and . . . constituting a nuisance prejudicial to the public health." [18]

The adoption of the two new technologies of piped-in water and the water closet, therefore, combined with higher urban densities to cause the breakdown of the cesspool–privy vault system of waste removal and to generate nuisances and hazards to health. Some cities permitted householders to connect their water closets with storm sewers, but because these were poorly designed, they became "sewers of deposit" and exacerbated the problem of waste collection. Many cities adopted the "odorless excavator," a vacuum pump that emptied the contents of cesspools into a horse-drawn tank truck for removal. But because it was both labor intensive and capital intensive, the excavator provided only a temporary or interim solution to the problem.[19] Faced by the incapacity of the existing system of waste disposal and stormwater removal, engineers, public health officials, and city authorities in the second half of the nineteenth century increasingly turned toward the water-carriage technology of waste removal.

The Construction and Distribution of Urban Sewerage Systems

The water-carriage system of waste removal (or sewerage) was essentially a system of waste removal that used the wastewater itself as a transporting medium and as a cleansing agent in the pipe. As Jon Peterson notes, it "represented a specialized form of urban planning [that] gradually supplanted long-accepted, piecemeal methods of waste removal, particularly the reliance upon privately built cesspools and privy vaults and the common municipal practice of constructing sewers without reference to a larger, city-wide plan."[20] In New York before 1860, for instance, sewers existed, but they did not necessarily constitute a sewerage system. That is, some sewers were public and others were private. They were often unconnected and made their own independent way to the river for disposal purposes. They were also of different shapes and material, some having elliptical forms while others were circular; some were made of stone and others of brick.[21] In the 1840s, the British sanitarian Edwin Chadwick had advocated the adoption of a system of earthenware small-pipe sewers that would take the place of London's cesspools and privy vaults. Chadwick's sewers would utilize the household water supply as a means to dispose of human wastes and keep the sewers free of deposits. A convinced anticontagionist, he believed that odors from decaying organic matter caused the spread of many fatal diseases and that fecal matter had to be swiftly transported from the household vicinity. The sewage from this "arterial-venous system," he maintained, could be utilized for agricultural purposes, thereby paying for the cost of the system. Water-carriage technology, therefore, would provide a system of self-financing health benefits.[22]

Chadwick's system was never implemented as he planned, but his vision stimulated debate in Great Britain about technology and health and strongly influenced American sanitarians concerning the benefits of systematic sewerage.[23] The engineers for the earliest sewerage systems in Brooklyn, Chicago, and Jersey City drew heavily on the English sanitary investigations and London debates of the 1840s and 1850s, as well as on the actual experience of London with a system of large, brick intercepting sewers constructed to solve the problem of sewage pollution of the Thames River. Throughout the remainder of the nineteenth century, visits to sewerage works in cities in Great Britain and Europe were almost mandatory for American engineers involved in planning new systems. Thus, water-carriage technology provides a good example of the interchange and transfer of ideas and experience concerning an urban technology from Great Britain and Europe to the United States, as well as the modification of this technology to meet American conditions.[24]

City councils, medical and engineering societies, and citizens held extensive debates and discussions concerning the advantages and disadvantages of water-carriage technology in the construction of new systems. These debates often lasted for years and involved the preparation of a number of engineering reports and extensive discussions about the advantages of one form or design of waste-disposal technology over another. Rather than review these debates in detail, I merely summarize them here so that the essay can focus on the effects of sewerage technology on the city and the larger society.

Among the critical factors discussed were economics, sanitation and health, and resource conservation. In regard to costs, proponents of the system often claimed that the capital and maintenance expenses of building sewerage systems would represent a savings for municipalities over the annual costs of collection under the cesspool–privy vault–scavenger system. In 1883 for instance, the influential sanitary engineer Rudolph Hering conducted a study of the sewerage needs of Wilmington, Delaware. He noted that while the annual cost of cleaning the city's privy vaults was approximately $15,000, this represented the interest at 5 percent on $300,000, or the exact sum he calculated as necessary to construct a sewerage system. Thus, he argued, sewers would provide a long-term savings for the city over continued use of the existing system.[25] In addition to these arguments, public officials estimated that sewerage systems would require only one-time connection costs and have minimal maintenance charges compared to the continual cost of emptying cesspools and privies.

Another powerful argument for sewerage was the claim that it would create improved sanitary conditions in cities and result in lowered morbidity and mortality from disease. Speaking in 1899, for example, Dr. William Osler of the Johns Hopkins Medical School confidently forecast that "with the completion of a good sewerage system, the present typhoid death rate of Baltimore, about 40 per 100,000," would fall to "from 4 to 8 per 100,000." Many sewerage advocates cited the health improvements of sewered European cities as evidence for their forecasts.[26] Sanitary experts often assigned a specific dollar value to the costs of sickness and death and calculated the benefits that would accrue from improved sanitation. The 1875 *Annual Report* of the Massachusetts Board of Health featured an essay entitled "The Value of Health to the State," which maintained that "the prosperity of a town, city, state, or country stands in immediate relation with its sanitary condition." The article calculated the savings that would accrue to the state's working population through the improved health that sewers would theoretically bring.[27] In 1876, the famous sanitarian George Waring, Jr., argued that the financial savings accomplished through health improvements rather than compassion for the sick should persuade municipalities to invest in sewerage systems.[28] Proponents of the technology also maintained that cities that built sewerage systems would become healthier and would thereby attract population and industry at a faster rate than those without the sanitary improvement. The sanitary engineer M. N.

Baker noted that "a village or town without waterworks and sewers is at great disadvantage as compared with communities having these conveniences and safeguards. Industries and population are not so quickly attracted to it."[29] Or, as the engineer for New London, Connecticut, observed, with the building of a sewerage system, the "good name" of the city appreciated, "thus attracting population and business, thereby increasing the value of real estate."[30] Improved sewerage was clearly an important element in the verbal arsenal of late-nineteenth-century city boosters.

The first municipal sewer systems intended for human wastes as well as stormwater were built in the 1850s (Brooklyn, 1855; Chicago, 1856; Jersey City, 1859), but extensive municipal construction did not begin until after the 1870s. The first date for which aggregate figures are available is 1890, and in that year the U.S. Census recorded 6,005 miles of all types of sewers in cities of over 25,000 population; by 1909, the mileage had increased to 24,972 for cities over 30,000 population, or from 1,832 persons per mile of sewer to 825 persons per mile. In the latter year, 85 percent of the population of cities with populations of over 300,000 were served by sewers; 71 percent of cities with populations between 100,000 and 300,000; 73 percent of cities with populations between 50,000 and 100,000; and 67 percent of cities with populations from 30,000 to 50,000 (see Table 8.1).[31] Small-diameter sewers were initially constructed almost entirely of vitrified clay pipe, while large sewers (42 inches or more) were made of brick. After approximately 1905, many sewers were constructed of reinforced concrete.[32]

The aggregate statistics conceal a number of critical stages and decisions involved in sewerage network construction. Before a municipality decided to construct a sewerage network, householders often made attempts to link their sinks and water closets, as well as their rainwater spouts, with the existing drainage sewers. In New York in 1844, for example, increasing use of water closets resulted in a demand by citizens that they be allowed to connect with existing stormwater sewers. Similar demands were made in other cities with stormwater sewers. While such connections were eventually permitted, storm sewers were poorly designed to serve as carriers of sewage and became "elongated cesspools" filled with putrescent material.[33] In the face of the difficulties encountered utilizing existing sewers, it was often necessary to construct an entirely new system. Such a decision raised questions about important matters such as materials and design. Among the materials considered, relatively high-cost preglazed clay pipes of small diameters had many advocates, as did materials such as brick, stone, and masonry.

TABLE 8.1.
Sewer Mileage by Type and Population Group, 1905, 1907, 1909

Population Group	Sanitary Sewers	Storm Sewers	Combined Sewers	Total Mileage by Group
1905				
300,000	335.2	157.0	8,229.9	9,422.6
100,000–300,000	809.0	120.5	2,961.0	4,101.5
50,000–100,000	965.8	242.8	2,491.1	3,709.7
30,000–50,000	1,313.4	326.6	1,507.4	3,147.7
Total by Sewer Type	3,756.2	846.9	14,856.5	20,381.5
1907				
300,000	554.8	352.0	9,242.3	10,149.1
100,000–300,000	1,300.0	262.9	3,690.5	5,253.4
50,000–100,000	1,097.1	181.6	2,627.1	3,905.8
30,000–50,000	1,611.3	383.9	1,562.9	3,558.1
Total by Sewer Type	4,563.2	1,180.4	17,122.8	22,866.4
1909				
300,000	789.5	349.9	9,834.3	10,973.7
100,000–300,000	1,404.4	284.2	4,405.8	6,094.4
50,000–100,000	1,831.5	384.2	2,615.5	4,831.2
30,000–50,000	1,232.9	333.8	1,505.9	3,072.6
Total by Sewer Type	5,258.3	1,352.1	18,361.5	24,971.6

Sources: U.S. Bureau of the Census, Statistics of Cities Having a Population of Over 30,000: 1905 (Washington, D.C., 1907), Appendix A, pp. 342–347; U.S. Bureau of the Census, Statistics of Cities Having a Population of Over 30,000: 1907 (Washington, D.C., 1910), pp. 458–463; and U.S. Bureau of the Census, General Statistics of Cities with Populations Greater Than 30,000: 1909 (Washington, D.C., 1913), pp. 88–93.

Small-diameter glazed pipes had self-cleansing characteristics as well as flexibility in installation, and it became common for municipalities to use them for household connections, while brick and masonry (and later concrete) were used for trunk lines and intercepting sewers.[34]

A critical design question faced by cities contemplating the construction of a sewerage system was whether to construct a separate or a combined system. In 1909, 18,361 miles, or 74 percent of the total mileage in cities with over 30,000 population, were combined sewers, and 5,238 miles (21 percent) were separate sanitary sewers, with only 1,352 miles of storm sewers. Separate sewers consisted of two sets of pipes—a small-pipe sanitary sewer to carry household wastes and a storm drain for rainwater; the combined sewer carried both kinds of wastewater in one pipe. Large cities that needed to remove stormwater from the streets as well as household wastewater generally installed combined sewers. Baltimore, the last major city to construct a sewerage system, however, built a separate system in 1911. Smaller cities and suburbs, with less need for below-ground storm drainage and with more limited funding abilities, often constructed sanitary sewers alone, leaving the stormwater to run off on the surface.[35]

During the 1880s and 1890s, decisions over whether to build a combined or a separate system were complicated by the anticontagionist argument that the separate system was more likely to prevent the formation of dangerous sewer gas than the combined method. The most famous advocate of this position was Colonel George E. Waring, Jr., who argued, as had Chadwick, that small-diameter separate sewers would speed the removal of fecal wastes from the vicinity of the household, thus preventing the formation of dangerous miasmas. Under Waring's direction and that of sanitary engineers who followed his doctrines, a number of smaller cities and towns installed separate sewers in the last decades of the nineteenth century. By 1900, however, largely because of the work of the sanitary engineer Rudolph Hering, most engineers understood that the two sewer designs had equal health benefits, and decisions regarding the implementation of one system rather than another were based primarily on the practical factors discussed above.[36]

Sewers and other infrastructure improvements were provided by both the public and private sectors, although they almost always were maintained by the municipality. As Christine Rosen has demonstrated, both methods of provision involved various kinds of frictions, making it difficult for urbanites to express their demands for services.[37] In some instances, the municipality itself provided the sewers, paying for them through general tax revenues. In such situations, city government often responded to business and commercial needs, especially for the removal of stormwater that might impede traffic. In Birmingham, Alabama, for instance, as Carl V. Harris has shown, limited resources resulted in sewers being first provided for the downtown business section and then for residential neighborhoods.[38] In residential areas, municipalities responded first to homeowners and real estate developers with economic and organizational power. Ann Durkin Keating has illustrated how residents of Chicago formed improvement associations to apply political pressure to secure services; Joseph Arnold has documented a similar process in Baltimore.[39] Poorer residents, who lacked strong voices to demand a share of services, often found themselves deprived.[40]

Betterment assessment laws that placed a special tax on abutting property owners in order to force them to contribute to the costs of sewers shifted the financial burden from the city proper to the home-owners. Those who could pay received first priority, often with the result that poorer areas of the cities lacked sewers for considerable periods of time. In Milwaukee, and Detroit, for instance, lower-income home-owners delayed the acquisition of sewers and other utilities in order to keep housing expenses low. In these situations, savings in housing costs were often replaced by health costs, as householders exposed themselves and their families to infectious disease as well as nuisance from inadequate sewerage facilities.[41]

Sewers were provided by developers mostly in newly developing middle-and upper-class urban subdivisions in the city and its suburbs. In a number of cities, builders would provide services before dwellings were constructed. In Milwaukee, for instance, developers often assumed the cost of installing sewers in the expectation of recouping the cost in the purchase price.[42] In contrast, in the Boston area suburban towns, authorities provided water and sewer lines after speculators had furnished rough-graded streets. Once the utilities were laid, the municipality would pave and maintain the streets. Such services were provided below cost as a subsidy to the development process and were paid for out of general taxes.[43] In Chicago, homeowners and developers in newly developing areas utilized the requirement that the city respond to their petitions for street paving (at the homeowners' cost) to secure water and sewerage services (provided at city expense). Once the street-paving orders were granted by the board of aldermen, the board of public works would be compelled to lay water and sewer pipe in order to save the cost of tearing up the streets in the future.[44]

Although central cities moved in the late nineteenth and early twentieth centuries to begin construction of sewerage systems and provide their inhabitants with other services, financial limitations forced some suburbs to forgo such luxuries. Faced by these deprivations, inhabitants of outlying communities often voted to merge with the central city to acquire superior services at a lower cost. City boosters concerned with municipal growth encouraged these mergers, and in many cases the cities offered services to the newly annexed areas at low cost in order to ensure a positive suburban voter response.[45]

Sewage Disposal, Sewage Treatment, and Health Externalities

All cities that constructed sewerage systems faced problems of sewage disposal, but it was an essentially more difficult and hence expensive question for cities with inland locations on rivers and lakes. If a water body of some size was available, municipal authorities would take advantage of it for disposal purposes. Such action might conceivably create a nuisance or pose a threat to the purity of the water supply of the city itself or downstream municipalities. For most of the nineteenth century, however, most municipal officials and engineers assumed that dumping raw sewage into streams was adequate treatment given the concept of the self-purifying nature of running water. In 1909, 88 percent of the wastewater of the sewered population was disposed of in waterways without treatment. When treatment was utilized at the beginning of the twentieth century, it was only to prevent local nuisance, not to avoid contamination of the drinking water of a downstream city.[46]

The number of municipalities that treated their sewage grew from about 28 in 1892 to 860 in 1920 (see Table 8.2). In the latter year, 9.5 million people, or 20 percent of the sewered population, had their sewage treated in some fashion, ranging, at the minimum, from the removal of solids through screening to what today would be called primary treatment. The remainder of municipalities continued to discharge raw sewage into waterways. The initial form of sewage treatment was on sewage farms and involved the use of sewage to fertilize crops, provide irrigation, or both. Sewage farms were land and labor intensive and presented environmental-control problems. Because of these factors, relatively few were utilized, and then often primarily for irrigation rather than sewage disposal.[47] From about 1890 to 1910, intermittent filtration, a sewage-treatment method that used the sand and soil as a filter medium without an attempt to grow crops, was widely implemented in Middle Atlantic and especially New England cities. This disposal technology required less land than sewage farms because of higher rates of sewage application, but worked well only under certain soil conditions.[48] In the first decades of the twentieth century, biological processes in man-made structures replaced land-treatment methods as the most popular approach to sewage treatment. These methods included techniques such as the septic tank, the Imhoff tank, the contract filter, the sprinkling filter, and the trickling filter. Eventually, the activated-sludge method of sewage treatment, first applied in Milwaukee in 1915, became the most commonly utilized technology-based municipal treatment process.[49] Not until 1940, however, did more than half of the sewered urban population have its sewage treated; dumping raw sewage into waterways remained the cheapest and most popular means of disposal.

The consequence of the disposal of untreated sewage in streams and lakes that supplied drinking water to other cities was a large increase in

TABLE 8.2.
Total U.S. Population, Urban Population, Water Treatment,
Sewers, and Sewage Treatment, 1880–1940

Year	Total Population	Urban Population	Water Treatment Population	Sewer Population	Sewage Treatment Population
1880	50,155,783	14,129,735	30,000	9,500,000	5,000
1890	62,947,714	22,106,265	310,000	16,100,000	100,000
1900	75,994,575	30,159,921	1,860,000	24,500,000	1,000,000
1910	91,972,266	41,998,932	13,264,140	34,700,000	4,455,117
1920	105,710,620	54,157,973	N.A.	47,500,000	9,500,000
1930	122,775,046	68,954,823	46,059,000[a]	60,000,000	18,000,000
1940	131,669,275	74,423,702	74,308,000	70,506,000	40,618,000

[a]1932 data.

mortality and morbidity from typhoid fever and other infectious water-borne diseases. In the 1890s, bacterial researchers, following the seminal work of Pasteur and Koch, identified the processes involved in water-borne disease. The work of William T. Sedgwick and other sanitary engineers, biologists, and chemists at the Massachusetts Board of Health Lawrence Experiment Station played a critical role in the clarification of the etiology of typhoid fever and confirmation of its relationship to sewage-polluted waterways.[50] Cities had built sewerage systems to improve the health of their residents, but disposal practices worsened health conditions (or produced "externalities") for downstream cities.[51] Sewerage technology, therefore, caused a sharp rise in societal health costs in addition to a number of improvements. Because these costs were often borne by second parties or downstream users, while the benefits accrued to upstream cities, cities continued to build sewerage systems and dispose of untreated wastes in adjacent waterways.

Policy Responses to Problems of Sewage Pollution

To attempt to cope with the nuisances and health risks of the cesspool–privy vault system and the supposed dangers of sewer gas and various forms of plumbing, late-nineteenth-century municipalities adopted sanitary and plumbing codes. In addition, they created local boards of health because of concern over epidemic disease.[52] Regulations and health boards were promoted primarily by a sanitary coalition composed of physicians, engineers, plumbers, and civic-minded citizens, and this coalition was instrumental in securing municipal adoption of the new sewerage technology.[53] The so-called public health movement in the United States was primarily an extension of the Sanitary Movement that began in Great Britain in the 1840s and 1850s with the work of Sir Edwin Chadwick. It was essentially a social movement by elites and professionals that aimed to change people's ideas about their personal habits of cleanliness, to create an enlarged role for government in areas related to health and sanitation, and to promote the construction of urban public works to achieve a healthful city. Chadwick's ideas greatly influenced the pioneer group of American sanitarians and public health reformers. Important among this group were men such as the public health reformer John H. Griscom and the sanitarian George E. Waring, Jr., who promoted sewerage as "not only the most economical, but the *only* mode in which the immense amounts of filth generated daily in [large cities] can be effectively removed."[54]

The sanitary coalition also pushed for laws and institutions to deal with the threats to health caused by municipal sewage-disposal practices. The transference of infectious disease carried in waste material from one user to another via sewerage technology, and then through the water medium used for disposal, necessitated an extralocal response. Legal redress for damages was possible in the case of nuisance, but difficulties in specifying the origins of waterborne disease prevented the affected individuals from seeking court relief.[55] The logical response was to create new institutions with regulating powers, such as the Massachusetts State Board of Health, formed in 1869, and to pass legislation to protect water quality. In 1905, the U.S. Geological Survey, in its *Review of the Laws Forbidding Pollution of Inland Waters in the United States*, listed thirty-six states with some legislation protecting drinking water and eight states with "unusual and stringent" laws. Supervision of water quality, whether through merely advisory powers or stricter enforcement provisions, was generally entrusted to state boards of health.[56]

By the early twentieth century, municipal and public health officials, and sanitary engineers, possessed several alternatives to deal with the threat of sewage to the water supplies of inland cities. One choice was to secure municipal water supplies from a distant and protected watershed, a course followed by cities such as Newark and Jersey City, New Jersey.[57] Sewage treatment, or "purification," provided a technological option, although it was more effective in preventing nuisance than in protecting drinking-water quality.[58] In addition, it provided benefits for the downstream city, but imposed the costs on the upstream community in an era when neither state nor federal programs subsidized waste treatment. A further technological option for protecting water supplies, water filtration at the intake, became available in the 1890s. Water filters had been utilized in Europe for some time, but it was not until this decade that experiments at the Lawrence Experiment Station and at Louisville, Kentucky, conclusively showed the effectiveness of both slow sand and mechanical filters in treating sewage-polluted waters. As a result of these successful demonstrations, many inland cities installed mechanical or sand filters in the years after 1897. The use of water-filtration technology resulted in an impressive decline in morbidity and mortality rates from typhoid fever and other diseases.[59]

Municipalities faced with both water pollution and poor access to pure supplies of water had to decide whether to filter their water *and* treat their sewage in order to protect their own water supply and that of downstream cities, or to filter the water alone, leaving downstream users with the responsibility of guarding the safety of their own water. In some cities, there was also a conflict over whether to filter the water from local polluted rivers in order to secure a water supply or to seek a pure source in a distant locality.[60] In the first years of the twentieth century, sanitary engineers took the position, as expressed editorially by the *Engineering Record*, that "it is often more equitable to all concerned for an upper riparian city to discharge its sewage into a stream and a lower riparian city to filter the water of the same stream for a domestic supply, than for the former city to be forced to put in sewage treatment works."[61] The sanitary engineer Allen Hazen expressed the rationale for this position in his 1907 book, *Clean Water and How to Get It*, by noting that "the discharge of crude sewage from the great majority of cities is not locally objectionable in any way to justify the cost of sewage purification." Instead, said Hazen, downstream cities should filter their water to protect the public health, and sewage purification should be utilized only in order to prevent nuisances such as odors and floating solids.[62]

While some public health officers and conservationists objected to this position on sewage disposal, most sanitary engineers and municipal officials espoused it.[63] Essentially, the engineering position was that the dilution power of streams should be utilized to its fullest for sewage disposal as long as no danger was posed to the public health and to property rights, and no nuisance was created. Where dilution was insufficient, sewage treatment would serve as a means of controlling specific water-quality problems rather than as a general means to achieve water purity. Water filtration and/or chlorination would serve as the main protection of the public health against waterborne disease.[64]

Statistics on sewage treatment and water filtration from 1910 to 1930 reflect this policy (see Table 8.2). During these decades, the newly sewered population rose by over 25 million, but the additional number whose sewage was treated increased by only 13.5 million. Simultaneously, the increase in the population receiving treated water was approximately 33 million. In 1930, not only did the great majority of the urban population dispose of its untreated sewage by dilution in waterways, but the number who were doing so was actually increasing over those who were treating their sewage before discharge. Because of the successes of water filtration and chlorination, however, waterborne infectious disease had greatly diminished and the earlier crisis atmosphere that had led to the first enforceable state legislation had disappeared.

Professional, Governmental, and Institutional Developments

The implementation of large-scale and capital-intensive networked technologies in cities necessitated a range of institutional developments. Jacques Ellul has suggested that the modern administrative organization derived from the application of the machine mindset, or "technique," into a social philosophy that promised progress through standardizing and rationalizing human behavior.[65] Public administration had other roots as well as technological ones, but Ellul's analysis still provides a link that relates the technological base of urban society to the growth of municipal administration, city planning, and various institutional innovations. Wastewater technology was extremely important as a networked technology that stimulated change in these governmental areas.[66]

Sewerage systems have characteristics of economies of continuous collection and of scale that often ignore municipal boundaries and require centralized administration. The same disregard of political boundaries is true of the health hazards created by waste disposal and water pollution. Ideally, then, wastewater collection and disposal should be dealt with on a regional basis, but many American urbanized areas were (and are) characterized by political fragmentation.[67] In order to secure cost and design advantages as well as efficiency and safety of disposal, sanitary engineers and public health officials pushed for unification of these fragmented districts. As early as the 1870s, the engineering press began urging regional cooperation in sewer and water services, and throughout the late nineteenth and early twentieth centuries it pushed for new regional administrative arrangements. The requirements of sewerage systems for efficient operation offered a powerful argument for overcoming the fragmentation produced by political boundaries that did not conform to environmental needs.[68]

Three means to achieve this unity actually came into use: inter-municipal and interstate cooperation, annexation or consolidation of suburban areas with a central city, and special district governments. The chief example of intermunicipal cooperation was the Passaic Valley Sewerage and Drainage Commission, where seven northern New Jersey municipalities joined to construct a joint outlet sewer. By 1927, the number of participants had increased to seventeen. There were, however, few examples of joint action between municipalities because of the difficulty in obtaining agreement on apportionment of responsibilities and costs.[69] An inability to agree on joint responsibilities also inhibited the development of interstate water pollution compacts before 1920; in the second quarter of the century, such pacts were instituted in important regions such as the Ohio River Valley.[70]

A more common method of solving sewerage problems involving several governmental jurisdictions was the annexation or consolidation of suburban territories by a central city. Sewerage and water supply are costly capital systems, and there was a financial incentive for suburban communities with weak tax bases to consolidate with central cities that could supply these services. As a further inducement, the annexed territories often received services at the regular city rate, even though the costs of installation exceeded revenues. Thus, as the historian Jon Teaford observes, the desire for improved service was a "countervailing force for unity" against the forces of fragmentation in the metropolis.[71]

The most readily adopted institutional means to handle sewerage and water supply projects has been the special district government. These governments are special state creations and are fiscally and administratively independent authorities of limited function and area. Some examples of special governments in regard to sewerage are the Chicago Sanitary District (1889), the Boston Metropolitan Sewerage Commission (1889), and the Washington Suburban Sanitary Commission (1918). Governmental reformers who pushed for these special authorities were motivated by the belief that metropolitan areas required a functional structure independent of political boundaries, by a desire to escape existing tax or debt limits, and by a wish to be free of municipal political control. Special district governments were also an alternative to central-city annexation, and were preferred by suburban authorities for this reason.[72]

Sewerage technology, therefore, with its characteristics of efficiency of continuous collection, scale economies of treatment and capital intensiveness, as well as its requisite for central administration, was an important factor in facilitating governmental integration. It encouraged consolidation of urban areas and promoted a new governmental form, the special district government. (It has also been argued that specialized districts have retarded full metropolitan integration.)[73] Such institutional innovations may have evolved without wastewater technology, but its requirements undoubtedly accelerated institutional adaptation that provided a model for further innovation in response to the needs of other urban systems.

When sewerage systems were constructed, engineers had to build to accommodate future urban population growth and changes in city functions in order to avoid the necessity of constant rebuilding. These requirements necessitated long-range planning and forecasts of population growth. The new technology required a permanent bureaucracy for day-to-day administration, for data collection, and for efficient planning. The massive costs of these public works, in addition, demanded fiscal planning by professionals. Ideally, the works could be constructed and maintained best by experts who could survey the topography scientifically, evaluate alternate materials, plan for population change probabilities, and keep the system functioning efficiently.

Before the city planning movement had fully begun, major expositions on sanitary reform argued the virtues of planned sewerage. By the 1870s and 1880s, this form of planning was firmly rooted as a major urban art, and sanitary engineering was established as an important subbranch of civil engineering concerned with sewerage and water supply. While the sanitary engineers' view of planning was restricted to sewage collection and disposal and water supply, it inherently contained a concept of the city as a physical container to be organized to provide more efficient delivery of services and disposal of wastes.[74] The engineering ideal was the comprehensively planned city, staffed and managed by disinterested experts such as themselves. Not surprisingly, sanitary engineers played important roles in the emergence of the city planning profession at the start of the twentieth century, with thirteen of the fifty-two charter members of the American Institute of Planners listed as engineers.[75]

Planning also involves making predictions and dealing with probabilities and uncertainties; sanitary engineering, more than other branches of engineering, had to deal with these factors. Writing in 1915 and 1916, the sanitary engineers Allen Hazen and George C. Whipple noted the special concern with variation that characterized sanitary engineering and placed it between the natural and the exact sciences.

They also pointed out the importance of probability theory to sanitary engineering problems.[76] Sanitary engineers were concerned with rainfall prediction, water use, and demographic change, all categories that shared a characteristic of high uncertainty. Unique among urban professionals, they were the first to attempt long-range urban population predictions, making such calculations considerably before city planners utilized the technique.[77]

The experience of sanitary engineers with population forecasting, social factors such as water use, and public health considerations, as well as their participation in planning large-scale capital works, prepared them for key roles in city government. They adhered to a set of values and procedures that stressed efficiency within a benefit–cost framework, and this appealed to late-nineteenth and early-twentieth-century reformers attempting to restructure municipal government in the direction of professionalism, efficiency, and bureaucratization. Sanitary engineers served in city government not only as municipal engineers but also as administrators, and were a principal group from which the majority of city managers were recruited before World War II.[78]

Conclusions

This article has discussed the development of sewerage systems as a critical element in the rise of the networked city. It is clear that sewerage technology played an important role in shaping the urban environment and in generating a set of institutional adaptations. Technological development and implementation, however, did not necessarily follow a linear path from cesspools and privy vaults to fully networked cities complete with sewage treatment and water-filtration plants. Adaptations were often gradual and included numerous efforts to retrofit old systems and avoid the large capital expenditures required by new system creation.

An important lesson from the sewerage history concerns the destabilizing effects of introducing new elements into a balanced system without attempting to calculate the impacts of the innovation. Technologies that appear to promise only benefits may have severe secondary costs, although these are often difficult to foresee. In the case of sewage disposal, problems of pollution were particularly difficult to deal with because they primarily affected downstream communities. Municipalities were quite willing to shift the burden of pollution if it meant improving their local environment.

The various characteristics of sewerage technology, such as its capital intensiveness and its planning requirements, as well as the fact that its effective operation bore little relationship to municipal boundaries, required a number of institutional adaptations and innovations. These involved planning innovations for large-scale public works, developing methods to raise capital funds and assess property owners, and measures to overcome the political fragmentation of urban areas. In some instances, suburbs actually courted annexation as a means of acquiring services, but increasingly special district governments were created to provide single functions without necessitating annexation or consolidation. In regard to professional training, the education of sanitary engineers and their experience with factors such as population forecasting, water use, and public health, as well as planning massive public works, fitted them for various important roles in city government.

By the end of the first third of the twentieth century, the modern networked metropolis was close to completion in terms of the technologies projected in the nineteenth century. Sewers, water systems, tracks, and electrical and telephone wires crisscrossed the city, providing services at a level that was inconceivable in the 1890s. Most urbanites took these technologies and their supply systems for granted, so intertwined had they become with their everyday lives. And yet system provision remained uneven and was often predicated on an ability-to-pay basis, meaning that poorer areas of the city were not well served. In addition, questions involving the basic scheduling of supply and maintenance tended to be inefficient in execution, resulting in the constant tearing up and rebuilding of city streets. Technological innovation had the capacity to improve the operations of existing infrastructure, for example, through sewage treatment and water filtration, but it could also cause disruptions, such as occurred with the automobile. Thus, in spite of the attempt of some urban planners to conceptualize the city as a smoothly operating, self-correcting technology, such a vision remained a "utopian dream." For, as Josef Konvitz has observed, "Vital cities are never finished."[79]

Notes

1. David P. Handlin, *The American Home: Architecture and Society, 1815–1915* (Boston, 1979), pp. 452–486.

2. See, for example, John Brotchie, Peter Newton, Peter Hall, and Peter Nijkamp, *The Future of Urban Form: The Impact of New Technology* (New York, 1985).

3. Nelson M. Blake, *Water for the Cities* (Syracuse, N.Y., 1956), pp. 12–13; Constance M. Green, *Washington: Village and Capital, 1800–1878* (Princeton, N.J., 1962), p. 95. The estimates on water usage are based on figures reported for cities without waterworks in John D. Bell, "Report on the Importance and Economy of Sanitary Measures to Cities," in *Proceedings and Debates of the Third National Quarantine and Sanitary Convention* (New York, 1859), pp. 576–577.

4. Jon Peterson calls the cesspool–privy vault system the "private-lot waste removal" system. See Jon Peterson, "The Impact of Sanitary Reform upon American Urban Planning," *Journal of Social History* 13 (Fall 1979): 85.

5. In 1829, it was estimated that each day New Yorkers deposited over 100 tons of excrement into the city's soil. See Eugene P. Moehring, *Public Works and the Patterns of Urban Real Estate Growth in Manhattan, 1835–1894* (New York, 1981), p. 15. For contemporary descriptions, see *Papers and Reports of the American Public Health Association* (hereafter cited as APHA), 3 (1876): 185–187; Mansfield Merriman, *Elements of Sanitary Engineering*, 3rd ed. (New York, 1906), pp. 139–142.

6. There is information on the "municipal cleansing" practices of more than a hundred cities in U.S. Department of the Interior, Census Office, *Tenth Census of the United States, 1880, Report of the Social Statistics of Cities* (George E. Waring, Jr., comp.), 2 vols. (Washington, D.C., 1887); "Report of Committee on Disposal of Waste and Garbage," APHA 17 (1891) 90–119.

7. Joel A. Tarr, "From City to Farm: Urban Wastes and the American Farmer," *Agricultural History* 49 (October 1975): 601–602. In 1880, the wastes of 103 of 222 U.S. cities were used on the land. See also Richard A. Wines, *Fertilizer in America: From Waste Recycling to Resource Exploitation* (Philadelphia, 1985), pp. 6–32.

8. See regulations cited in Waring, *Social Statistics of Cities* (1887). See, for instance, descriptions of sewers in Julius W. Adams, *Report of the Engineers to the Commissioners of Drainage* (Brooklyn, 1857); Henry I. Bowditch, *Public Hygiene in America* (Boston, 1877), pp. 103–109; Leonard Metcalf and Harrison P. Eddy, *American Sewerage Practice*, 2nd ed. (New York, 1928), 1:15–19. For private sewers, see Peterson, "Impact of Sanitary Reform," p. 85; Geoffrey Giglierno, "The City and the System: Developing a Municipal Service, 1800-1915," *Cincinnati Historical Society Bulletin* 35 (Winter 1977): 223–224.

9. U.S. Department of Commerce, Bureau of the Census, *Historical Statistics of the United States . . . to 1970*, (Washington, D.C., 1975), 1:11–12.

10. George E. Waring, Jr., "The Sanitary Drainage of Houses and Towns," *Atlantic Monthly* 36 (October 1875): 434; Clay McShane, "Transforming the Use of Urban Space: A Look at the Revolutions in Street Pavements, 1880–1924," *Journal of Urban History* 5 (May 1979): 288.

11. Blake, *Water for the Cities*, pp. 3–17; Ellis L. Armstrong, Michael C. Robinson, and Suellen M. Hoy, eds., *History of Public Works in the United States, 1776–1976* (Chicago, 1976), pp. 217–235; and Moehring, *Public Works and Patterns of Urban Real Estate Growth*, pp. 23–51.

12. J. T. Fanning, *A Practical Treatise on Hydraulic and Water-Supply Engineering* (New York, 1886), p. 625.

13. Town of Pawtucket, Committee on Sewers, *Report, 1885* (Pawtucket, R. I., 1885), p. 15; E. S. Chesbrough, "The Drainage and Sewerage of Chicago," APHA 4 (1878): 18–19; Town Improvement Society of East Orange, *The Sewerage of East Orange* (East Orange, N.J., 1884).

14. May N. Stone, "The Plumbing Paradox: American Attitudes Towards Late Nineteenth-Century Domestic Sanitary Arrangements," *Winterthur Portfolio* 14 (1979): 284–285. See Reginald Reynolds, *Cleanliness and Godliness* (New York, 1974), for an amusing description of the evolution of the water closet; see also Lawrence Wright, *Clean and Decent: The Fascinating History of the Bathroom and the Water Closet* (Toronto, 1972).

15. Boston, Cochituate Water Board, *Annual Report for 1863* (Boston, 1864), p. 43; City of Buffalo, *Sixth Annual Report of the City Water Works, 1874* (Buffalo, 1875), p. 47. The estimate for 1880 is based upon information in Waring, *Social Statistics of Cities.*

16. William H. Bent, George H. Rhodes, and William Tinkham, *Report of Special Committee on Sewerage for City of Taunton* (Taunton, 1878), pp. 25–26; E. S. Chesbrough, *Report on Plan of Sewerage for the City of Newport* (Newport, 1880), pp. 5–6; Rudolph Hering, *Report on a System of Sewerage for the City of Wilmington, Delaware* (Wilmington, 1883), pp. 5–6; Maryland State Board of Health, "The Sanitation of Cities and Towns and the Agricultural Utilization of Excremental Matter," *Annual Report, 1887* (Baltimore, 1887), pp. 229–230.

17. Bell, "Importance and Economy of Sanitary Measures to Cities," pp. 479–575; Charles E. Rosenberg, *The Cholera Years* (Chicago, 1962), pp. 75–81, 117, 202; Charles V. Chapin, "The End of the Filth Theory of Disease," *Popular Science Monthly* 60 (January 1902): 234–239; and George E. Waring, Jr., "The Sanitary Drainage of Houses and Towns," *Atlantic Monthly* 36 (November 1875): 535–551.

18. Benjamin Lee, "The Cart Before the Horse," APHA 20 (1895): 34–36. Lee was concerned with the bacterial danger presented by the fecal pollution.

19. Azel Ames, "The Removal and Utilization of Domestic Excreta," APHA 4 (1877): 65–70. *Social Statistics of Cities* listed eleven cities using the odorless excavator in 1880, but this is probably an underestimate.

20. Peterson, "Impact of Sanitary Reform," p. 84.

21. Moehring, *Public Works and Patterns of Urban Real Estate Growth*, pp. 87–95.

22. Francis Sheppard, *London 1808–1870: The Infernal Wen* (Berkeley, 1971), pp. 250–278.

23. Peterson, "Impact of Sanitary Reform," pp. 86–87.

24. Rudolph Hering, "Report of the Results of an Examination Made in 1880 of Several Sewage Works in Europe," *Annual Report of the National Board of Health, 1881* (Washington, D.C., 1882), was probably the most influential report about European sewerage systems written by an American engineer. For the background and influence of this report, see Joel A. Tarr, "The Separate vs.

Combined Sewer Problem: A Case Study in Urban Technology Design Choice," *Journal of Urban History* 5 (May 1979): 308–333.

25. Hering, *System of Sewerage for Wilmington, Delaware*, p. 6; Joseph E. Nute, "The Sewerage of Malden [Mass.]," B.S. thesis, MIT, 1884; Baltimore, Sewerage Commission, *Second Report* (Baltimore, 1899), p. 30.

26. Baltimore, Sewerage Commission, *Second Report*, pp. 14, 30.

27. Massachusetts State Board of Health, "The Value of Health to the State," *Annual Report*, 1875, pp. 57–75.

28. George E. Waring, Jr., *Draining for Health and Draining for Profit* (New York, 1867), pp. 222–223; Henry E. Sigerist, ed., "The Value of Health to a City: Two Lectures, Delivered in 1873, by Max Von Pettenkofer," *Bulletin of the History of Medicine* 10 (October 1941): 473–503, 593–613.

29. M. N. Baker, *Sewerage and Sewage Purification* (New York, 1896), p. 11.

30. New London Board of Sewer Commissioners, *First Annual Report* (New London, 1887), p. 4.

31. U. S. Bureau of the Census, "Sewers," *Report on the Social Statistics of Cities, Eleventh Census* (Washington, D.C., 1895), pp. 29–32; and "Sewers and Sewer Service," *General Statistics of Cities: 1909* (Washington, D.C., 1913), pp. 20–23.

32. Harold E. Babbitt, *Sewerage and Sewage Treatment* (New York, 1932), pp. 132–133; H. F. Peckworth, *Concrete Pipe Handbook* (Chicago, 1959).

33. John Duffy, *A History of Public Health in New York City, 1625–1866* (New York, 1968), pp. 409–411; C. C. Clarke, *Main Drainage Works of Boston* (Boston, 1885), p. 10. Boston permitted fecal matter in its sewers in 1833, St. Louis in 1842, Washington in 1858, and Cincinnati in 1863. For discussions of the problems with such sewers, see H. I. Bowdith, *Public Hygiene in America* (Boston, 1877), pp. 103–109; J. S. Billings, "Sewage Disposal in Cities," *Harper's New Monthly Magazine* 71 (1885): 579–580; and Waring, "Sanitary Drainage of Houses and Towns," pp. 537–553.

34. A good discussion of this question is in Giglierano, "City and System," 225–237; see also Metacalf and Eddy, *American Sewerage Practice*, vol. 1 *Design of Sewers*, pp. 328–382.

35. Tarr, "Separate vs. Combined Sewers," pp. 308–328, 332.

36. Ibid., pp. 313–329. For an appreciation of Waring as an environmentalist, see Martin V. Melosi, *Pragmatic Environmentalist: Sanitary Engineer, Col. George E. Waring, Jr.*, Essays in Public Works History 4 (Washington, D.C., April 1977). For a more critical appraisal of Waring, see James H. Cassedy, "The Flamboyant Colonel Waring: An Anti-Contagionist Holds the American Stage in the Age of Pasteur and Koch," *Bulletin of the History of Medicine* 36 (March–April 1962): 163–176.

37. Christine Meisner Rosen, "Infrastructural Improvement in Nineteenth-Century Cities: A Conceptual Framework and Cases," *Journal of Urban History* 12 (May 1986): 211–256.

38. Carl V. Harris, *Political Power in Birmingham, 1871–1921* (Knoxville, Tenn., 1977), pp. 149–153.

39. Ann Durkin Keating, "From City to Metropolis: Infrastructure and Residential Growth in Urban Chicago," in *Infrastructure and Urban Growth in the Nineteenth Century*, Essays in Public Works History 14 (Chicago, 1985), pp. 3–27; Joseph Arnold, "The Neighborhood and City Hall: The Origin of the Neighborhood Associations in Baltimore, 1890–1911," *Journal of Urban History* 6 (November 1979): 3–30.

40. Roger D. Simon, *The City-Building Process: Housing and Services in New Milwaukee Neighborhoods 1880–1910* (Philadelphia, 1978), pp. 245–247.

41. Ibid., pp. 40–41; Olivier Zunz, *The Changing Face of Inequality: Urbanization, Industrial Development, and Immigrants in Detroit 1880–1920* (Chicago, 1982), p. 174.

42. Simon, *The City Building Process*, p. 40.

43. Sam Bass Warner, Jr., *Streetcar Suburbs: The Process of Growth in Boston 1870–1900* (Cambridge, Mass., 1962), pp. 154–155.

44. Rosen, "Infrastructure Improvement in Nineteenth Century Cities," pp. 241–242.

45. Jon C. Teaford, *City and Suburb: The Political Fragmentation of Metropolitan America, 1850–1970* (Baltimore, 1979), pp. 32–63.

46. "Sewage Purification and Storm and Ground Water," *Engineering News* 28 (August 25, 1892): 180–181; Sedgwick, *Principles of Sanitary Science*, pp. 231–237; Rudolph Hering, "Notes on the Pollution of Streams," APHA 13 (1888): 272–279; Moses N. Baker, "Sewerage and Sewage Disposal," in *Statistics of Cities Having a Population of over 30,000: 1905* (Washington, D.C. 1907), pp. 103–106.

47. Tarr, "From City to Farm: Urban Wastes and the American Farmer," pp. 607–611.

48. Metcalf and Eddy, *American Sewerage Practice*, vol. 3, *Disposal of Sewage* (New York, 1935), pp. 4–5.

49. For a summary of the various steps leading to the construction of the first full-sized activated-sludge working plant in the United States, see Babbitt, *Sewerage and Sewage Treatment*, pp. 434–435.

50. Rosenkrantz, *Public Health and the State*, pp. 97–107.

51. George C. Whipple, *Typhoid Fever* (New York, 1908).

52. Howard D. Kramer, "Agitation for Public Health Reform in the 1870s, pt. 1," *Journal of the History of Medicine* 3 (Autumn 1948): 474–476; ibid., pts. 1, 2, 3, 4 (Autumn 1948 and Winter 1949): 473–488, 75–89.

53. Barbara G. Rosenkrantz, "Cart Before Horse: Theory, Practice and Professional Image in American Public Health, 1870–1920," *Journal of the History of Medicine and Allied Sciences* 29 (January 1974): 55–56; idem, *Public Health and the State*, pp. 1–127; Stephen Smith, "The History of Public Health, 1871–1921," in *A Half Century of Public Health*, ed. Mazyck P. Ravenel (New York, 1921), pp. 1–12; George Rosen, *A History of Public Health* (New York, 1958), pp. 233–250.

54. The statement is by John H. Griscom and is quoted in Peterson, "Impact of Sanitary Reform," p. 86.

55. "Sewage Purification and Water Pollution in the United States," *Engineering News* 47 (April 3, 1902): 276; "Sewage Pollution of Water Supplies," *Engineering Record* 28 (August 1, 1903): 117. After 1910, the courts awarded damages against municipalities in cases where negligence in the operation of public waterworks resulted in individuals contracting typhoid fever. See James A. Tobey, *Public Health Law*, 2nd ed. (New York, 1939), pp. 277–280.

56. Edwin B. Goodell, *Review of Laws Forbidding Pollution of Inland Waters in the United States*, U.S. Geological Survey Water Supply and Irrigation Paper 152 (Washington, D.C. 1906). A useful summary of Goodell is "Pollution of Streams," *Municipal Journal and Engineer* 21 (October 3, 10, 17, 1906): 333–334, 364–365, 384.

57. Stuart Galishoff, "Triumph and Failure: The American Response to the Urban Water Supply Problem, 1860–1923," in *Pollution and Reform in American Cities, 1880–1930*, ed. Martin V. Melosi (Austin, 1980), pp. 46–47.

58. Metcalf and Eddy, *Disposal of Sewage*, pp. 190–231; "Sewage Purification and Storm and Ground Water," pp. 180–181.

59. Allen Hazen, *Clean Water and How to Get It* (New York, 1907), pp. 68–75; George C. Whipple, "Fifty Years of Water Purification," in Ravenel, ed., *Half Century of Public Health*, pp. 161–180. See George A. Johnson, "Present Day Water Filtration Practice," *Journal of the American Water Works Association* 1 (March 1914): 31–80, for figures on typhoid death rates for leading cities before and after filtration.

60. For a discussion of this dispute in Boston, see Fern L. Nesson, *Great Waters: A History of Boston's Water Supply* (Hanover, N. H., 1983).

61. "Sewage Pollution of Water Supplies," p. 117. See also, "Water Supply of Large Cities," p. 73.

62. Hazen, *Clean Water*, pp. 34–37.

63. For a discussion of the conflict over this issue, see Joel A. Tarr, Terry Yosie, and James McCurley III, "Disputes over Water Quality Policy: Professional Cultures in Conflict, 1900–1917," *American Journal of Public Health* 70 (April 1980): 427–435.

64. George W. Fuller, "Relations between Sewage Disposal and Water Supply Are Changing," *Engineering News Record* 28 (April 5, 1917): 11–12. See also George Fuller, "Is It Practicable to Discontinue the Emptying of Sewage into Streams?" *American City* 7 (1912): 43–45.

65. Jacques Ellul, *The Technological Society* (New York, 1964), pp. 3–13, 171–183, 229.

66. For discussions of different aspects of this question, see Stanley K. Schultz and Clay McShane, "To Engineer the Metropolis: Sewers, Sanitation, and City Planning in Late-Nineteenth-Century America," *Journal of American History* 65 (September 1978): 389–411; and Martin J. Schiesl, *The Politics of Efficiency: Municipal Administration and Reform in America: 1880–1920* (Berkeley, 1977).

67. Paul Studenski, *The Government of Metropolitan Areas in the United States* (New York, 1930), p. 18.

68. Stanley K. Schultz and Clay McShane, "Pollution and Political Reform in Urban America: The Role of Municipal Engineers, 1840–1920," in Melosi, ed., *Pollution and Reform in American Cities*, pp. 160–167.

69. "Municipal Cooperation as a Possible Substitute for Consolidation," *Engineering News* 41 (February 16, 1899): 104–106; "Sewerage of the Passaic River Valley," *Engineering Record* 44 (December 28, 1901): 60; and Studenski, *Government of Metropolitan Areas*, pp. 47–48. States such as Ohio and California passed legislation providing for intergovernmental contractual relations. See Teaford, *City and Suburb*, p. 81.

70. Edward J. Cleary, *The ORSANCO Story: Water Quality Management in the Ohio Valley under an Interstate Compact* (Baltimore, 1967).

71. Teaford, *City and Suburb*, pp. 39–40, 59–60; Studenski, *Government of Metropolitan Areas*, 166–167.

72. Robert B. Hawkins, Jr., *Self-Government by District: Myth or Reality* (Stanford, 1976), p. 25; Studenski, *Government of Metropolitan Areas*, pp. 256–262; Teaford, *City and Suburb*, pp. 79–81, 173–174; and Louis P. Cain, *The Search for an Optimum Sanitation Jurisdiction: The Metropolitan Sanitary District of Greater Chicago, a Case Study*, Essays in Public Works History 10 (Chicago, July 1980), pp. 1–5.

73. For a discussion of this literature, see Cain, *Search for an Optimum Sanitation Jurisdiction*, p. 31, note 4.

74. Peterson, "Impact of Sanitary Reform," p. 89.

75. Mel Scott, *American City Planning Since 1890* (Berkeley, 1969), pp. 163–164; Nelson P. Lewis, *The Planning of the Modern City* (New York, 1916), esp. chap. 21, "The Opportunities and Responsibilities of the Municipal Engineers"; and Jeffrey K. Stine, *Nelson P. Lewis and the City Efficient: The Municipal Engineering City Planning during the Progressive Era*, Essays in Public Works History 11 (Chicago, April 1981), pp. 13–17. William Paul Gerhard was notable among sanitary engineers for directly addressing the question of city planning. See "The Laying Out of Cities and Towns," *Journal of the Franklin Institute* 140 (August 1895): 90–99.

76. George C. Whipple, "The Element of Chance in Sanitation," *Journal of the Franklin Institute* 182 (July–August 1916): 37–59, 205–227.

77. For examples of population forecasting by sanitary engineers, see Henry N. Ogden, *Sewer Design* (New York, 1899), pp. 93–101; and George W. Rafter and M. N. Baker, *Sewage Disposal in the United States* (New York, 1894), pp. 129–131. Most of the early population forecasts were based on straight-line extrapolation of past trends. They were often faulty. In 1895, for instance, a Boston civil engineer, Frederic P. Stearns, made 186 population forecasts for 27 cities and towns within 10 miles of Boston. A later check on the accuracy of his predictions showed that he had underestimated growth in 27 cases, overestimated in 156, and made 3 accurate predictions. Writing in 1928, sanitary engineers Leonard Metcalf and Harrison P. Eddy noted that "forecasts of population based upon experience of the past are likely to prove somewhat too high in about 85 percent of the cases and too low in the remainder." For the Stearns

estimates, see Paul M. Berthouex, "Some Historical Statistics Related to Future Standards," *Journal of the Environmental Engineering Division, Proceedings of the American Society of Civil Engineers* 100 (April 1974): 423–424; Metcalf and Eddy, *Design of Sewers*, pp. 191–192.

78. Schultz and McShane, "To Engineer the Metropolis," pp. 409–410; Schiesl, *The Politics of Efficiency*, pp. 171–188. For a discussion of the role of engineers in regard to city government in an earlier period, see Raymond H. Merritt, *Engineering in American Society, 1850–1875* (Lexington, Ky., 1969), pp. 136–176.

79. Josef W. Konvitz, *The Urban Millennium: The City-Building Process From the Early Middle Ages to the Present* (Carbondale, Ill., 1985), p. 195.

9 Historical Origins and Development of a Sewerage System in a German City: Bielefeld, 1850–1904

Georges Knaebel

THIS essay traces, very briefly, the origins of the sewerage networks of Bielefeld in the Federal Republic of Germany. It deals primarily with changes in sanitation techniques and examines elements in the society related to them.

The period covered stretches from 1850 to 1904, the year in which the major part of the network was finished. The principal sources of information used are the *Annual Reports* on the administration and the state of local affairs. These reports were special documents produced by those involved with local politics. They were edited by the local council, under the supervision of the magistrate, and were intended for distribution to the local elected assembly and higher state authorities. They were also read by the local bourgeoisie. They contained information about activities carried on in the town as well as policies adopted by the magistrate in dealing with administrative problems.

The Evolution of the Town

From 1850 onward, Bielefeld's history is mainly marked by the upheavals brought about by the movement toward industrialization. The town of Bielefeld, in North Rhine–Westphalia, 38 miles east of Münster, is spread out at the foot of Teutoburg Forest, at the northern tip of a gap through which passed the important trade routes of the Middle Ages. The routes led from the areas around the Rhine to those around the Weser and from central Germany to Flanders and the Low Countries. In 1787, 1,100 families lived in the town of Bielefeld, and its population numbered 5,310 inhabitants, of which 1,994 belonged to the garrison; the population of the town itself was not much more than 3,300.[1] From 1683 to 1798, the population hardly altered, but at the beginning of the nineteenth century it slowly began to increase, rising from 6,077 inhabitants in 1811 to around 12,000 in 1857.

Situated with its back up against Sparrenberg Castle and surrounded by ramparts and ditches, the town communicated with the outside world through five gates. At first, these gave onto gardens and later on, onto the Feldmark, land held in common by the inhabitants of Bielefeld. Accompanying the farming, market gardening, and the sort of artisan work found in most small German towns was the town's principal activity, the manufacture and trade of linen cloth. The click-clack of spinning and weaving looms could be heard in many houses. "Around here, everyone is either a spinner or a manufacturer. In a certain fashion, the whole area can be considered a factory in which both young and old, from the seven-year-old child to the old man, are occupied at one stage or another of the manufacturing process."[2] Bielefeld's tradesmen bought directly from the region's weavers, and their businesses had been conducted on an international level for some time.

This, then, was the small, bourgeois town that was to be shaken by industrialization. Since the beginning of the nineteenth century, the local industry had been in increasingly one-sided competition with English and Irish factories where mechanization was advancing rapidly. The first reaction of the town tradesmen was to improve the methods of production, particularly by trying to strengthen the spinners' and weavers' work ethic.[3] This proved in vain, and finally, at the end of November 1854, despite opposition to mechanization from all social classes, several merchants from the town and the surrounding area founded the Ravensberg spinning mill, a joint-stock company. With 20,000 spindles in operation from 1861, this factory was to remain the largest linen spinning mill in Germany until 1914.

A rapid process of industrialization took place after 1854. Some of the industries complemented existing ones, while others were more diversified. Often, economic crises gave impetus to these developments. A linen-weaving mill, the Mechanische Weberei, was founded in 1862. Silk weaving also was introduced, and this kept the home looms busy. But it too would be mechanized toward the end of the century, as would the production of plush. A metallurgical industry would appear, and factories that produced textile machinery, particularly sewing machines. One factory began the production of bicycles in the 1880s. Finally, in 1900, a certain Dr. Oetker set up a food-production factory that rapidly became the largest factory in the town. To all this must be added the small craftsmen who serviced and furnished the parts required by large-scale production. The metallurgical industry, which was at first a simple contributor to the textile industry, would supersede it by the end of the century.

The town grew rapidly. At first, workers had to be brought in from the western provinces of Prussia because local workers refused to work at the Ravensberg spinning mill. The influx of workers grew, and by March 1905, there were 69,343 inhabitants of Bielefeld. The town pushed beyond the ramparts, and urbanization was carried into the Feldmark. It was here that factories were built. Urbanization in this area was relatively extensive; at first, it occurred along the existing roads and then, partly because of speculation, new districts sprang up.

Sanitary Arrangements

What were sanitary arrangements like in 1850 at the time when we begin to consider them? One answer to this question can be found in the town's roadway regulations that were in force over a considerable period of time. For example, article 36 of the regulations prohibited anyone from causing "tainted" water to flow during the hours of daylight; this could be done only after dark, and the street had to be rinsed with clean water afterward.[4]

In 1746, the king of Prussia issued a decree, "Regulations Concerning the Streets of the Town of Bielefeld," the aim of which was to improve the cleanliness of the town. The town's inhabitants were required, within a period of two weeks, to remove rubbish from around their houses and yards. They were also to fill in and pave over the ditches. Each property owner and inhabitant had to sweep his part of the street in order to clear it of refuse and excrement. This regulation was also extended to cover the landlords of drinking houses. In winter, on Wednesday and Saturday mornings, those in charge of the water fountains had to clear the ice away from around them in order to make the streets "passable." Two beadles went about the town to ensure that the regulations were being observed. If necessary, they could order the removal of neglected rubbish and assess those responsible with the cost. Offenders could also be fined by the bishop. In order to prevent possible corruption, the beadles were themselves checked on by the police.

In 1830, Mayor Delius laid down other regulations concerning the public thoroughfares.[5]

1. On Wednesdays and on Saturday afternoons before nightfall, as well as on the day before public holidays and on any other occasion stipulated by the police, people living along the roadways were to keep

the surface, pavement, and gutter clear and sweep them clear of any refuse deposited there.

2. The same conditions applied to the removal of weeds.

3. Waste should not be thrown into the street and the gutter.

4. Public conveniences, and tanners' and glue-makers' pits, could be cleaned out only between 10 P.M. and 4A.M., and nothing was to be deposited in the street itself; liquid manure and "all other foul liquids" could be emptied into the gutters, but these had to be rinsed at once with water.

5. Butchers, tanners, dyers, and glue makers were banned from throwing used water onto the street. It was forbidden to throw any liquid or other object out the window.

6. Vehicles carrying manure, straw, earth, and debris had to be fitted with impervious sides to prevent substances from falling on the road surface. Beasts of burden could not be fed in the street or other public places unless they were hitched temporarily in the course of the removal or loading of goods.

7. Water from kitchen sinks that emptied into the street had to flow through pipes or "protective boxes." In the same way, gutters around the roof had to be equipped with drainpipes that reached pavement level.

Rainwater ran off the surface of the streets into a few old underground drainpipes. Rapidity of disposal depended on factors such as slope and permeability, and townspeople often encountered problems. From the repeated prescriptions of the *Strassenordungen*, it can be inferred that the process was not without its difficulties. The outlets of the ducts in the town itself were the Graben (a defensive ditch) and the Lutterbach River. The Lutterbach, which flowed from the south through the gap between Johannesberg and Sparrenberg and whose natural course ran through the town, dividing it in two (Alstadt to the north and Newstaat to the south), had been partially diverted so that it might feed the Graben.[6] At Aammator, a windmill functioned with its own dam and head-race; the Stadtgraben also was put to different uses, including the bleaching of linen. Finally, tanks for firefighting and washhouses had been set up on each place. The ditch waters, especially, were stagnant and gave off a nauseating stench. The Bach and Stadtgraben also collected the water that ran off the land lying by the Feldmark. The Feldmark itself, covered by meadows, grazing land, and fields, was furrowed with small streams, the source of which lay at the foot of Teutoburg Forest.

Sanitary improvements in the coordinated removal of rainwater, various kinds of dirty water, and excrement, which occurred during the second half of the century, can be divided into three parts.

1. Until 1871–1872, the work of the town's administrators was aimed only at *rainwater and the development of the Graben and the Lutterbach within the city walls.*

Within the town, certain streets caused particular concern: the Ritterstrasse, which was flooded as soon as heavy rain fell for any length of time; and the Breitestrasse, where the water level rose in small springs beside or even under the carriageway on the road, causing damage to its surface whenever there was frost, were among the worst. In 1861, after a disagreement among interest groups in the town regarding the cost of its upkeep had been settled, the sewer in the Renteistrasse, which channeled waters draining off the Obernstrasse to the Graben, was reopened. In 1863, the sewer in the Niedernstrasse was renovated and made larger. In 1866–1867, an overflow pipe was installed from the sewer in the Renteistrasse to the Graben, and work was carried out on the adjacent sewers (Breiten, Renteistrasse, Ritterstrasse, and Niedernstrasse). The magistrate's concern was to clear water from the streets and make the plots of land more sanitary. Low-lying parts of the town, like the Ritterstrasse, caused particular problems, as did areas situated at the foot of Sparrenberg Castle, where water sprang from the ground in many places. In general, however, the desire to prevent flooding was linked to problems along the highways. Work on drainage lines most often took place when roads were being resurfaced or when a new road was opened. In this last category, after the installation of the "new market" (Neumarkt), a communicating road with the Gehrenberg was opened and piping was laid that changed the direction of the previous drainage flow and that of the few existing sewers. In the medium term, the town, and therefore the sanitation system, changed little by little, in an unplanned manner.

A similar process took place in the Feldmark, which was beginning to assume more and more importance in the magistrate's mind. Here the development of the town was carried on in an extensive manner. There were two areas of development in the immediate vicinity of the town walls—the Ravensberg spinning mill to the east and the railway station to the north, which saw the establishment of road links of a sufficient level to cater to the growing traffic to and from the city center. This explains the construction of the Bahnhofsstrasse (1864) and the Viktoriastrasse (1856–1870), where the installation of the roadway went hand in hand with the installation of the necessary drainage by means of underground sewers. Farther out in the Feldmark, the draining of surface water to protect the carriageway was without exception carried out by lateral ditches along the main roads that led to the neighboring areas of Werther, Schildesche, Herford, Heepen, and Dekmold. The Wertherstrasse, in particular, required serious repair work to be done each year, as well as work to drain off surface and spring waters. Both in the Feldmark and within the town, work was carried on as and when the magistrate deemed necessary.

The town was concerned with conditions in the Graben, but energetic action was prevented by the many user rights associated with the ditch. In 1853–1855, the cleaning out of the ditch between Obernthor and Niedernthor was rendered impossible by a dispute with a Dutch bleaching company. The Ravensberg spinning mill had also set up a reservoir on the ditch. Thus, action on the ditch was prevented by these private arrangements. As time went on, the ditch, as well as the town's walls and gates, came to be regarded more and more as obstacles blocking better relations between the town and the Feldmark.

Little by little, these obstacles would be overcome. In 1856, demolition work was begun on the walls, and the materials recovered were used to fill in the ditch; work also began on that part of the Lutterbach that was within the town walls. In 1856, the covering over of the ditch on its western side was planned. The vaulting work was begun in 1866–1867, together with the establishment of the Oberwall- and Niedenwall-strassen. Work progressed slowly, by disjointed sections, and depending on the success of negotiations carried on with those who used the ditch or had property alongside it.

2. In the town's 1872–1873 report, mention is made for the first time of the need for a complete clean-up of the urban area, "in relation to the installment of a water-supply network to provide the town with abundant, healthy drinking water." This marks a reversal of the policy of previous decades, when work was carried out largely according to the needs and requests of private individuals. Now, operations would be carried out according to a general, planned strategy.

The new policy was not a complete break with the old, as can be seen in previous reports. By this decade, the installation of sewers had become an almost routine action. Indeed, the construction of a sewer was elevated to the level of a piece of civil engineering, with all the technical characteristics of the work made clear in the annual report. Before 1871, sanitation works were classified under the heading *Road Surfacing* or *Public Roads* (for the town and the Feldmark, respectively), but in that year a separate heading, *Sewers*, made its appearance. Sewer installations were now found to be necessary "for the improvement of the state of sanitation in the town." In other words, even if they still related to highway problems, sewers had acquired a specificity and their own reason for existence. From this point on, the practice of removing not only surface water but also dirty water along these underground routes came into use.

Nevertheless, more than twenty years would pass before a complete and unified "canalization" in the town would become a reality. In 1874–1875, the elected assembly rejected the town administration's request to be allowed to collect funds to set up a unified system of sanitation and water distribution. The request was rejected on the grounds that the town had other, more pressing needs, and because the elected assembly doubted the feasibility of the plan, "although this year's rainfall has clearly demonstrated the inadequacy of the town's sewers and despite the fact that the quality of the local drinking water is deteriorating more and more." A search for an improved water supply began, but no action took place in regard to the construction of a planned sewerage system. In 1875–1876, the administration made another try, but this also failed, and a solution for what had become "a question of a general improvement in sanitation" was not attempted again until 1890–1891. Work on the sewers, however, continued during this period. In the Feldmark, side by side with the drainage equipment installed when new roads were opened, the old side ditches were vaulted (Kesselbrink in 1882–1883 and Burgerweg in 1887–1888). In the town, new sewers were installed and older ones renovated or extended, but the outlet was always the ditch or, above all, the Lutterbach. In 1879–1880, in order to drain off crossroads in the upper part of the Obernstrasse and Welle, a conduit was carried to the river. At first, the pipe was underground, but because of the water level in the reservoir in front of the windmill, it was made into a stone gutter and covered by metal sheeting. In 1881–1882, a circular sewer was installed "with the participation of the residents" in the Brinkstrasse in order to ease the strain on the gutters, which did not slope very steeply, and to reduce the amount of ice that formed on the road, making it hazardous to pedestrians.

In this period, although work was still closely linked to highway problems, questions of house drainage became much more important. The volume of these waters increased (probably due to new paving, especially in the older courtyards) and outlet problems were felt more and more acutely. Not only was total outlet capacity not increasing, it even showed signs of decreasing because many natural drainage lines (ditches, streams) had been diverted or simply blocked up or done away with during building work; this was particularly true in the Feldmark. Although drainage and sewer work had formerly been carried out as the need arose, henceforth they would be constructed as parts of a unified network that one day would be built. (In 1873–1874, it had already been said that a large sewer installed in the Breitestrasse would be used as a main sewer.) This idea of an integrated network was clearly demonstrated in 1884–1885 when an engineer from Frankfurt on Hain named Schmick designed a project for a separate system. In this project, rainwater would be carried by the shortest route to the Lutterbach, the gradient of which would be regulated by the removal of the reservoir ponds used by the windmills. The old sewers would also be assigned to the discharge of rainwater. Wastewater would be carried along the Heepenstrasse toward a water-purification plant and then released into the Lutterbach, on the outskirts of town (near Heepen). "This project for the improvement of sanitation will not be able to be carried through on a large scale until a water supply network is set up," said the annual report for that year. "Until this happens, it will offer a solid base for the construction of the fragments of the sanitation system, which is progressing each year."

3. In 1889–1890, the municipality reported that "after the finishing of work on the water supply, the municipal administration is still concerned about the progressive installation of an organized sanitation system and preliminary work in this direction has begun again." Thus a period was initiated that would lead to the rapid installation of an integrated network beginning in 1899, with the whole decade devoted to resolving questions regarding the type of network to be installed, the site of the outlet pipe, and a definition of the way in which wastewater would be removed.[7]

Regarding choice of system, Schmick's project for a separate system was submitted to an "eminent specialist," Dr. Hobrecht, who had come from Berlin in August 1891. Hobrecht objected to Schmick's plan, arguing that "up to now, no town in Europe, or at least no town of any size or which does not present special topographical characteristics, has installed a separate system and, in cases where this system has been used, the results have always been unfavorable . . . the conduit pipes of the separate system, reduced to the smallest possible dimensions, cause obstructions and require excavation, cleaning and repairwork of the said pipes . . . the separate system lacks ventilation since it does not connect with the rainwater pipes." Because of Hobrecht's objections, engineers prepared plans for a combined system. The project defined the functions required of the town's "drainage" as follows: (1) draining domestic wastewater from kitchens, washing areas, etc.; (2) the removal of wastewater from industrial areas "insofar as this water is in a state which does not endanger the maintenance of the sewers"; (3) draining rainwater from the area where the direction of natural drainage is toward the town; (4) the removal and maintenance at a certain depth of underground water in order to protect cellars from flooding and to make possible better land use in built-up areas situated on low-lying land; (5) the removal of human waste where appropriate; and (6) the regulation of the Lutterbach. The system itself was to be self-cleaning.[8]

Once the decision was made for a combined system, a site had to be found for the outlet pipe. Initially it was believed that the Lutterbach River could be used for disposal. In April 1893, however, the president of the city of Minden, on the River Weser, whose approval was required if the project was to proceed, appointed a commission on the sewage-disposal question that expressed its disapproval of the discharge of effluent, even if treated, into the Lutterbach. The commission argued that "the rate of flow of the Lutterbach is too weak, considering the dirty water that will be discharged therein" on the Senne, on the other side of Teutoburg Forest. The land here was divided into many small plots, and its owners demanded high prices for it. The cost of the operation was so high that the municipality backed down and once again sought approval for its first plan. Finally, in 1896, a plan for mechanical purification of the sewage, with a limited spreading of the effluent, received the approval of the authorities at Minden.

The necessary land was immediately purchased in the neighboring parish of Heepen, and during the winter of 1897–1898, contracts were awarded for the construction of decantation tanks for the main outlet and the main sewers. The inhabitants of the towns of Herford and Heepen attempted to block construction, but a ministerial order, issued September 8, was favorable to the town of Bielefeld, and work progressed rapidly. By 1904, virtually the whole of the central part of the urban area was served except for an area where new roads were still being built and for areas located away from the center, such as the district of Godderbaum, which had just been integrated into the municipality. In addition, although they still constituted a very important element of the countryside of old Bielefeld, the Lutterbach and the Stadtgraben had been rerouted round the town center, not to reappear for more than a kilometer to the northeast.

Background

There is nothing original about what has been sketched out in the previous pages. About this time, many towns in Germany acquired sewerage networks. It could not even be said that the installation of the Bielefeld network occurred late in the day in relation to the western part of the empire. Admittedly, large towns had had networks for some considerable time (work was begun in 1867 at Frankfort on the Main, in 1874 at Stuttgart and Nuremberg, in 1881 at Cologne, Dortmund, Freiburg in Breisgau, and Munich), but the 1890s and 1900s witnessed intense activity in more than a hundred areas, many of which were of a comparable size to Bielefeld. It is also impossible to say that Bielefeld was representative, since the only comparative statistics involve technical details of existing networks and provide no information about the events that preceded them.

What we can explore is the context of Bielefeld's development. On the one hand, the installation of the network at Bielefeld can be regarded as the result of a progressive elimination of former practices because they were becoming anachronistic in a changing environment. The result of the decision to build the new system was the simultaneous treatment of rainwater, various wastewaters, and human excrement; previously, the three had entailed the use of three separate treatments.[9] For a time, piecemeal attempts were made to respond to urban problems caused by the disposal of these three substances, but a threshold developed beyond which this type of response was no longer tenable. At that point, it became necessary to change the way in which sewerage problems were solved.

This is very clear with reference to drainage outlets. At the start of the period, the outlets for the areas within the ramparts and the new buildings on the outskirts of the town were the Stadtgraben and Lutterbach rivers. Farther out, in the Feldmark, were scattered the *Kotten*, the homes of poor peasant workers. These peasants threw all their wastewater into ditches, and so affected the many small rills and streams that ran across the slope of Teutoburg Forest. This zone began to build up more and more rapidly in the 1850s. As a result, the development of new roads and their ditches, and of houses with paved courtyards, caused the amount of waste that had to be removed to increase rapidly at the same time that it became more and more difficult to find a disposal site. Eventually the ditches formed a labyrinth whose capacity grew more and more uncertain. The streams, where they had not disappeared altogether, were subject to frequent overflowing, a situation that was especially intolerable because they were now situated in built-up areas. When the urban agglomeration grew beyond a certain size, it was no longer possible to act only when the need arose; a global approach to the problem had to be taken.

The transformations that the town underwent were not only quantitative but also qualitative. The town came into the grip of a new economy, which shook up its old ways of existence. The appearance of factories and the considerable decline in the amount of work done at home meant that the highway system became an essential element in the mobility of both loom workers and goods.[10] As was illustrated earlier, during the whole of the first period, sanitation work was almost always the result of a desire to maintain the roads in a passable state. In order to guarantee free movement on the roads, it became necessary to put an end to many former uses made of these collective spaces. Similarly, it was important to establish better traffic flow by separating pedestrians and vehicles. This gave rise to the construction of pavements and gutters, to the disappearance of furrow drains and the filling in of holes, to the removal or transformation of old fountains, and to the passing of decrees relating to roads and prohibitions regarding what might be termed "deviant" behavior.

These comments can also be applied to the water supply. Two methods of supply had existed side by side in the town: wells and three systems of pipes. The latter, installed at the beginning of the seventeenth century, had served some eighty houses at that time.[11] They were kept up until the nineteenth century, the wood having been replaced by cast iron in 1826 in one case and in 1871 in another. Their maintenance and administration were shared on a common basis and seem to have been a frequent cause of dispute. Water from these "pipe works" was also used by the town, particularly for firefighting. Here and there, basins holding reserves of water were dug, and public fountains were equipped with attachments that fitted the firemen's water carts. For this reason, the town contributed financially to the three associations that operated the pipe works. The failure of householders in the association to agree on the upkeep of the pipe works resulted in an administration proposal in 1868–1869 to integrate the two networks of pipes into a single system, which would then be incorporated into the public domain and governed by a specific set of police regulations. Because of the private interests involved, the negotiations proved unfruitful. Rising urban demand and their innate limitations (the works were spring fed), however, made it doubtful that the pipe works would have been adequate to supply the town even if an agreement had been reached.

Once again, as with the sewers, initial measures were fragmentary. Later, the magistrate called for a general water supply that would be administered by the town. Hence, the development of the town (the fact of its becoming a social complex) led to the appearance of a single

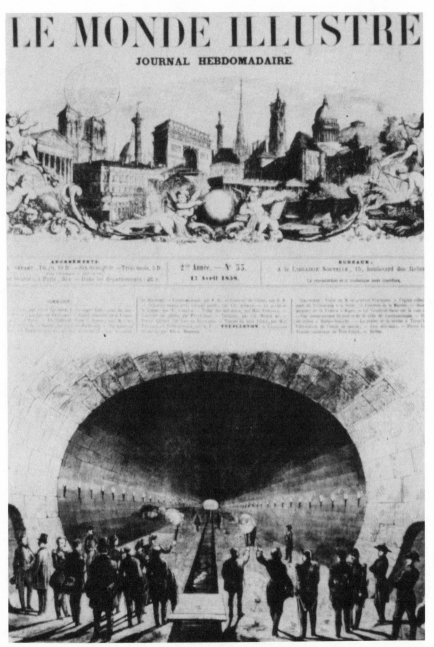

The grand opening of the collecting sewer at the Strasbourg gate, 1858, as pictured in *Le Monde Illustré*. Courtesy Gabriel Dupuy.

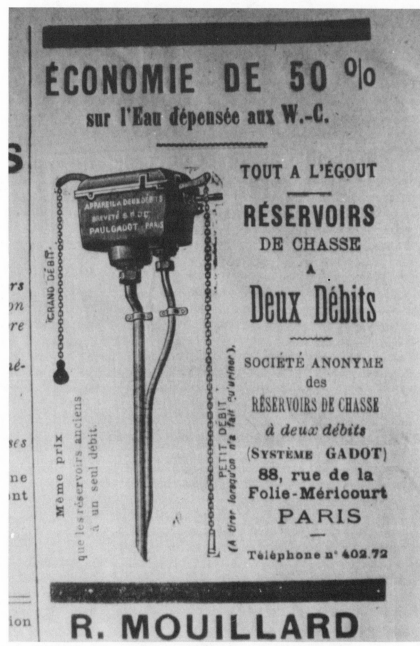

An advertisement in a Parisian newspaper around the turn of the century for a water closet with a two-part flush tank to save water. Courtesy Gabriel Dupuy.

The careless disposal of household wastes was common in many nineteenth-century households. Photo from *Harper's Weekly*, 1873.

Constructing sewers in Milwaukee in the early twentieth century. Photo from Harold E. Babbitt, *Sewerage and Sewage Treatment* (New York, 1932).

Privy vaults and water spigot outside a Pittsburgh tenement, 1907. Photo from Margaret Byington, *Homestead: The Households of a Mill Town* (New York, 1910).

The Destructor at Milwaukee built in 1909. At the time of construction it was the largest incinerator in the nation. Photo from Walter F. Goodrich, *Modern Disposal Practices* (New York, 1912), page 205.

decision-making authority that was henceforth to introduce economic criteria of management into the planning and setting up of infrastructures (general conditions for the functioning of an urban area). It is remarkable to see the explosion of municipal "technical services" that took place in less than two decades: first, the water service; then, almost simultaneously, the sanitation service and the bringing under the control of the administration the removal of household refuse and street cleaning; finally, the electricity plant and the establishment of a tram network. Considered from this structural point of view, it can be understood why a water supply had to be laid down as a preliminary condition to the establishment of a sanitation network.

So, the centralized supply of water was laid down as a necessary first move toward the establishment of a general, unified sanitation system. The question could be posed whether the municipality's action could have stopped there. Logically, it is true that a concern to supply houses and industry with water—a big event for the "parish" administration—implies that thought had been given at least to what would become of these waters after they had been used and, hence, to their removal. On the one hand, the water supply was organized and systematized; on the other, thought was given to how to remove wastewater in a systematic way. The need for organization and systemization obviated piecemeal solutions and demanded the setting up of a central apparatus. Having said that, organizing and systematizing the removal of water did not imply that this should be done exclusively through underground galleries, which worked through gravitation from upstream to downstream.

In Bielefeld, the influence of other models played a key role. What happened there could be placed in the general context of problems specific to industrialized Europe, as they slowly spread to other towns in Westphalia. The town called on men of "experience" who had already participated in similar enterprises in the big towns of the Reich to design the water supply system, then to direct the works, and finally to administer the apparatus that was set up. Schmick left behind him in Bielefeld an engineer named Bock, one of his assistants from Frankfort on the Main, who would come to be in charge of the infrastructures department. Dr. Hobrecht was consulted when it became necessary to have an expert's opinion on conflicts arising from the choice of the Senne as a wastewater-disposal zone. Because Hobrecht came from Berlin, where he had operated the sanitary works, and because he enjoyed great prestige, his opinions had the force of commands. This explains why the combined system was chosen when it could have been supposed that Schmick had the advantage of long experience of a separate network and better local knowledge of conditions peculiar to Bielefeld. It was a clash between two different schools in which the "winner" had no more experience than the "loser" but simply drew on his greater prestige from having worked in the empire's capital. Moreover, when a new section of the district was equipped (from 1902 to 1904), it was necessary to fall back on the separate system, which was the only economically viable solution. In the project that he drew up in 1892, under Hobrecht's direction in favor of the combined system, Bock seemed at first to regret that a separate system had been rejected. Later, however, he accepted Hobrecht's line of reasoning, which is particularly significant in terms of power. Given continued growth, and supposing that the town was equipped completely as a separate system, it would have been virtually impossible, despite all possible policing measures, to avoid a long-term increase in the number of clandestine branches of rainwater conduit pipes fitted into those built to carry away wastewater.

The laying of pipes underground stemmed from yet another problem than that of the simple removal of traffic obstructions. In Bielefeld, laying pipes underground went hand in hand with the emergence of a sense of urban "beauty." For example, toward the end of the century, an effort was made to provide the town with a unique identity by designating specific "monuments." This showed itself first in the purchase of the Sparrenberg, an ancient, abandoned castle, and then by the erection of various monuments, particularly the statue of the Linen Weaver. A primary concern of the magistrate, once the water distribution system was completed, was to add to the town's beauty by the construction of fountains. This process of adding to the town's beauty also manifested itself in the establishment of "promenades" on the sides of the Sparrenberg—but also over certain parts of the channeled ditch. "Going for a stroll" seemed to be a new activity. It is not known which groups in society were the most directly concerned, but it can be supposed that the working classes were pretty much excluded. For example, the self-celebration of the local bourgeoisie was typified by the construction of the municipal theater; but if one reads the passages of annual reports devoted to the administration of the poor, nothing is found but moralizing and invectives against places selling alcohol.

This new moral tone was clearly revealed in material manifestations: If one considers the ditch, above it distinguished people took their "promenade," taking great care of their appearance, their gestures, and what they said; beneath them floated, in underground shadow and humidity, their own excrement, which was considered shameful and unspeakable. Underground installation was not simply the result of a desire to reduce traffic obstructions, of a calculated efficiency, of the fear of epidemic;[12] it also corresponded to the moral outlook. Within the limited perspective of the Bielefeld of that time, the same people who dominated the municipal council, either directly or indirectly through their representatives, responded to economic crises by making wholesale reductions in the workforce. They were also the people who preached to the "poor," who decided to move the cemetery to the outskirts of town at the same time as they contributed to the installation of districts reserved for workers; who freed credit for the construction of the theater, and the formation of a local orchestra; who contributed to state memorials and met in their own closed circles. These were the people who had a dominant position in the arena of social struggles, those who administered and made decisions about municipal technical services, including sanitation. They were merchants and factory owners, for the most part, and it is no surprise that, as elected representatives, they employed the same reasoning and value systems in municipal enterprises as guided them in their private and professional lives.

Relations between the town and private property owners underwent considerable evolution. At the start of the period under study, property owners participated greatly in urban development; the magistrate seemed to do no more than embody the organization of Bielefeld property owners (proprietors of ground used for production and commerce, as a form of investment, or for their own residences). It was virtually only property owners who voted for or could be elected to the town council.[13] The annual reports are studded with references to reciprocal payments made by the town and the owner-occupiers—for pavements, sewer installations, the opening up of roads in the Feldmark. Admittedly, these relations were not without conflict, but the principal means of settling problems was through negotiation. This situation was to change in the course of time. Thus, notes in the reports reveal a split between the newcomers—property owners in the Feldmark—and the bourgeois of the center—the old families. As described by Gustav Engel, the local bourgeois class was becoming diversified, with factions appearing in the new rise of the urban economy.[14] From the time when the distribution of water was set in place, and as the town progressively changed into a supplier of services, relations between bourgeoisie and town changed from one based on negotiation to one of pure administration. The list of local bylaws grew longer each year. In regard to water, the policy followed seemed to be to charge the cost price for the service. At the outset, a double system of billing was introduced—paying by set fee or by meter—and industries were the first to have meters installed.

Then, very quickly, the system of fee payment vanished. In regard to sanitation, relations between the administration and private property owners were closely regulated, virtually from the moment the service started operation. A police ordinance of July 23, 1900, made it obligatory for all houses equipped with a drain that bordered roads to have a branch pipe. Other regulations stipulated the installations required inside the houses. The system, then, was established as much in its technical parts as in the legal conditions governing its functioning.

This study suggests that we would gain much from the study of urban infrastructure not only as a public, collective device but also, and above all, as a machine. The observable changes in this regard derive from the same circumstances as those in force in the domain of merchandise production: "Technological revolutions" reflect the "progress" of science and technology less than the determination, through the concurrence of capital and the contradiction of capital and work, of the financial means available for the undertaking. In effect, the machine is by definition an apparatus that sets up agents according to the relationships of power. It is the same for machines dealing with urban wastewater.

Notes

1. "Consbruch. Topographisch-Statistiche Beschreibung der Stadt Bielefeld aus dem February 1787," in *Jahresbericht des Historischen Vereins fur die Grafschaft Ravensberg zu Bielefeld* 19 (Bielefeld, 1905): 32–46.

2. Heinrich Wilhelm Schubart, *Topographisch-historich-statische Beschreibung der Stadt Bielefeld* (Bielefeld, 1835).

3. Wolfgang Hofman, *Die Bielefelder Stadtverordneten Ein Beitrag zu burgerlicher Selbstverwaltung und sozialem Wandel. 1850–1914* (Lubeck and Hamburg, 1964).

4. *Jus Statuarium civitatis Bilefeldiensis oder Bilefeldisches Stadt-Recht und Burger-Sprache samt beygefugter Policey Ordnung wie auch einige von hoher Lands-Obrigkeit gnadigst confirmirten Privilegien und Freyheiten* (Bielefeld, 1685).

5. *Strassenordnung fur die Stadt Bielefeld* (Minden, November 1830).

6. Dr. H. Tumpel, "Die Ableitung der Lutter nach Bielefeld im Jahre 1452 und das Kloster Marienfeld," in *Jahresbericht des Historischen Vereins fur die Grafschaft Ravensberg zu Bielefeld* 12 (1898): 100–102.

7. Indecision and controversy regarding the resolution of these questions prevented work on the total system, but some important preliminary steps were taken in the 1890s. The Lutterbach, or windmill reservoir, ponds situated in the center of the town, for instance, were bought between 1891 and 1893, and that

section of the river was vaulted between 1893 and 1895. The work of covering over the ditch was completed in 1895.

8. A small amount of work done on the sewers during these years bore the stamp of these divergent approaches. In 1891–1892, for instance, the sewers in Am Damm Street were laid out in such a way that they could be connected to a future separate system; in 1893, the sewers in the Teich- and Freidrichstrassen were designed according to the combined system.

9. Actually, there was a small degree of mixing in some places.

10. Henne, *Bielefeld-Stadterweiterung im Industrialisierungs prozess* (Bielefeld, 1974).

11. Konrad, Balke, "Die Entwicklung der Trinkwasserversorgung und Abwasserbeseitigung der Stadt Bielefeld," Ph.D. dissertation, Maschinenschr, Frankfurt/Main, 1950.

12. Except for a small epidemic recorded in 1866–1867, Bielefeld seems to have been safe from cholera in the second half of the nineteenth century.

13. Jürgen Reulecke, *Die deutsche Stadt im Industrie-Zeitalter* (Wuppertal, 1978).

14. Gustav Engel, "Die alten Wasserleitungen der Stadt Bielefeld," in *Ravensberger Blatter fur Gerschichte, Volks, und Heimatkunde* (1936), pp. 49–51.

10 Technology Diffusion and Refuse Disposal: The Case of the British Destructor

Martin V. Melosi

TECHNOLOGY diffusion, or transfer, has played a critical role in the spread of the Industrial Revolution. In the last hundred and fifty years, the scale and impact of technological activities has accelerated greatly.[1] Those who have studied technology diffusion in relation to the Industrial Revolution caution not to separate technology transfer from the economic process. Nathan Rosenberg has argued "that the diffusion of invention is an essentially economic phenomenon, the timing of which can be largely explained by expected profits, is by now well established."[2] Several historians of technology have stated the relationship between technology transfer and the economic process in broader terms. Peter Mathias, for example, has asserted: "For too long, in fact, the history of technology was separated from the more general analysis of economic development in the eighteenth and nineteenth centuries; pursued as the antiquarian study of actual techniques in their own right rather than viewed in relation to the dynamics of economic change as a whole."[3]

Is the intimacy between technology and broad-scale economic change the key variable in all cases of technology diffusion? Few scholars have claimed such, and most are quick to assert the complex nature of technology transfer. Nevertheless, in trying to determine the factors that produce technology diffusion and its various impacts, the focus of historical study in the past two or three decades—discounting the current interest in transfer from developed nations to the Third World—has been limited to the diffusion of agricultural and industrial technologies in the private sector.[4]

In several instances, the general suppositions advanced by students of technology diffusion apply equally to the public sector, but some do not. Exploring the transfer of public works technology offers an effective way to examine the process of diffusion in the public sector. In large measure, historians and economists have ignored this topic, which is important in understanding the expansion of city services. The diffusion of sewerage technology, water-purification equipment, firefighting hardware, street-cleaning and refuse-disposal systems, and expertise on road construction and maintenance deserve more attention. In particular, questions of "public interest," the role of government officials, budgetary priorities and constraints, and the input from technical consultants are crucial.[5]

As a departure point in examining the public dimension of technology diffusion, this essay presents one important case: the transfer of the British Destructor in the late nineteenth and early twentieth centuries, especially to the United States. While one case cannot address all the significant issues, it can be suggestive about the nature of technology diffusion in the public arena in general, and about the spread of public works technology in particular.

The British Destructor: Stages of Development

Disposing of wastes by fire is an age-old practice. Yet the first systematic cremation of refuse was not attempted until the late nineteenth century. "Cremation," "incineration," or "destruction" of refuse is a relatively new practice in towns and cities, not only because of the technical requirements but also because modern industrial society, not ancient agrarian society, demanded large-scale disposal systems to alleviate its massive urban refuse problem.[6]

Industrialized, urbanized Great Britain led the way in the systematic disposal of waste by fire. Various environmental and economic factors motivated British engineers and technicians to develop the "destructor." Alternative methods of disposal posed serious problems. Land dumping was not possible in a country with limited open space. Sea dumping threatened the English livelihood linked to maritime pursuits, and was a source of potential animosity with England's neighbors across the narrow channel. Also, the English population was sufficiently dense to make centralized systems of disposal, such as cremation, practical. But any centralized system had to operate efficiently—it had to avoid threats to public health—and economically—it could not be a financial burden on the citizenry. Many engineers believed that destructors could meet these criteria, and if they were designed to burn waste without utilizing additional fuel, they could help preserve coal and other costly sources of energy.

Between the late 1860s and 1910, the British Destructor moved through three stages of development: (1) low-temperature and slow-combustion furnaces; (2) destructors with artificial drafts, which operated at higher temperatures and greater incinerating capacity, capable of producing steam for various work purposes; and (3) destructors capable of providing power for the generation of electricity or for pumping various liquids.

First-generation furnaces were designed simply to burn waste, with little attention to efficiency of operation or secondary features, such as utilizing the heat generated for other purposes. As early as 1865, a furnace was constructed on Gibraltar to dispose of refuse at the British garrison located there. About 1870, the first municipal furnace was erected in Paddington, a suburb of London. It employed a natural draft for the burning process, but operated poorly and produced a great amount of noisome smoke. The first systematic cremation of refuse on the municipal level was tested in Nottingham in 1874. In 1874–1875, a series of furnaces—dubbed "destructors"—was built in Manchester. The group of eight furnaces consumed about 5 tons of refuse each in a twenty-four-hour period.[7] Elaborating on the experience obtained at Manchester, Albert Fryer in 1876 constructed an improved destructor, which consisted of two cells, at the Water Street depot. Soon destructors were being erected in Ealing, Birmingham, Leeds, Bradford, Derby, and elsewhere.[8]

The poor operation of the first-generation destructors undermined public confidence in cremation as the best method of waste disposal for English towns and cities. Aside from the inefficiency of the burning, the major criticism was the "noxious fumes" billowing from destructors and blanketing nearby residences. Efforts to disguise the pollution by constructing high smokestacks failed to alleviate the public outrage. In 1885, an Ealing engineer, Charles Jones, made the first serious attempt to eliminate the smoke by patenting the "fume cremator." Jones placed a secondary fire in the flue leading from the destructor to the chimney that was supposed to burn the gases emanating from the cells before they escaped into the atmosphere. This first attempt to make destructors sanitary proved largely unsuccessful, not simply because the technology was inappropriate but also because it was expensive. Experiments in Edinburgh demonstrated that the need to use high-grade coal or coke in the secondary fire to reduce the smoke doomed the prospect of economical operation. Furthermore, the low-temperature destructors of the first generation could not produce enough heat to generate steam or sufficiently good residium (clinker) to help offset the cost of cremation with commercial by-products.[9]

Despite its failings, the fume cremator was a bridge to the second generation of destructors. To alleviate the nuisance quality of the destructors, British engineers experimented with new designs. The goal was complete combustion. Forced combustion and high-temperature burning replaced low-temperature burning through the addition of the artificial draft. High temperatures were attained by the introduction of powerful currents of steam or heated air under the ashpits of the furnace. The old destructors rarely produced temperatures in excess of 1000° F; in 1887, temperatures ranged only from 500° to 750°. By 1902, however, a destructor achieved 3000° F. High temperatures appeared to remedy the emissions problem. They also allowed the destroying capacity to increase without enlarging the plant—an economic as well as a technical benefit. The high-temperature process yielded an additional boon: potentially usable by-products, especially steam and serviceable clinker.[10]

Several new destructors quickly appeared on the market in the late 1880s and 1890s. The Horsfall model was completed in 1887–1888. It utilized a forced draft over and under the firebars that increased temperatures and permitted the use of a water-tube boiler close to the destructor itself. It was the first destructor to be used in connection with electricity works (at Oldham).[11] In 1898, Meldrum Brothers of Manchester introduced the Simplex model. It was supposed to combine efficiency, inoffensive cremation of refuse, and the production of a maximum amount of steam power. It had one long cell fed through several openings. The Simplex was touted as a highly economical model that developed intense heat through the use of the poorest kinds of fuel: coal or coke dust, ashpit refuse, or anything with caloric value. Like Jones's fume cremator, the Simplex was a transition technology. It was designed to destroy refuse, but also to produce large quantities of usable power.[12]

From the 1890s to about 1905, the development of the high-temperature destructor was attended by wild claims but slow progress. Manufacturers were sometimes the culprits. One engineer noted in 1900 that the "disastrous consequences" of the working of some destructor companies "discredited the subject of the utilisation of town's refuse to such an extent that it has been very difficult of late years to persuade the vestries and local authorities to treat the matter seriously at all."[13] But sometimes engineers and other professionals also exaggerated the technical achievements and economic benefits of the new destructors. As one British engineer, W. Francis Goodrich, asserted in a book published in 1904:

> The harm done by enthusiastic professional men at this time has had its effects ever since. Results in power production were prophesied which have not yet been attained, and which never will be attained. This is a candid admission, but no apology is needed. The modern Destructor has an excellent record, but it has its limits, and had this been recognized at its advent, much misunderstanding might have been avoided, and as the result greater progress might have been recorded.[14]

Public confidence in destructors began sinking once again, largely because claims were not matched by tangible results. In 1888, Colonel Thomas Codrington was instructed to conduct a study of destructors for the Local Government Board. His report concluded that heat as a by-product of cremation varied greatly from one location to the other.[15] No one produced figures or evidence to prove that the new, experimental high-temperature destructors could produce power and other by-products using only refuse as fuel. New evidence would have to be acquired or expectations lowered if the new destructors were to proliferate after the turn of the century.[16·]

(It is helpful to note here that when a British municipality decided to install a refuse destructor, it normally went through a rigorous review process. First, it referred the matter to a standing or special committee or council. The city engineer was then called upon to evaluate and test the destructor. The Local Government Board—a central body with offices in London—then held an inquiry and sanctioned the necessary loans for the project. Taxpayers had an opportunity to speak at the inquiry. After public advertisement, contracts were awarded, with provisions for an efficiency test before the final acceptance of the destructor.[17])

The third generation of destructors—beginning with the Simplex and others like it—were meant to be adjuncts to electrical works (or other "work purposes"), that is, they had functions beyond the simple cremation of waste. Before sufficient improvements in design and function were achieved—especially before 1902—many engineers and technicians assumed that the destructors could burn waste and produce energy with little or no fuel other than the refuse itself. The Shoreditch destructors and electrical plant (1897), for example, originally were installed to operate by burning only refuse. To generate sufficient power, however, coal had to be added. Ironically, the Shoreditch destructors may have become the largest consumer of coal of any destructors in England.[18]

By 1905, it was widely accepted that to generate electrical power effectively, additional fuel had to be added to the refuse. Engineers began to modify their expectations about destructor-generated power. As early as 1902, Goodrich noted:

> The real value of a destructor as an adjunct to an electricity works is now more clearly understood than it has been in the past, and experience has shown exactly how far and under what conditions the destructor may be relied upon as a power producer. Experience has also shown its weaknesses, and if we are to gain knowledge we must be even more ready to detect weaknesses than to congratulate ourselves upon success.[19]

Technical improvements soon followed. Larger steam boilers, for example, were directly linked to the destructors to utilize a larger portion of the generated heat. Efforts at raising heat to the highest point possible were criticized by some engineers, however; they argued that temperatures between 1250° and 2000° F were desirable for economic as well as sanitary operation of the destructors. The debate over the optimum temperature range persisted for several years,[20] as did the debate over the "fuel value" of waste.[21]

Cost factors also were monitored more carefully than before. It was estimated that the construction of a power-producing plant cost from one-third to one-half more than the standard destructor. Cells covered a considerable area by themselves. By adding boilers and very large storage space for rubbish, construction expenses and the cost for land rose sharply.[22] Nonetheless, it was believed that the return from the sale of by-products made destructors relatively economical to run. A 1906 estimate placed the average cost per ton of disposal of refuse in English cities using destructors at 26 cents, with very few exceeding 40 cents. (This does not include capital costs.) A 1908 report set the cost per ton at 30 cents.[23] While the evidence is meager, the cost of incineration, based on these figures, made cremation at least competitive with other means of disposal available at the time.

Although the technical and economic feasibility of using destructors in conjunction with energy-producing apparatuses was inconclusive in the third generation, the new destructor "systems" spread rapidly. In Nottingham, enough steam was produced by destructors to provide electricity for one-third of the needs of the tramway system.[24] In 1899, 81 cities and towns in Great Britain employed cremation as their chief disposal method; of these, 76 utilized the generated heat in some fashion.[25] In 1906, there were from 140 to 180 towns and cities with destructors (some sources claimed as many as 250), of which more than half utilized the power generated for various purposes. Destructors combined with electrical works ranged from 45 to 65.[26] The first destructor combined with sewage pumps was constructed in Hereford in 1897. By 1902, 26 towns constructed similar facilities; by 1912, there were 36 or more.[27]

International Diffusion of the Destructor

The publicity and notoriety that the British Destructor received led to its proliferation throughout the world. By 1906, the destructor spread across the European continent, most notably to Brussels, Copenhagen, Rotterdam, Paris, Monaco, Gibraltar, Zurich, Berlin, Hamburg, and St. Petersburg. Seven cities in South America, including Buenos Aires, adopted it; seven in Australia and New Zealand; four in South Africa; and four in India. Destructors could also be found in Cairo, Singapore, Shanghai, and across North America.[28]

German engineers were not satisfied simply to adopt the new technology, but began refining it. As is typical in many instances of technology transfer, technicians (firemen) were imported from the nation of origin (England) to the adopting nation (Germany) in 1896 to train local technicians in the operation of the destructors. To test furnaces built in Berlin and Hamburg, refuse was transported from remote locations.[29] German engineers then gave particular attention to the furnace design, resulting in a destructor differing in detail and operation from the British version. The German design utilized a single cell and combustion chamber in place of the multiple cell and continuous grate used in many parts of Great Britain. It also increased the pressure and the temperature of the air for the forced draft in the furnaces.[30] In 1899, the Hamburg works was the largest plant in existence, and the German design was rivaling the British in adoptions.[31]

Conditions in France were substantially different from those in Germany. Between 1886 and 1898, French engineers carefully studied incineration systems. After several trips to England, they built a trial oven and operated it through 1895. Parisian garbage proved to be self-burning, according to the tests. But the French were more ambivalent about the technology than the Germans. They believed that no savings were to be achieved because of the high content of ashes and cinders in the refuse, which had little commercial value in France. Also, a market for the generated heat was not available at the time. Economy, rather than hygiene, was the crucial factor.[32] Although incineration of waste was adopted in Paris, it had little success in outlying areas where disposal through programs of agricultural utilization seemed more appropriate.[33]

Transfer to the United States

Despite its widespread diffusion, the British Destructor was not transferred to other countries without difficulty and some modification. In the United States, the transfer resulted in as many problems as it hoped to resolve. Actually, the British Destructor was adopted in the United States in two stages: The first, beginning in 1885, introduced the low-temperature furnaces that were under criticism in Great Britain at the time and ultimately proved to be poorly adapted to the needs of European municipalities. The second, primarily after the turn of the century, introduced the high-temperature destructor with its adjunct boilers and power-generating capability. Adoption of the low-temperature furnaces was impulsive and widespread in the first stage. In the second stage, adoption of the high-temperature destructors was more cautious with much more attention to effective operation. Nonetheless, widespread adoption failed to materialize. The reasons for the dual failures arose from a combination of several factors, not the least of which was a lack of appreciation for the unique physical, economic, and social context into which the British technology was being introduced.

The American experience with incineration began with a small garbage furnace built by an army lieutenant, H. J. Reilly, on Governor's Island, New York, in 1885.[34] The miniature crematory—burning 10 to 30 cubic feet of waste a day—handled the refuse of the post located on the island. This modest device became the early model for American crematories.[35] In 1886, a small garbage crematory was built in Allegheny City, Pennsylvania—the first municipal furnace—using natural gas to burn refuse. A gas furnace was also constructed in Pittsburgh the following year. Between 1886 and 1888, furnaces were built in several more cities, including Detroit, Chicago, Milwaukee, Wheeling (West Virginia), and Des Moines (Iowa).[36]

Municipal officials across the country quickly took note of these installations, and orders for incinerators were brisk. The Engle Sanitary and Cremation Company of Des Moines was a pioneer in the field. By 1894, it had installed incinerators in cities from Portland, Oregon, to New York (at Coney Island), and from Milwaukee to St. Augustine, Florida. It also constructed an incinerator for the Chicago World's Fair and built several systems in Latin America.[37] Although some engineers and scientists cautioned against expecting too much from the cremation method—or any single method of disposal—regulated disposal by fire was hailed as a panacea.[38]

City officials and many engineers marveled at the way the British had apparently solved their waste problems efficiently and "scientifically" through the use of fire as a disinfectant. Burning waste appeared to be the perfect disposal method: There were no stench-ridden dumps, no pollution of streams and other watercourses, no unsanitary landfills. One physician called the cremation of garbage "a great sanitary device." A doctor from Wheeling, West Virginia, stated that "at last we have secured a means of entirely destroying these substances and their power to do evil."[39] In 1888, Dr. S. S. Kilvington told an audience attending the annual meeting of the American Public Health Association (APHA):

> Everywhere interest in the question of cremation is awakening, and the present points to the future—a near future—in which every city, large or small, upon the American continent will consider the crematory a necessary part of its municipal outfit; forward to a time when our cities will be redeemed from the curse of accumulating waste, . . . when the age of filth-formation will be superseded by the era of filth-destruction, when fire will purify alike the refuse of the living and the remains of the dead —but also is it allotted to each one of us to help to bring in the coming of this sanitary consummation.[40]

The widespread adoption of this "ultimate" disposal method led the APHA's refuse committee to give its provisional endorsement to incineration in 1897.[41]

The high hopes for the adaption of British destructor technology to American needs were soon dashed, however. Although many engineers and city officials believed that incineration was a theoretically perfect disposal method, they soon expressed dissatisfaction with the performance of the first generation of American furnaces and crematories installed in the late 1880s and early 1890s. Of the 180 furnaces erected between 1885 and 1908, 102 were abandoned or dismantled by 1909.[42]

Many reasons were advanced for the failure of the American crematories. Some engineers claimed that faulty design of the furnaces was the problem—large grates, too much reliance on stoking, and so forth. Of particular note was low-temperature burning, which posed the same problems as it had in England: incomplete combustion of the waste and the creation of billowing smoke.[43] Some critics argued that British engineering expertise had not been taken into account in building crematories in America, or that the firemen and stokers who operated the plants were incompetent.[44] Others criticized the use of excessive amounts of coal and other fuels to augment the burning of refuse, which sharply increased the cost of operation.[45] Still others noted the difference in English and American refuse, speculating that the latter may contain a higher water content and thus require higher temperatures to burn.[46] Leaders in the field, such as Colonel William F. Morse, believed all these factors contributed to the failure.[47]

In a larger sense, the American experience with cremation in its first stage was unsuccessful because American engineers did not apply the technology with a realistic appreciation of its capabilities. In their enthusiasm for the new method, few bothered to consider whether British-designed equipment would meet American needs without substantial modification. Circumstances of space and demography were much different in the United States. The American population was dispersed over a larger area and, in many regions, was substantially less dense, making cremation less practical. Dumping refuse on land was an inexpensive alternative to incineration because of the availability of abundant, cheap tracts in or near most American cities. Moreover, energy sources were more plentiful and cheaper in America, which meant that waste could be hauled greater distances for dumping or fuels might be used to augment burning refuse without adequate attention to efficient operation of low-temperature furnaces.[48]

The unique spatial and demographic characteristics of the American urban society, the availability of alternative disposal methods, and the inherent limitations of the low-temperature furnaces kept the first-generation incinerators from becoming a competitive or reliable disposal method in the United States. Also, the earliest American furnaces were intended primarily for burning organic wastes, unlike the British practice of burning mixed refuse. In major American cities, primary separation, or source separation, was popular because inorganic materials such as ashes and rubbish could be utilized as fill or disposed of more conveniently by inexpensive dumping practices. Thus, incineration acquired a more specialized function in the United States than in England. Using incinerators to burn materials high in water content required greater temperatures than the first-generation crematories could provide or led to the use of additional fuels.[49]

Disappointment with the poor performance of the American crematories went beyond faulty designs and improper operation. Some American crematory manufacturers built and promoted poor-quality equipment simply to exploit the prevailing high expectations for the technology. Frustrated sanitary engineers faulted city officials who adopted equipment without consulting them first. They strenuously objected to having their expertise ignored and their authority usurped by bureaucrats untrained in the public works field. Colonel Morse, a sanitary engineer who was also a promoter of incineration equipment, led the charge against the "extravagant claims" of the peddlers of the first-generation crematories. He charged that the "sharp competition of opposing interests—builders of crematories and city authorities—developed mutual misrepresentation and recrimination. Contracts were obtained by personal and political favor, by influential pull, by manipulation and graft, with little regard to the interests of the city or town."[50] Perhaps Morse was so openly critical of the equipment companies because of his own ties to the Engle Sanitary and Cremation Company and then the Morse–Boulger Destructor Company.[51]

But other engineers echoed Morse's sentiments that American municipalities had too quickly adopted the new method without adequate investigation and without the sound advice of experts. The 1901 report of the APHA's Committee on Disposal of Refuse stated: "No city in America has yet undertaken a systematic comparative test of the several types of destructors that are recommended."[52] The sanitary engineer Rudolph Hering charged that "in but a few instances did city officials take the initiative in devising proper methods of disposal; hardly one American city can be found where exhaustive preliminary investigations were made and where the solutions suggested and practiced have been as yet entirely satisfactory and final."[53] *Engineering News* asked: "Why do so many of our cities persist in building discredited types of garbage and refuse furnaces?"[54] While doling out criticism, most sanitary engineers sought to exonerate themselves for the failures. They tenaciously guarded the territory of their expertise, but were not willing to declare any responsibility for what was, in part at least, an engineering problem.

In a broader context, the adoption of the low-temperature furnaces in the United States was a typical transfer of a single invention or device from a more technically advanced country to a developing country. As the historian Brooke Hindle persuasively argued:

> The acquisition of a single invention can be a relatively simple process, most easily accomplished when the receiving country imports the finished machine and immigrant artisans with the skills required to install and maintain the device. Yet, this sort of transfer may be helpful, it may fail completely, or it may quickly die. At best, it will remain a static, limited,

and terminal movement until it can be absorbed with the technological systems of the receiving country.[55]

Using Peter Mathias's terms, the British Destructor might be considered a "showpiece" innovation, that is, a best-practice technology that dramatically changes prevailing techniques in its field.[56]

Nevertheless, the transfer of the destructor was not altogether typical of the transfer of many industrial and agricultural technologies in the private sector. For one thing, American designers tried to copy the destructor rather than reassemble a gigantic piece of equipment built in England. As such, British artisans and British capital played a limited role in the construction and implementation of the device in the United States. Unlike the Germans, who imported British firemen to train their technicians in the operation of the destructor, the Americans modified the low-temperature destructor without a firsthand knowledge of the original intention of the technology or the operations it served.

The adoption of the British Destructor in the United States was subject to several nontechnical variables because of its nature as a piece of public works technology. Nathan Rosenberg has made a useful distinction between the transfer of "industrial" technology and "agricultural" technology:

> With all of the difficulties attaching to the transfer of industrial technology, such technology is nevertheless much easier to transfer than agricultural technology because industrial technology is at least very much self-contained. It tends to be a relatively closed system. Even its dependence upon human inputs can be reduced by making the technology more capital-intensive. But where ecological relationships are involved, as they are in agriculture, everything is quite different. Here there are important interactions between the human enterprise and specific features of the natural environment. Here natural phenomena participate in a much more active way, and productive activity must be more highly responsive to even minor variations and peculiarities of the environment. Agricultural activity is much more closely enmeshed with the natural environment, and this, in turn, has important consequences.[57]

In a sense, the transfer of the destructor was a hybrid of "industrial" and "agricultural" technologies. Like most industrial technology, the destructor was "self-contained." In its earliest form, at least, it was a single device that performed a particular function. But, like agricultural technology, the destructor—and public works technology in general— also had to respond to conditions within the natural environment, albeit the physical setting of the city. Considerations of health and sanitation, siting, and access to sources of fuel and supplies of refuse were crucial to the adoption of the destructor.

Public works technology, more than industrial and agricultural technology, is particularly sensitive to political and budgetary constraints, which it inevitably faces. It must stand the test of adoption, which may have little to do with its technical merits or how it enhances profitability or efficiency. City officials and municipal engineers, of course, do not want to invest in useless equipment. But they must justify a purchase of this kind in terms of the city's priorities and fixed budgetary limits. In the adoption of the destructor, city officials and engineers were seduced by the "idea" of efficient and cheap disposal by cremation, rather than careful consideration of the specific limits of the low-temperature furnaces. Expectations for the destructors ran ahead of careful investigation, and that was sufficient cause to place it high on the priority list of capital purchases. Unfortunately, these decisions were being made in the United States at a time when the British themselves were rethinking the design and functions of the destructor and slowly moving toward modifications that made the low-temperature furnace obsolete.

About the time the first-generation American crematories were being discredited, the British were well into the second stage of development, that is, the construction of high-temperature destructors. After 1902, the serious criticisms of the engineering community in the United States and the mediocre to poor performance of the first-generation crematories led to more careful scrutiny of the practice of burning refuse. In that year, experts in the field conducted the first major investigation of American crematories. The resulting debate over the merits of incineration lasted fifteen or twenty years.[58] In 1906, engineers made the first successful adaptation of an English-style destructor in Westmount, Quebec. This project was followed by similar ones in Vancouver, Seattle, Milwaukee, and West New Brighton, New York.[59] By 1910, many engineers were claiming that a new generation of incinerators had finally arrived.[60] By 1914, approximately 300 incinerating plants were in operation in the United States and Canada, 88 of them built between 1908 and 1914. About half the plants constructed after 1908 were built in the South.[61]

English and German projects to convert waste into steam and electrical power eventually prompted similar experiments in the United States. The writings of Goodrich and Joseph G. Branch introduced to American engineers the English method of using high-temperature destructors to produce power. Colonel Morse called Goodrich's first book on the subject—*Refuse Disposal and Power Production* (1904)—"a revelation to American engineers and town officials."[62] In 1905, New York City began a project to combine a rubbish incinerator and an electric light plant. The steam was to run dynamos and provide power to light several city structures, including an East River bridge. Before World War I, there was talk of several incineration plants to be coupled in a similar manner, but only a few were completed.[63] Electric generators were installed in the huge Milwaukee incinerator in December 1913, but, as Samuel Greeley noted: "With this exception no new record of actual steam utilization on a full output basis has been made in this country."[64] A few small-scale projects were developed, such as one in Minneapolis to light and heat a hospital and to light some city streets. In Seattle, power from waste was used for some manufacturing functions; Miami generated steam for a municipal hospital to help operate the kitchen and laundry; and Savannah, Georgia, sent power to a steam heater in a nearby waterworks plant.[65]

Cost was a major deterrent to large-scale projects in heat and power generation in the United States. A report of the U.S. Chamber of Commerce concluded that "only a small number of American incinerators develop steam, the more common practice being to erect plants of cheaper initial cost which are neither designed nor equipped for steam production."[66] To obtain sufficient value from a steam-producing incinerator required utilizing the generated steam day and night, effectively regulating the collection of wastes to ensure proper content for the burning process, calculating the requirements for auxiliary fuel to create the steam, and, of course, acquiring a market for the energy produced. Although proponents of the method enumerated the advantages of energy-producing incinerators, it was difficult to overcome the objection that the costs of construction were higher than those of traditional crematories. Colonel Morse calculated that the initial cost of a steam-producing plant, which used 30 to 75 tons of waste a day, was 15 percent higher than the cost of a simple incinerator. A strong proponent of the new technology, Morse tried to persuade his peers that the investment was worthwhile. But many city officials and engineers were convinced that higher construction costs and potentially higher operating costs—a hotly debated point—made steam-generating incinerators hardly worth the effort.[67]

Utilization of power failed to make its anticipated impact because it was out of step with certain realities of industrial America. The availability of cheap energy sources made conversion of waste into heat and light unnecessary—or at least undesirable. Private power companies frustrated efforts at extensive experimentation with European innovations. Furthermore, the United States had so many disposal methods available to it that utilization was inevitably compared with them on a cost–benefit basis. In fact, by World War I, burning or incinerating refuse was used as the major disposal method in only about 25 percent of American cities.[68] The advent of the sanitary landfill, in particular, undercut several utilization programs, since it appeared to offer both economical and efficient means of disposal.

Ironically, at the point where American engineering expertise was capable of translating the new British high-temperature destructors into the American setting, other factors intervened to frustrate effective transfer. Relative to Great Britain and other European nations, implementation of the high-temperature destructor technology was retarded in the United States because city authorities were unwilling to spend money on a combined system of disposal and power generation. In this instance, the high-temperature destructor was viewed as a means to overcome the problem of incomplete burning of organic materials, with less interest in the production of salable by-products. The British had moved toward high-temperature destructors not simply to consume waste more effectively but because they were keen on developing a better method to utilize heat generated by the process. As Hindle noted: "The transfer of technological systems represents an achievement of a different order of magnitude from the transfer of a single invention."[69] And contrary to the reception of the low-temperature furnaces, city authorities wanted to apply the technology of high-temperature burning without accepting the idea of the new destructor technology as the British devised it. Therefore, they now regarded the cost of incineration as excessive; they were thinking in terms of a *single device* rather than a *system*.

In the transfer of the British Destructor to the United States, technical feasibility was evaluated with little appreciation of the context in which the device was introduced. Furthermore, nontechnical issues and economic considerations peculiar to municipalities influenced the decision to implement or not to implement the new technology. In the first stage, economic issues (budgetary considerations) were subordinated to unrealistic expectations that the destructor would be a panacea —a "technical fix"—to the disposal problem. But the low-temperature furnaces were inherently faulty and were adopted with little attention to proper testing in the American setting. In the second stage, economic issues were crucial, not because city authorities had learned a lesson from previous experience, but because their expectations were limited to implementing a single device meant to accomplish what its predecessor failed to accomplish. The major change in the technology—more effective generation of steam—had less appeal. Consequently, the new systems appeared to be prohibitively expensive or, at least, more complex than necessary.

The transfer of the British Destructor to the United States exhibits many of the general characteristics of the transfer of industrial and agricultural technologies. But it is most interesting and most enlightening for what it demonstrates about the transfer of technology in the public sector. To understand the significance of public works technology diffusion is to pay particular attention to the urban environment and the way in which new services evolve or are changed by new technical devices and systems. One hopes that examination of other instances of technology transfer in the public sector will shed more light on the general topic, and will enhance our understanding of the diffusion process.

Notes

1. Nathan Rosenberg, *Inside the Black Box: Technology and Economics* (Cambridge, Mass., 1982), p. 246.

2. Nathan Rosenberg, "Factors Affecting the Diffusion of Technology," *Explorations in Economic History* 10 (Fall 1972): 6.

3. Peter Mathias, "Skills and the Diffusion of Innovations from Britain in the Eighteenth Century," *Transactions of the Royal Historical Society* 25 (1974): 113.

4. For a good, brief introduction to the theory, historiography, and methodology of technology transfer, see David J. Jeremy, *Transatlantic Industrial Revolution* (Cambridge, Mass., 1981), pp. 1–5.

5. For the United States, at least, a good bibliography of public works is Suellen M. Hoy and Michael C. Robinson, eds., *Public Works History in the United States: A Guide to the Literature* (Nashville, 1982).

6. See Martin V. Melosi, *Garbage in the Cities: Refuse, Reform, and the Environment, 1880–1980* (College Station, Tex., 1981).

7. William F. Morse, "The Utilization and Disposal of Municipal Waste, *Journal of the Franklin Institute* 157 (June 1904): 411–414.

8. S. S. Kilvington, "Garbage Furnaces and the Destruction of Organic Matter by Fire," in American Public Health Association (APHA), *Public Health: Papers and Reports* 14 (1889): 158–159; Mazyck P. Ravenel, ed., *APHA, A Half Century of Public Health* (New York, 1970; orig. pub., 1921), p. 189.

9. W. Francis Goodrich, *Refuse Disposal and Power Production* (Westminster, England, 1904), pp. 1–3; "Refuse Destruction," *Engineering* 67 (April 7, 1899): 459; "The Town's Refuse Problem," *Electrical Review* 46 (June 1, 1900): 957.

10. Goodrich, *Refuse Disposal*, pp. 3–10; Morse, "Utilization and Disposal of Municipal Waste," p. 420.

11. "The Horsfall Refuse Destructor and Forced Draught System," *Electrical Review* 53 (July 31, 1903): 168–169; "The Modified Horsfall Crematory," *Engineering Record* 41 (February 3, 1900): 111.

12. "Refuse Destructors," *Engineering* 66 (September 9, 1898): 342–344; Kilvington, "Garbage Crematories and the Destruction of Organic Matter," pp. 159–160; "Refuse Cremators," *Engineering Record* 26 (October 8, 1892): 297.

13. C. De Segundo, ed., "Refuse and Its Calorific Power," *Electrical Review* 46 (January 12, 1900): 41.

14. Goodrich, *Refuse Disposal*, p. 5.

15. De Segundo, "Refuse and Its Calorific Power," p. 41.

16. "Town's Refuse Problem," pp. 945, 958; "Refuse Disposal in Edinburgh," *Engineering Record* 38 (August 6, 1898): 199; Henry A. Garrett, "Refuse Disposal," *Engineering* 68 (August 18, 1899): 215.

17. See M. N. Baker, "Refuse Destructors," *Municipal Engineering* 27 (December 1904): 448.

18. A. S. Atkinson, "The Economy of the Modern Garbage Destructor," *Western Electrician* 36 (March 25, 1905): 236.

19. W. Francis Goodrich, "Electricity from Refuse—The Case for the Modern Destructor," *Electrician* 50 (November 28, 1902): 220.

20. "Municipal Refuse Disposal," *Transactions of the American Society of Civil Engineers* 60 (June 1908): 378; William F. Morse, *The Collection and Disposal of Municipal Waste* (New York, 1908), pp. 216–261.

21. See, for example, Frank Broadbent, "The 'Fuel' Value of Town Refuse: A Rejoinder," *Electrical Review* 52 (January 23, 1903): 130–132; W. H. Booth, "The Fuel Value of Refuse," *Electrical Review* 52 (February 6, 1903): 245–246.

22. William F. Morse, "The Next Step in the Work of Refuse and Garbage Disposal," in *Public Health: Papers and Reports* 25 (1900): 325–326; Ernest Kilburn Scott, "Combined Destructor and Electric Light Plants," *Electrical Review* 43 (December 9, 1898): 857.

23. Joseph G. Branch, *Heat and Light from Municipal and Other Waste* (St. Louis, 1906), p. 11; J. T. Fetherston, "Municipal Refuse Disposal, An Investigation," *Transactions of the ASCE* 60 (1908): 348.

24. "Running Municipal Trolley Cars with Garbage and Refuse," *Scientific American* 96 (June 1, 1907): 446.

25. Morse, "The Next Step," p. 322.

26. See Branch, *Heat and Light from Municipal and Other Waste*, pp. 11, 17–24; Morse, "Disposal of Municipal Waste," *Municipal Journal and Engineer* 21 (August 1, 1906): 111; Morse, "Utilization and Disposal of Municipal Waste," p. 421.

27. W. Francis Goodrich, *Modern Destructor Practice* (London, 1912), pp. 47–52; Goodrich, "Electricity from Refuse," p. 222.

28. Branch, *Heat and Light from Municipal and Other Waste*, pp. 25–26;

William F. Morse, "The Disposal of Municipal Refuse," *Municipal Journal and Engineer* 21 (October 3, 1906): 346–348; ibid. (September 5, 1906): 239.

29. "Garbage Cremation Experiments in Germany," *Engineering Record* 34 (June 20, 1896): 102–103.

30. American Society of Municipal Improvements, *Proceedings of the 21st Annual Convention, 1914* (Boston, 1915), pp. 51–52.

31. Rudolph Hering, "Construction, Cost and Operation of the Hamburg, Germany, Works," *Engineering Record* 36 (October 23, 1897): 446; Morse, "The Next Step," p. 320.

32. P. Tur, "Disposal of Municipal Refuse: Note on the Removal and Utilization of Municipal Refuse in French Cities," *Transactions of the ASCE* 54 (1905): 313–316.

33. See "Desiccation et Transformation des Matières Putrescibles," *Le Genie Civil* 6 (November 1, 1884): 12–15; "La Combustion des Ordures Ménagères," *Le Genie Civil* 27 (October 26, 1895): 415–416; "La Destruction des Ordures Menageres," *Le Genie Civil* 41 (March 3, 1902): 7–10.

34. Americans used the terms *garbage furnace, cremator,* and *incinerator,* instead of *destructor.*

35. Kilvington, "Garbage Furnaces and Destruction of Organic Matter," p. 159; Goodrich, *Refuse Disposal and Power Production,* pp. 3, 9–10; William F. Morse, "The Disposal of the City's Waste," *American City* 2 (May 1910): 23.

36. Charles T. Bayless and Arthur E. Merkel, "Garbage Cremation in America," *Engineering News* 32 (August 30, 1894): 167–168; Kilvington, "Garbage Furnaces and the Destruction of Organic Matter," p. 159.

37. William Mayo Venable, *Garbage Crematories in America* (New York, 1906), p. 88; "Garbage-cremation," *Science* 12 (December 7, 1888): 265–266; Morse, "Utilization and Disposal of Municipal Waste," pp. 25–42.

38. "City Refuse Disposal," *Engineering News* 28 (October 6, 1892): 325; George H. Rohe, "Recent Advances in Preventive Medicine," *Journal of the AMA* 9 (July 2, 1887): 5–6.

39. J. Berrien Lindsley, "On the Cremation of Garbage," *Journal of the AMA* 11 (October 13, 1888): 514; George Baird, "Destruction of Night-Soil and Garbage by Fire," in *Public Health: Papers and Reports* 12 (1886): 120.

40. Kilvington, "Garbage Furnaces and the Destruction of Organic Matter," p. 170. See also Thomas B. Carpenter, "Garbage Disposal," *Sanitarian* 40 (May 1898): 402.

41. "Report of the Committee on the Disposal of Garbage and Refuse," in *Public Health: Papers and Reports* 22 (April 1897): 108.

42. "Refuse Disposal in California," *Municipal Journal and Engineer* 42 (January 25, 1917): 101; U.S. Chamber of Commerce, Construction and Civic Development Department, *Refuse Disposal in American Cities: A Report* (Washington, D.C., 1931), pp. 15–16.

43. Harry deB. Parsons, *The Disposal of Municipal Refuse* (New York, 1906), p. 108; "Report of the Committee on Street Cleaning and Garbage Disposal," *Proceedings of the American Society of Municipal Improvements, 1910,* p. 61.

44. "Why American Crematories Fail," *Municipal Journal and Engineer* 13 (November 1902): 234; *Engineering News* 64 (August 11, 1910): 153.

45. "British Refuse Destructors and American Garbage Furnaces," *Engineering News* 53 (April 13, 1905): 388; W. Howard White, "European Garbage Removal and Sewage Disposal," *Transactions of the ASCE* 15 (December 1886): 872.

46. Howard G. Bayles, "Incineration of Municipal Waste," *Municipal Engineering* 29 (October 1905): 255.

47. William F. Morse, "The Disposal of Municipal Wastes," *Municipal Journal and Engineer* 22 (March 6, 1907): 232–235. See also Goodrich, *Refuse Disposal and Power Production*, pp. 10–13.

48. See Rudolph Hering and Samuel A. Greeley, *Collection and Disposal of Municipal Refuse* (New York, 1921), pp. 311–312.

49. See Melosi, *Garbage in the Cities*, for a detailed discussion of American disposal methods.

50. Morse, *Collection and Disposal of Municipal Waste*, p. 98.

51. Ibid., pp. 148–149, 161–163, 191–193.

52. "Report of the Committee on Disposal of Refuse Materials," in *Public Health: Papers and Reports* 27 (1901): 184.

53. Rudolph Hering, "Disposal of Municipal Refuse: Review of General Practice," *Transactions of the ASCE* 54 (1904): 266. See also *Proceedings of the ASMI Convention, 1916*, p. 245.

54. *Engineering News* 64 (August 11, 1910): 153.

55. Brooke Hindle, "The Transfer of Power and Metallurgical Technologies to the United States, 1800–1880," in *L'Acquisition des Techniques par les Pays Non-initiateurs* (1970), p. 408.

56. Mathias, "Skills and the Diffusion of Innovations from Britain," p. 102.

57. Nathan Rosenberg, *Perspectives on Technology* (Cambridge, Mass., 1976), p. 168.

58. Morse, *Collection and Disposal of Municipal Waste*, pp. 98–99; J. T. Fetherston, "Incineration of Refuse," *American Journal of Public Health* 2 (December 1912): 943–945.

59. "Modern Refuse Disposal Plants," *Municipal Journal and Engineer* 32 (May 30, 1912): 832; U.S. Chamber of Commerce, *Refuse Disposal in American Cities*, p. 16.

60. *Engineering News* 63 (June 23, 1910): 729; Hering and Greeley, *Collection and Disposal of Municipal Refuse*, p. 314.

61. "Recent Refuse Disposal Practice," *Municipal Journal and Engineer* 37 (December 17, 1914): 849–850.

62. William F. Morse, "Refuse Disposal and Power Production," *Municipal Journal and Engineer* 17 (September 1904): 107–109.

63. "New York Light from Rubbish," *Bulletin of the League of American Municipalities* 4 (December 1905): 190; *Engineering News* 53 (January 5, 1905): 17.

64. Samuel A. Greeley, "Refuse Disposal and Street Cleaning," *Engineering Record* 69 (January 3, 1914): 15.

65. Charles Zueblin, *American Municipal Progress* (New York, 1916), p. 80; Greeley, "Refuse Disposal and Street Cleaning," p. 15; Robert H. Wyld, "Modern Methods of Municipal Refuse Disposal," *American City* 5 (October 1911): 208–209.

66. U.S. Chamber of Commerce, *Refuse Disposal in American Cities*, p. 17.

67. Morse, *Collection and Disposal of Municipal Refuse*, pp. 431–432; P. M. Hall, "Report of the Committee on City Wastes," *American Journal of Public Health* 5 (November 1915): 1164–1167.

68. See Melosi, *Garbage in the Cities*, p. 163.

69. Hindle, "Transfer of Power and Metallurgical Technologies to the United States," p. 408. See also Mathias, "Skills and Diffusion of Innovations from Britain," 110; Jeremy, *Transatlantic Industrial Revolution*.

PART IV

Energy, Heat, and Power

CITIES cannot exist without water or food; neither can they survive without adequate sources of energy. Increasingly, throughout the nineteenth century, energy for purposes such as illumination and power was supplied through networks of wires or pipes. Electricity was most significant, coming to serve a wide variety of urban needs for improved lighting, electric traction, and industrial and domestic purposes. Natural gas also came into use around the turn of the century in areas where it was readily available.

In the first two essays in this part, Mark H. Rose and Harold L. Platt explore the development and meaning of electrical and natural gas networks for three American cities: Denver, Kansas City, and Chicago. Rose is particularly concerned with the urban spatial distribution of electrical and gas service and shows how it was linked to income patterns. Platt focuses on the spread of electrical power throughout the Chicago region and particularly on the role of Samuel Insull in making this energy source a driving factor in suburbanization. Roselyne Messager deals with so-called heat networks, or district heating, in Germany, providing informative comparisons with district heating in France.[1] She emphasizes the extent to which a historical institutional context can shape policy choices in areas such as energy supply. In his case study, Gabriel Dupuy provides us with insights into the interrelated regional effects of utility networks by examining the history of the town of Andresy, located near Paris. Andresy is unusual in that, for many years, a single concessionaire built and controlled its gas, electricity, and water networks and used them as the base from which to extend his regional influence.

227

Note

1. District heating is much more widely used in Europe today than in the United States, although a number of systems were installed in American cities around the turn-of-the-century. They were largely abandoned until the energy crisis in the 1970s stimulated a renewal of interest. The history of district heating in the United States, however, remains to be written.

11 Urban Gas and Electric Systems and Social Change, 1900–1940

Mark H. Rose

THE relationship between technological systems, group experiences, and social change within the urban context is a complex one and has engendered considerable study by urban scholars.[1] Questions focusing on these relationships, for instance, were first explored by members of the Chicago School of Sociology in the 1920s when they were attempting to determine the nature of city life.[2] More recently, scholars from different disciplines have studied the interaction of technology, the group, and society in an urban context from three lines of inquiry: the effects of transportation systems, with the streetcar or trolley enjoying a remarkable amount of scholarly attention;[3] an examination of the proliferation of domestic appliances for the urban household, particularly in terms of the experiences of women;[4] and a consideration of the concept of sanitation and its links to changes in personal hygiene and political action. One must add that although science, engineering, and technical know-how shaped the urban technologies such as electrical and sanitary systems as well as domestic appliances such as irons, all were known popularly as the products of remarkable science.[5]

Curiously, scholars have not probed the interrelationships among cities, popular notions about science, and the household and industrial technologies that, all agree, were so critical in facilitating and conditioning social change on so vast a scale. Gas and electric systems, in part because of their ubiquity and their importance to the functioning of modern cities, provide a point of entry to the analysis of the experiences of urban residents who were purchasing gas and electric service and enjoying brilliant illumination at work and school and the advantages of gas cooking in their kitchens. Those residents were participants in a pro-

229

cess of diffusing equipment and technological knowledge. That diffusion, in turn, rested on ideas of personal convenience, cleanliness, and hygiene informed by science. Indeed, between 1890 and 1940, higher-income households in every city demanded electric and gas service as part of a search for comfort and cleanliness.[6] The availability of gas and electric service came to underlie the creation of the interior, built environment. The net result of this interaction of gas and electric systems, urban culture and spatial ordering, and popular notions of science was the creation of overlapping social and technological ecologies.

Beginning late in the 1870s, electric lighting began to replace gas, kerosene, and candles. Between 1880 and 1883, numerous businesses, usually retailers, installed arc and incandescent lights. Publicity and sales were their principal motives, although many also sought to advertise their city as part of the race for urban greatness. By the early 1880s, electrical manufacturers were providing outdoor lighting as part of an effort to promote the sale of equipment, whether for use in manufacturing and retail firms or in a central station serving numerous customers. Either way, electric lighting enchanted urbanites, who remarked both about its naturalness and about the awesome and mysterious powers of science. As an example, during the evening of March 23, 1881, the G. Y. Smith Dry Goods Company of Kansas City turned on sixteen carbon-arc lights, the first located indoors in the city. According to a reporter for the *Kansas City Evening Star*, "Almost every inch of available space in street, sidewalk and even gutter was covered by a jostling, crowding, pushing, hurrying mass of people," each seeking admission and a glimpse of the lights. The store's gas lamps, compared to this brilliant illumination, appeared "yellow, ghastly and ashamed of themselves." The arc lights, he thought, represented "the splendid triumph of science."[7]

Over the next three decades, residents of every city began the process of adapting electricity and manufactured gas to households, office buildings, stores, and plants. It was a slow process. During the 1890s, only the wealthiest families and most exclusive hotels and office buildings featured gas and electric service. After 1900, political pressures and improved methods of gas and electric manufacture and distribution led to lower rates and more customers. As late as 1910, however, most urban residents knew of gas and electric service by virtue of their rapid introduction in central business districts for street lighting. As an indoor service, for heating, lighting, and cooking, they remained too expensive and complicated, still the triumph of splendid and mysterious science.[8]

One factor retarding the diffusion of electric and gas service for general household and commercial purposes was a lack of popular knowledge about their use. Beginning late in the 1890s, public and private agencies in many cities institutionalized mechanisms for the diffusion of knowledge regarding gas and electric service. Both general and technological high schools functioned, in part, with a view toward teaching the techniques of gas and electric operations to men and to alert women to their potential in the home. As early as 1898, the physics course required of all juniors at Kansas City's Central High School included instruction in the manufacture of electrical instruments such as electrodynamometers. By 1915, utility firms had made substantial donations of electrical, cooking, and laundry equipment to high schools in their city, enabling large numbers of men and women to become familiar with units of measure and with safe and routine use of power tools and domestic appliances. Adequate lighting was an important component in school design, and by 1916 administrators were adopting standards of measure—footcandles at each desk as determined by a photometer. A conviction that illumination eased learning and improved deportment guided them.[9]

After 1900, then, thousands of students graduated yearly from the nation's general and technological high schools. As part of their courses of study, many had been exposed to the rudiments of gas and electric systems. They had acquired the methods of routine investigation, sharp calculation, construction with any eye toward efficient operations, and mass sales based on a trained demand. Often, these graduates took jobs in gas and electric firms or in electric shops, repairing and selling equipment or monitoring utility operations. No doubt, many adopted a higher level of lighting for their homes on the basis of illumination standards in the public schools. Those who had studied gas and electricity (or simply experienced it in the form of improved lighting) helped transform them from the realm of mysterious science, for the elite, to an area of general application—a technological system serving thousands and yet still popularly conceived as science.

The creation of licensing agencies by city governments supplemented efforts of the public schools to provide knowledge regarding the safe and economical use of gas and electric appliances. During the 1890s, most cities created gas and electric codes. These documents represented the outcomes of negotiations among contractors, journeymen, and city officials. By 1910, the licensing of plumbers and electricians, based on knowledge of gas and electric characteristics, was a common feature of urban policy and building practice. After 1910, national building codes, created by experts in gas and electric operations, were adjusted to local circumstances, again following negotiations between electricians, plumbers, employers, and the city's representative.[10] By World War I, then, building codes and licensing requirements, along with instruction in the public schools, set the framework within which builders of houses, factories, shops, and offices would install electric and gas equipment. The law and a growing body of customs supplemented one another in guaranteeing that diffusion would occur along standardized lines.

After 1900, sales representatives employed by gas and electric utilities in many cities began to distribute information regarding the use of appliances for home, office, and shop. Salesmen marketed each new appliance by informing the customer of its personal advantages, which were either economic, aesthetic, or simple convenience. The gas house-heating campaigns of the 1920s and 1930s illustrate these methods as well as the general paths of diffusion taken by gas and electric appliances adapted for domestic purposes.[11]

Natural gas, for house heating, represented the most innovative sales effort undertaken by gas companies during the interwar period. Up to that point, distributors had been reluctant to sell gas for house heating, fearing demand would exceed capacity during chilly winter evenings, leading to the expense of providing an increase in distributional capacity to serve only an occasional peak in demand. If operators risked a peak beyond capacity, they feared service interruptions and dissatisfied customers, leading perhaps to placing the franchise in jeopardy. (Indeed, by the 1920s, the idea of service, especially uninterrupted service, was a regular fixture guiding policymaking and the maintenance of operations.) Yet by 1926, executives at the Public Service Company of Colorado (PSCC) had decided to take advantage of the diversity in Denver, assuming that they could balance their domestic and industrial loads against one another, thus creating a large potential market without expensive improvements in capacity.[12]

As with earlier innovations at gas and electric firms, such as rate changes and the introduction of new equipment, executives educated employees regarding the basic concepts and routine applications of gas for house heating. For instance, salesmen were enrolled in a correspondence course with the American Gas Association. Materials sent for study took a practical turn, focusing on combustion principles and on the quantities of gas consumed by different appliances, especially the furnace. Each morning, salesmen met for thirty minutes in order to polish methods "to overcome sales resistance in an intelligent manner." On Tuesday evenings, moreover, executives conducted classes on the "theory and practice" of gas distribution and utilization. Salesmen were tested at the conclusion of each session, and cash prizes were awarded for the highest scores.[13]

Advertisements and home presentations for gas house heating revolved around several themes. First, executives stressed the pushing ahead of inevitable progress and similarity with products already known to householders. Officials at gas firms recognized that residents lacked familiarity with a gas-fired furnace. After all, in appearance and operation, it was different from a coal-burning unit. The designers of advertising literature for the Denver Gas and Electric Company (subsequently the Public Service Company of Colorado) thus spoke of the replacement during the past two decades of horsecar lines, kerosene lamps, and water pumps by "more efficient methods of public service." A new gas unit, in this approach, was called "the invisible furnaceman," reminiscent of the stoking activity performed in every household where heating took place "by the old fashioned method of solid fuel furnaces." Consequently, the gas furnace was a natural, inevitable, and certainly "progressive step" in the evolution of household appliances, much in the same fashion as the "Hoover has overthrown the broom."[14]

Control of the household environment constituted the second theme in these promotions. Gas heating, it appeared, offered a level of household amenity unavailable to those heating with coal. Gas heating was automatic. It promised additional leisure time with family members, "instant . . . regulat[ion] [of] the heat to suit our changeable climate," and protection of health, because the heat was regulated, against "numerous wintertime coughs and colds." Coal, as before, represented the major competitor for the house-heating market, and rather than discuss price (coal was cheaper), the company emphasized cleanliness and convenience.[15]

In short, the introduction of gas heating resonated with several of the intellectual and social currents of the preceding decades. For years, aesthetic considerations had caused sensitive and knowledgeable urbanites to judge coal smoke a miserable problem. "The smoke is shutting out the mountains . . . besides doing other damage," reported a visitor to Denver in 1906. Observers routinely reported that soft coal and ignorance as to proper methods of stoking fires were the principal factors leading to excessive discharges of smoke. Between 1900 and 1920, businessmen and civic reformers secured the imposition of smoke-control legislation. During the 1920s, though, smoke remained a major social and health problem, particularly in areas around railroad yards and adjoining industrial districts. "Denver plants," according to the city's smoke inspector in 1920, "pour forth smoke to hide the mountains from the city's vision." Leaders of business and civic affairs still resided and worked near the smoke, encouraging many to agitate for improved enforcement and the education of firemen as to proper stoking. Yet public policy and the self-interest of manufacturers, railroads, and other coal consumers in achieving economical methods of firing failed to reduce the filth and ugliness of excessive smoke. One reason for moving to the periphery was distance from the major sources of smoke. "There is less smoke in the district," reported the editor of the *Country Club District Bulletin*, a periodical aimed at residents of Kansas City's most fashionable area. As it always had, responsibility for clean air rested on the ability of each person to select a homesite in a neighborhood miles from major sources of coal smoke. Gas heating, in a house located far from downtown amid other dwellings burning gas, helped break the exterior connection with coal and its inevitable smoke.[16]

Domestic hygiene and environment had always been the special responsibility of householders. By the 1920s, moreover, the lexicons of public health, engineering, and medical science coalesced to become the rhetorical framework within which decisions about personal and domestic well-being were made. Part of the intellectual apparatus for this emphasis on the personal and familial drew from a vast body of literature and practice focused on cleanliness in public schools and the hygiene of children. As one illustration, E. S. Hallett reported to his professional society in 1920 the results of his work in a school where teachers had complained of the "odoriferous condition" caused by children who both ate garlic and who "were from families of foreigners who never bathed." He installed ozone equipment "under such medical supervision as would inspire confidence," and soon thereafter deportment improved, teachers "were as fresh at the close of the day as in the morning," and "colds and coughs nearly disappeared." During the 1920s, builders of houses for the upper middle classes had adopted this approach, installing heating equipment and other appliances that appealed to buyers seeking cleanliness, convenience, and environmental control. One new home in Kansas City featured "a ventilating system in the kitchen that will remove every cookery odor"; another had a "specially designed gas boiler" as part of the effort to "secure [an] even temperature." No doubt, the ideologies of science and self masked the paradox that personal responsibility for control of the interior environment was accomplished routinely through a commercial purchase.[17] Either way, householders found in gas heating and cooking an escape from the obvious problems of dirt and smoke and their association with disease.

TABLE 11.1.
Indices of the Cost of Electricity, Gas, and Coal for
Household Use in Denver and Kansas City, 1919–1936
(1923 = 100)

	1919	1920	1925	1930	1935	1936
Denver						
Coal (bituminous)	79.5	88.9	92.3	101.7	82.0	73.9
Electricity			100.0	84.4	78.8	78.8
Gas			100.0	94.7	94.7	94.7
Kansas City						
Coal (bituminous)	92.2	118.9	100.1		75.9	76.5
Electricity			100.0	86.2	86.2	86.2
Gas			100.0	100.0	100.0	100.0

Source: M. Ada Beney, Cost of Living in the United States, 1914–1936, National Industrial Conference Board Studies 228 (New York, 1936), pp. 74–77, 80–83.

Attitudes toward the care of women and children served as still a third factor helping the gas house-heating campaigns resonate with popular ideology and common practices. Especially among those who could afford the cost of installing a gas-fired furnace and the higher price of gas over coal as a fuel, the ideal of the companionate marriage loomed alongside the reality of patriarchy in the family setting. By the 1920s, several generations of reformers had sought to use government and moral suasion to protect women and children from danger and abuse. They had achieved considerable results. Yet responsibility for protecting women and children from coal dust, smoke, chilly rooms, and the drudgery of starting a fire rested with husbands and fathers, or so ran the reasoning. Utility executives bridged the ambiguity of companionate and patriarchal notions by marketing furnaces in terms of guilt. This "sense of guilt," as the gas company in Denver labeled it, "rests upon all men as they sit in their comfortable downtown offices knowing that their mothers or wives are doing janitor or stoker work."[18] By the late 1920s, then, utility executives were able to combine modern advertising with the extant coalescence of ideologies and habits to give a powerful boost to the demand for gas house-heating equipment.

As much as utility operators distributed advertising literature and as often as salesmen made calls, still, by the late 1920s, urban marketplaces were saturated with cheap coal and urban households knowledgeable as to its use (see Table 11.1). Despite the higher price of gas (over coal), the early gas house-heating campaigns achieved remarkable results. During

1926, the Kansas City Gas Company sold 2,130 gas-designed furnaces and conversions. Between June 1928 and November 1929, the gas company in Denver sold 6,349 house-heating units, including 1,000 during one period of thirty days. By May 1932, executives in Denver reported 10,000 installations. Falling coal prices and declining incomes slowed the diffusion process, even reversing it in many households, and by August 1934, only 8,160 households in Denver burned gas for house heating. Thereafter, gas prices were slashed, and coupled with rising incomes, the number of installations increased rapidly once again.[19]

The successes achieved in these early campaigns set the framework for the diffusion of gas heating and other appliances on a citywide basis following World War II and the return of prosperity. But up to this point, technology diffusion, at the household level, had rested on a social base grounded in popular notions about hygiene, comfort, convenience, and science. Households earning modest incomes followed the well-off, first by overcoming the inertia associated with knowledge of an existing fuel, eventually leading to the abandonment of large investments in coal equipment.

The scholarly problem at this juncture is to explain several consequences of the interaction of these heterogeneous and changing cities (socially and spatially) and their technological systems, particularly gas and electric systems. At first glance, appliances and income intersected, forming neighborhoods of material plenty and others of uncooked dinners, chilly baths, and cold winters. Indeed, as much as income, transport, and urban morphology were (and are) contingently linked, still I want to probe technology–urban relationships in another, I think more penetrating, fashion.

The most straightforward manner to conduct this phase of the study is to create a sense of the location of gas and electric service across urban space. After 1900, the Denver Country Club district emerged as one of the most fashionable in the city. Located about 1.5 miles southeast of the central business district and accessible on broad boulevards and mass transit, the area developed around the Denver Country Club as a distinctive community for the directors of Denver's commercial and industrial firms. Householders in this district purchased gas and electric appliances before most urbanites did. During 1909 and 1910, Alfred A. Fisher, an architect for many of Denver's leading families, installed electric lighting, two gas ranges, and a washing machine in his new home. During the 1930s, younger families moving into the Country Club district continued the tradition of installing dense quantities of modern appliances. Gas and electric service were virtually universal in the district. The home of Hudson Moore, Jr., a PSCC executive, was typical of the homes near the Denver Country Club. In 1931, Moore ordered gas service for a cooking stove as well as a furnace along with the usual assortment, for well-to-do households, of electric lighting and convenience outlets controlled by switches. In addition, Moore installed an electric line, for a range, next to the gas stove.[20] By financing the installation of more than one appliance or utility, the owners of homes and office buildings regularly hedged as to fuel availability and price and the direction of technological innovation. So important had appliances such as refrigerators and radios (and automobiles) become, an observer of the real estate scene in Denver noted, that many families earning large incomes were constrained nonetheless "to minimize expenditures for housing."[21]

In districts nearby, housing lower-income residents, fewer were enjoying gas and electric appliances. Because these households did not patronize architects and conduct elaborate correspondence about interior decorations, we have only survey data on which to base a sketch in the aggregate of their tastes and technologies. By the 1930s, gas and electric service were available everywhere in Denver (although executives of the local utility could refuse service in areas where the number of consumers fell below a legal minimum). Even so, gas and electric salesmen routinely canvassed each household. At the conclusion of the depression decade, still, the areal distribution of appliances remained uneven. In many districts, fewer than 10 percent of the dwelling units featured central heating; on a citywide basis, 41.6 percent of the residential units included electric refrigerators, but in nineteen out of forty-four census tracts, the level of saturation was less than 25 percent. Unlike wealthier urbanites, moreover, poorer residents of Denver were not prepared "to minimize expenditures for housing" in order to enjoy modern appliances. This same observer found that "dwelling units provided with inexpensive equipment for heating, cooking, and refrigeration are more completely occupied than are those more expensively equipped."[22] Such in brief was the social–spatial distribution of gas and electric service, starting from a highly articulated demand and near saturation in several districts to a lower demand—in part based on lower income—and fewer installed appliances across the cityscape.

The experiences of appliance users, women in particular, add another dimension to the emerging picture of cities and technologies. The historian Ruth Schwartz Cowan describes the diffusion of appliances in terms of the material and working conditions of women in two groups, the modestly well-to-do and those who struggled "to make ends meet."

Among housewives located in the higher-income groups, Cowan finds a curious anomaly. These women purchased household appliances such as vacuum cleaners and washing machines, substituting machinery and their own labor for expensive helpers. Among the women of upper-income households, by about the late 1920s, the prospect of "two washes in the morning and a bridge party at night" held a broad appeal. Thus began the process, according to Cowan, "which might properly be called the 'proletarianization' of the work of economically comfortable housewives." But among working-class women, these same appliances facilitated a marked improvement in the health of their families and the conditions of their homes. Less dust, cleaner sheets and clothing, greater variety in meals, hot water, and a warmer home during the winter and better ventilated during the summer were brought about by purchasing appliances, although the length of her workday remained about the same as that of her mother. The experiences of appliance use, then, varied with family income.[23]

Because women and their households resided in different sections of the city essentially according to income, their experiences with gas and electric service also coincided with the diffusion of appliances across urban space. In higher-income areas, many housewives experienced incipient proletarianization, redeploying Cowan's elegant conceptualization. Experiences varied throughout the remaining sections of the city, ranging as far as the near-poor for whom an electric sewing machine and iron meant a substantial increase in the housewife's productivity, to borrow again from Cowan.[24] The net result was a gradation of experiences with gas and electric service across the landscape.

The varying experiences of householders, again especially women, with technological systems such as gas and electric service were related to levels of education in technological and general schools. Education played an important part in shaping outlooks on household technologies. The lengthier the stay in school, the greater the likelihood of exposure to brilliant illumination and to instruction in housekeeping, physics, and cooking. Educators did not advise on the purchase of appliances, but served as agents for the diffusion of skills and enthusiasm—habits of mind and an outlook—about ways of doing things at home and office. In Denver, as elsewhere, a higher income and a lengthier period of education generally overlapped, and these households clustered near one another in neighborhoods such as the Country Club district. In 1933, as the depression reached its low point, households in that area earned among the highest incomes in the city and had attended school the longest, an average of 12.9 years. More than half of those households owned a mechanical refrigerator; nine of ten reported ownership of a radio; and about one in three had a washing machine.[25] In this analytic framework, then, a map of educational levels in each section of a city ought to provide a rough index of technological attitudes and knowledge, as well as excitement regarding the benefits of gas and electricity applied in the home.

In 1934, the PSCC installed a battery of appliances in the Denver home of Dr. R. G. Howlett. Altogether this equipment promised a year-round temperature of 72°F and a humidity level of at least 25 percent during the coldest winter days. Residents of every city, the wealthy first, could begin to plan for a house that was free of stagnation and dryness, without pollutants, and capable of providing a uniform temperature in each room every day. Urbanites applying gas and electric service toward the perfection of their interior environments had as counterparts during the 1930s those who perceived hydropower as a form of energy at once clean, abundant, and capable of leading to the establishment of a utopian society.[26] Such, at any rate, were the nostrums and notions attached to energy choices and social and environmental change through the application of technological systems.

On the local urban scene, householders and utility officials had begun to reshape the domestic and city environments along these lines. By the 1920s, neighborhoods formed along income and occupational groupings, further separating according to race and perhaps religious orientation and national origin. Jobs, housing, and transport formed the "building blocks" of the city.[27] The addition of new energy forms—gas and electric service, as technological systems—added another dimension to each neighborhood and to the experiences of its inhabitants.

Knowledge about gas and electricity were concentrated in a couple of neighborhoods of the city, brought by persons seeking to control domestic hygiene, ease the burdens placed on women, and achieve a cleaner environment. Women located in higher-income households accepted additional tasks such as cleaning, cooking, and laundering, a curious process in which machinery and technological knowledge created a labor-intensive household. Subsequently, this environment nourished the feminine mystique, a set of attitudes and practices under way during the 1920s that glorified housework and childrearing.[28]

As wealthier households moved to the urban periphery, they left behind domestic and neighborhood environments that were less smoky and more packed with appliances, wires, and pipes. Lower-income households replaced the well-to-do in these districts, inheriting the accouterments of gas and electric service. But poorer urbanites possessed firsthand experience with the drudgery of housework and cooking on a limited budget. Consequently, they interpreted the presence of appliances to mean that domestic life would prove less burdensome and more productive.[29]

Gas and electric service thus dispersed across the city, following the migratory paths of the affluent. Utility operators facilitated the process by offering uniform rates citywide and education as to the use of appliances. The newer technologies of gas and electric service were thus part of the "ceaseless temporal and spatial interweaving of . . . physical components, . . . individual inhabitants, and the concrete practices of . . . families."[30] The net result on the urban scene of this interaction was a rough confluence of social and technological ecologies in the form of knowledge, appliances, experiences, and systems. Just as much, this emerging ecological system coincided with a more basic urge to use technologies (in the popular lexicon known as science) to achieve privacy and foster selfishness.

Notes

Acknowledgments
I am happy to acknowledge the following sources of funding in support of research: the Albert J. Beveridge Award Committee of the American Historical Association; the Herbert Hoover Library Association; the Eleanor Roosevelt Institute; the Harry S. Truman Library Institute; and the National Science Foundation. At Michigan Tech I received support from the Dean of Sciences and Arts; the Office of Research; and the Program in Science, Technology, and Society with funds granted by the Sohio Corporation. W. Bernard Carlson read an earlier version of the manuscript and made a number of valuable suggestions. Any opinions, findings, conclusions, and recommendations expressed herein are those of the author and do not necessarily reflect the views of those who supported my research.

1. See George H. Daniels and Mark H. Rose, "Introduction," in their *Energy and Transport: Historical Perspectives on Policy Issues* (Beverly Hills, Calif., 1982), pp. 15, 23.

2. Richard Sennett, "An Introduction," in his *Classic Essays on the Culture of Cities* (New York, 1969), pp. 14–15; Brian J. L. Berry, *The Human Consequences of Urbanization: Divergent Paths in the Urban Experience of the Twentieth Century* (New York, 1973), pp. 30–36; James Borchert, *Alley Life in Washington: Family, Community, Religion, and Folklife in the City, 1850–1970* (Urbana, Ill., 1980), pp. ix–x; and see Kenneth L. Kusmer, "Industrialization and Ethnicity: A New Model," *Reviews in American History* 11 (December 1983): 561–562.

3. Charles W. Cheape, *Moving the Masses: Urban Public Transit in New York, Boston, and Philadelphia, 1880–1912,* Harvard Studies in Business History 31 (Cambridge, Mass., 1980); Paul Barrett, *The Automobile and Urban Transit: The Formation of Public Policy in Chicago, 1900–1930,* Technology and Urban Growth Series (Philadelphia, 1983).

4. Charles A. Thrall, "The Conservative Use of Modern Household Technology," *Technology and Culture* 23 (April 1982): 175–194. See, especially, Ruth Schwartz Cowan, *More Work for Mother: The Ironies of Household Technology from the Open Hearth to the Microwave* (New York, 1983), pp. 151–191; and Mary Corbin Sie's, "The City Transformed: Nature, Technology, and the Suburban Ideal, 1877–1917," *Journal of Urban History* 14 (November 1987): 81–111.

5. Mark H. Rose, "Science as an Idiom in the Domain of Technology," *Science and Technology Studies* 5 (Spring 1987): 3–11. For the popular appeal of the idea of science, if not for its contents and methods, see John C. Burnham, *How Superstition Won and Science Lost: Popularizing Science and Health in the United States* (New Brunswick and London, 1987). Martin V. Melosi, *Garbage in the Cities: Refuse, Reform, and the Environment, 1880–1980* (College Station, Texas, 1981), explores the growing consciousness among civic reformers and the educated public of garbage as one of several environmental pollutants.

6. Although I speak often in this essay of cities, the bulk of my data is drawn from materials gathered in Kansas City and Denver. In turn, they were selected for study on several counts, including rough comparability of their dates of founding, rates of population increase, and the composition of their industrial and commercial sectors, principally meatpacking, railroading, printing, and jobbing for their respective hinterlands. Social and political developments in these two cities are carefully explored by Gunther Barth, *Instant Cities: Urbanization and the Rise of San Francisco and Denver* (New York, 1975); Lyle W. Dorsett, *The Queen City: A History of Denver* (Boulder, 1977); Dorsett and A. Theodore Brown, *K.C.: A History of Kansas City, Missouri* (Boulder, 1978); and Lawrence H. Larsen, *The Urban West at the End of the Frontier* (Lawrence, Kan., 1978). For a superb account of politics and culture in the formation of electric power systems, see Thomas P. Hughes, *Networks of Power: Electrification in Western Society, 1880–1930* (Baltimore and London, 1983).

In previous articles dealing with energy in cities, I focused on professional tastes, the mechanisms of diffusion, and the social bases of technological innovation. This study assesses the interaction of cities and technological systems with a view toward adding to our understanding of the development of the social–spatial order.

7. W. Bernard Carlson, "Invention, Science, and Business: The Professional Career of Elihu Thomson, 1870–1900," Ph.D. dissertation, University of Pennsylvania, 1984; W. H. Preece, "Electric Lighting in America," *Journal of the Society of Arts* 33 (December 5, 1884): 75; *Kansas City Evening Star,* March 24, 1881, in Kansas City Public Library, Missouri Valley Room, clippings file.

8. High prices also impeded the diffusion process in Great Britain. See Leslie R. Hannah, *Electricity Before Nationalisation: A Study of the Devel-*

opment of the Electricity Supply Industry in Britain to 1948 (Baltimore and London, 1979), p. 186.

9. Kansas City, Missouri, Central High School, Catalogue and Revised Courses of Study of the Central High School of Kansas City, Mo., 1898–1899 (1898), pp. 31–32; Denver, Manual Training High School, School District Number One, Arapahoe County, Colorado, Courses of Study, Requirements for Admission, General and Special Information (1900), pp. 14–15, 25; see "Denver's Technical High School," City of Denver 2 (April 11, 1914): 11–12, for a brief review of a similar course of study in the city's vocational high school. Lewis M. Terman, Report of the School Survey of School District Number One in the City and County of Denver, p. 5, The Building Situation and Medical Inspection (Denver, 1916), pp. 13–17, in State Historical Society of Colorado, Denver, Special Collections, Box 11. By the 1920s, electric firms provided educational materials to the public schools regarding the advantages of adequate home lighting. See G. W. Weston, "Better Home Lighting Contest," Tie 4 (December 1, 1924): 3.

10. "Electrical Department Active," March 28, 1914, p. 11, and "Electricians Must Pass Examination," April 25, 1914, p. 7, both in City of Denver 2; and "Recodification of Denver Electrical Ordinances," Electrical World 61 (April 26, 1913): 889.

11. In the easygoing regulatory environments of most cities up to the 1920s, government created few impediments to sales promotions. For an account of early sales activities in Detroit, see Raymond C. Miller, Kilowatts at Work: A History of the Detroit Edison Company (Detroit, 1957), p. 146. See Burnham, "New Psychology," for a review of popular psychology and sales appeals during the 1920s; see also Cowan, More Work for Mother, p. 187, for an explanation of advertisements in women's magazines.

12. Roy G. Munroe, "Selective Load Building Makes Even Load Curve," Gas Age-Record, October 8, 1927, pp. 545–546, 550; George Wehrle, "Load Factor and Its Bearing upon Gas Distribution Investment," American Gas Journal 135 (October 1931): 44–45.

13. Business Training Corporation, under the direction and supervision of the American Gas Association, "Building the Gas Load," Unit 5 [c. 1927], in Public Service Company of Colorado, Denver, Roy G. Munroe Files, loc. B16C4, trans. 24653 (cited hereafter as PSCC, Munroe Files); Thomas R. Thompson, "Denver Breaks All House-Heating Sales Records," Gas Age-Record 63 (March 9, 1929): 334.

14. PSCC, The Story of the Invisible Furnaceman, Being a Short Statement of Another Modern Convenience That Will Make Your Home a Healthier and Happier Place in Which to Live (Denver, 1923), in PSCC, Munroe Files, loc. B16C4, trans. 24653.

15. Ibid. O. J. Horrom, "Gas Heat," and V. E. Philpott, "Gas Heat Gives You More for Your Dollars," were awarded prizes for best house-heating sales presentation to be written within twenty minutes and to consist of not more

than 200 words, August 10, 1931, both in PSCC, Munroe Files, loc. B16C4, trans. 24653. Comfort and convenience were traditional themes in marketing natural gas. See Robert S. and Helen M. Lynd, *Middletown: A Study in Modern American Culture* (New York, 1929, 1956), p. 98, for a note that Muncie's press announced "the millennium of comfort and convenience" following the arrival of natural gas in 1895.

16. Charles M. Robinson, *Proposed Plans for the Improvement of the City of Denver* (Denver, 1906), p. 21. See also "Manufacturers and City to Co-operate in Steady Effort to Reduce Smoke Pall," November 11, 1920, p. 5; "City Inspector Ladd Says Way Discovered to Abate Smoke Pall," March 31, 1921, p. 1; and "Model Smoke Abatement Equipment Suggested for C. of C. Building," April 14, 1921, p. 1, all in *Denver Commercial* 12. And see "There Is Less Smoke in the District," *Country Club District Bulletin* 5 (February 1925): 2. For the problem of smoke caused by railroad locomotives and the limited success in controlling it, see Joel A. Tarr and Kenneth E. Koons, "Railroad Smoke Control: The Regulation of a Mobile Pollution Source," in *Energy and Transport: Historical Perspectives on Policy Issues*, ed. George H. Daniels and Mark H. Rose (Beverly Hills, Calif., 1982), pp. 71–92. For the perception that the smoke problem was both worse and yet susceptible to a technical–political fix, see Joel A. Tarr, "Changing Fuel Use Behavior and Energy Transitions: The Pittsburgh Smoke Control Movement, 1940–1950," *Journal of Social History* 14 (Summer 1981): 561–588.

17. E. S. Hallett, "A New Method of Air Conditioning in School Buildings," April 1920, pp. 423, 425, and J. H. Dunlap, "Common Sense, Science and Drinking Fountains," May 1919, pp. 470–472, both in *American City*. For one of the permutations of these ideas in a later period, see "Bacterial Rays Eliminate Familiar Health Hazard"; *Architectural Record Combined with American Architect and Architecture* 85 (April 1939): 73, advertising a method of sterilizing the family's toilet seat to the level "that 98.52% of the bacteria had been killed by ultraviolet radiation." Details regarding the design of homes for middle- and upper-income markets in Kansas City during the 1920s are in the *Country Club District Bulletin*. The Missouri Valley Room of the Kansas City Public Library holds nearly a full run. On the ideology of science and the increasing self-centeredness of urbanites, see Burnham, "Essay," p. 19, and his "New Psychology," pp. 352–353. Finally, see Cowan, *More Work for Mother*, pp. 176–188, for an explanation of the special responsibility of women to achieve domestic purity.

18. See Joan Jacobs Brumberg, "Zenanas and Girlless Villages: The Ethnology of American Evangelical Women, 1870–1910," *Journal of American History* 69 (September 1982): 348, 358, 367–369, for a description of ethnologies prepared by missionaries and the meaning of that body of literature in the American mind in terms of the desirability of protecting women and children. PSCC, *Story of the Invisible Furnaceman*; see also Cowan, *More Work for Mother*, p. 187, for women's magazines and the use of guilt in advertising.

19. Willis Parker, "More Than 1000 House-Heating Sales in 30 Days," *Gas Age-Record* 64 (November 16, 1929): 741–742; B. J. Strickler, "Natural Gas and the Public," paper presented to the Oklahoma Utilities Association, Oklahoma City, March 9, 1926. See also Roy G. Munroe to V. L. Board, July 26, 1933; Roy G. Munroe to G. B. Buck, August 28, 1933, and August 30, 1933, all in PSCC, Munroe Files, loc. B16C4, trans. 24653.

20. Invoices dated May 14, 1909, November 1909, April 1910, June 3, 1911; W. E. Fisher to Hudson Moore, Jr., March 13, 1931; Arthur A. Fisher to Brown-Schrepferman and Co., April 27, 1931; all in Fisher Architectural Records Collection, 1897–1978, Denver Public Library, Western History Department.

21. F. L. Carmichael, "Housing in Denver: A Report on the Real Property and Low Income Housing Survey," *University of Denver Reports* 17 (June 1941): 30. Carmichael's observation was based on aggregate data and thus not specific to the Country Club district.

22. Ibid., pp. 32, 43–44.

23. Ruth Schwartz Cowan, "Two Washes in the Morning and a Bridge Party at Night: The American Housewife between the Wars," *Women's Studies* 3 (1976): 149; idem, *More Work for Mother*, pp. 174, 178, 180, 183, 189.

24. Cowan, *More Work for Mother*, p. 190.

25. F. L. Carmichael, "Employment and Earnings of Heads of Families in Denver," *University of Denver Reports* 10 (September 1934): 6.

26. Service to the home of R. G. Howlett is reported in the *Gas Pilot* 17 (March 22, 1934): 1, in PSCC, Munroe Files, loc. B16C4, trans. 24652. Cooling was achieved through dehumidification rather than the removal of heat. Units supplying air-conditioning in the modern sense of the term did not appear in substantial numbers in Denver until about 1937. Compare "Electric Appliances Sold," (1930–1941), in ibid., loc. B16C4, trans. 24653; and "Air Conditioning Summary: 1936," *Automatic Heat and Air Conditioning* 8 (March 1937): 43. The utopian dimensions of energy enthusiasms during the twentieth century are explored by George Basalla, "Some Persistent Energy Myths," in Daniels and Rose, eds., *Energy and Transport*, pp. 27–38.

27. See Theodore Hershberg, ed., *Philadelphia: Work, Space, Family, and Group Experience in the 19th Century* (New York, 1981), pp. 28–29.

28. Cowan, "Two Washes in the Morning," pp. 147–149.

29. For a review of the patterns of urban deconcentration from the 1920s through the late 1950s, see Homer Hoyt, "Recent Distortions of the Classical Models of Urban Structure," in *Internal Structure of the City: Readings on Space and Environment*, ed. Larry S. Bourne (New York, 1971), pp. 84–103. See also Susan J. Kleinberg, "Technology and Women's Work: The Lives of Working Class Women in Pittsburgh, 1870–1900," *Labor History* 17 (Winter 1976): 65–70, for a review of household labor among working-class women prior to the diffusion of appliances, however limited. Finally, compare ibid., and Cowan, *More Work for Mother*, pp. 173–181, as to the significance of appliances for the domestic routines of women in higher-income households.

30. The quote is from Allan Pred's review of Hershberg, ed., *Philadelphia*, in *American Historical Review* 87 (February 1982): 268–269. See also George H. Daniels, "The Big Questions in the History of American Technology," *Technology and Culture* 11 (January 1970): 1–21, for the argument, which I have adopted, that "the preferences of people have a lot to do with the development of their technology."

12   City Lights: The Electrification of the Chicago Region, 1880–1930

Harold L. Platt

CHICAGO's World Fair of 1893 showed the potential of electricity to improve urban life for the first time. Without question, the "wondrous enchantment of the night illumination" was the most spectacular sight of an exposition devoted to dazzling displays from all over the world. In fact, the Dream City's 100,000 incandescent bulbs, 20,000 decorative "glow lights," and 5,000 brilliant arc lamps represented more than double the amount of lighting supplied to the real city by the Chicago Edison Company. Visitors to the fair were amazed by seemingly endless displays of electricity: whirring dynamos and motors, dancing fountains, futuristic kitchens, and an elevated railway.[1] Electricity made such a singular impression because it best symbolized the promise of American life through the inexorable triumph of technology over nature. Like the computer today, electrical devices a century ago held a mystical power that evoked visions of a future world free of drudgery. Abundance and leisure would spread among all the people, fulfilling our democratic ideals.[2]

By 1893, Americans were well prepared—even anxious—to base their optimistic faith in progress on such a technological foundation. On the one hand, Chicagoans and other city dwellers had been witness for over fifteen years to numerous experiments, demonstrations, and commercial applications of electrical lighting in exclusive downtown establishments. They had also received almost daily news reports about the exciting work of the American Wizard, Thomas Edison. The inventor had already achieved the status of a cultural hero. He personified traditional beliefs both in the Horatio Alger myth of the self-made man and in technology as an engine of progress.

On the other hand, Americans thought that contemporary society was coming apart at the seams. During the troubled 1870s and 1880s, the rise of big business, among other forces, had thrown workers, farmers, and capitalists into violent struggle. With the nation sinking into a deep depression on the eve of the fair, the frightening Haymarket Riot of 1887 took on an even more ominous meaning as the first tremor of an impending revolution. The White City embodied the search for the restoration of social cohesion and civic order. Many Americans believed that environmental improvements inspired moral uplift and good citizenship. As they toured the fairgrounds, electricity and other technological innovations appeared to offer the best hope for upgrading the quality of urban life.[3]

Yet even the most ardent technocrat would have to wait another ten years before a second generation of large-scale electrical systems could begin to supply energy to the real city on a par with the Dream City. It took a quarter of a century, from 1878 to 1903, for central-station operators to deliver an ample amount of energy at a reasonable price to the metropolis. In comparison to this period of technological incubation, the next twenty-five years represented a period of building momentum in the diffusion and use of electricity. In large part, the process of electrification had a built-in accelerator. As utility companies gained more customers, the unit cost of electricity fell, encouraging even higher levels of consumption. By the mid-1920s, virtually all households in the Chicago metropolitan area were enjoying central-station service, while industrial demand skyrocketed. The electrical motor caused a revolution in the factory by speeding up the work process with mechanized, assembly-line methods of production.

The diffusion of modern utility services occurred at a highly uneven pace among various consumer groups, social classes, and geographical areas. A history of these diverse patterns of energy use in the Chicago region provides important insights into the impact of technology on the city. From 1880 to 1930, steady increases in the consumption of energy affected the growth of Chicago and the daily lives of its inhabitants. The use of electricity exercised a pervasive influence on urban life and culture, extending from elite institutions and public places to the homes and workshops of ordinary city dwellers. Electrical technologies also played an integral role in suburbanization and urban deconcentration. In Chicago, a unified electrical grid eventually spanned a 10,000-square-mile territory of northern Illinois. Although most Chicagoans remained unaware of their membership in this regional community of energy consumers, each customer helped to determine the availability and the cost of service to the others. The emergence of this unwitting community underscores the historic lag between the momentum of technological change and the formation of public policies designed to harness innovation for the promotion of the general welfare.

The unique characteristics of electrical technology suggest that a supply–demand model of development best explains the timing and spatial patterns of energy use in Chicago. Unlike any other public service, electrical utilities must stand ready to meet every period of peak demand or face a breakdown. Other utilities can deny or delay services at times of peak load: the telephone's busy signal, the packed transit car's stranded commuters, and the waterworks' dry mains, for example. With gas and water, moreover, utility operators can store their products in tanks or reservoirs to meet anticipated periods of heavy demand. The storage battery supposedly promised an equivalent capacity but never proved practical for electrical suppliers.

Instead, they had to discover new technologies to keep supply and demand in constant equilibrium. They also had to create corresponding principles of economics to make electricity available at a price consumers could afford. In Chicago and other places, pent-up demand for better lighting initially outstripped the ability of utility operators to supply it. In spite of luxury prices, brownouts and blowups caused by overloaded circuits were the most common problems plaguing central-station operators in the early days.

More than any other individual, Samuel Insull was responsible for putting the new technology on a sound economic footing. During the first stage of technological incubation, from 1878 to 1898, Insull and other novice electrical men were preoccupied with overcoming the problems of supplying the demand for better lighting at a reasonable price. They tried a wide variety of approaches to loosen a seemingly endless train of supply-side constraints such as faulty equipment, limited service areas, and noncompetitive rates. In 1898, however, Insull combined several promising technological innovations with novel economic concepts to announce a gospel of consumption. Insull proposed that a coordinated marketing plan of building and diversifying the use of energy would set into motion a self-perpetuating cycle of rising consumption and falling rates. Over the next decade, Insull led the industry in removing supply-side constraints. During this second stage of economic maturity, he turned Chicago's Commonwealth Edison Company into the world's leading central-station utility. Electric rates fell dramatically as Insull installed a new generation of large-scale technology on a metropolitan scale. At the same time, he mastered the new economics of load management, rate structures, and marketing electricity to an enlarging community of energy consumers.

In 1911, the resulting drop in rates inaugurated a third era of rapid diffusion of electrical services into the home. More efficient light bulbs and improved household appliances helped Insull to spread a web of electrical lines throughout Chicago. He also expanded the scope of his operations to encompass a 6,000-square-mile region by consolidating several suburban and rural utilities into a single holding company. World War I slowed the pace of home electrification, but severe coal shortages during the conflict turned industry irreversibly toward central-station service. Scrapping their power plants and steam engines, manufacturers converted their factories to assembly-line methods of production during the postwar decade.

The addition of these big power consumers marked a final stage in the electrification of the Chicago region. Falling rates during the 1920s further accelerated the rising demand for energy by every type of consumer. Householders not only burned the lights longer but also added more and more appliances. Refrigeration machines in restaurants and grocery stores helped establish new standards of nutrition. The icebox became standard equipment in the home and the radio was an instant success. To meet the ever-growing demands of a regional community of energy consumers, Insull built a third generation of "superpower" generating stations and transmission lines. By the time of the Great Depression and his fall from grace, Chicago had become a society dependent on an intensive use of energy to sustain everyday life.

Sam Insull was the proverbial right man at the right time and place to emerge in the late 1890s as a leading spokesman of the electrical industry. Born in London, England, in 1859, Insull would always maintain close ties to Europe where inventors made many of the most important advances in electrical technology. Starting as an office boy, the ambitious young man learned shorthand, got a job at London's first (Edison) telephone exchange, and in 1880 won a job as Edison's personal secretary. The immigrant landed in New York City just as the Wizard was perfecting his incandescent lighting system. At the very center of the whirlwind for the next five years, Insull learned every phase of the new industry, from laying cables in the streets to making deals in the boardrooms of Edison's financial backers. In 1886, the rising executive was sent to Schenectady, New York, where he set up manufacturing facilities for Edison lighting equipment. But, six years later, a consolidation of the Edison and the rival Thomson–Houston companies left Insull as the number-two man of the resulting General Electric Company (GE). Armed with his technical, financial, and manufacturing experience, Insull put his own name forward as the most promising candidate for the position of chief executive of the fledgling Chicago Edison Company. Arriving at the time of the World's Fair, he soon put the dramatic lessons of the White City to work in the real one.[4]

Electrical services in Chicago reflected the industry's uncertain state of technological incubation. Supply-side constraints kept prices at prohibitive levels and restricted distribution areas to 3–4 square miles.[5] Edison had built a direct current (DC) system of low voltage that could not be transmitted at a reasonable cost for more than a half-mile from the generating station. To make a profit, Edison companies had to supply only the most densely packed commercial districts where the high cost of electric lighting could be absorbed as a business expense. In contrast, George Westinghouse's alternating current (AC) system used high voltages and transformers to solve the problem of long-distance transmission. But before the late 1890s, AC technologies lacked reliable appliances, including a motor for the trolley car, the single largest consumer of electricity. Moreover, both systems were based on small-scale generating equipment that individual consumers such as hotels, factories, office buildings, and theaters could buy as attachments to their existing steam engines. Central stations—themselves mere agglomerations of small-scale machines—consequently faced tough competition from these self-contained, "isolated plants." Cheaper lighting from coal gas and kerosene lamps also confronted utility operators with strong market pressures to reduce their rates.[6]

When Insull reached Chicago in 1893, the demand for better lighting was encouraging a rapid proliferation of the new, small-scale technology. Central-station companies had to compete not only against one another but also against the self-contained system. In the Loop, for example, the new head of the Edison venture confronted a potentially fatal rivalry with the Chicago Arc Light and Power Company (CALPC). It had been formed out of eight small companies in 1887, the same year that Chicago Edison began central-station service. The onset of this competition was more than coincidental, since the owners of the CALPC were major shareholders of the gas company. Members of Chicago's elite, they were hedging their bets against the new technology. In addition, Insull had to cope with the discouraging fact that many of the largest and most prestigious users of electric lighting had already installed their own Edison equipment, including the Marshall Field store, the McVickers theater, the Auditorium, the Rookery, and the Union League Club. At the same time, real estate developers and other entrepreneurs were hooking up homes and shops in outlying neighborhoods such as Hyde Park, Lakeview, and Edgewater.[7] The polygenesis of first-generation technology meant that central-station service remained a luxury twenty years after its introduction.

From the beginning of commercial applications in the late 1870s, electric lighting was associated with downtown areas and notions of class status. In most respects, this imagery of urban elitism represented a continuing tradition of big-city leadership in cultural and fashion trends. Regardless of cost, the most exclusive hotels, shops, and homes were the first to obtain the most up-to-date amenities. Certainly, it was only men like the merchant princes and Edison company directors John Doane and Marshall Field who could afford to buy a complete self-contained system in order to light up their neighboring homes on Prairie Avenue. The fancy downtown hotels were especially important as pace setters in interior design and housekeeping appliances. Chicago's high-class hotels had been among the first to install indoor plumbing and steam heat in the antebellum period as well as mechanized cleaning and laundry equipment in the Gilded Age. The coming of electric lighting proved no different; it became the new status symbol of the elite. Images of the "Great White Ways" and glittering city lights reinforced the magnetic appeal of America's urban centers.[8]

Notions of modernity and class status go a long way in explaining the origins of electrical services in the suburban and rural areas of the Chicago region. In a process similar to the polygenesis of electric lighting in the city's outlying neighborhoods, local businessmen started utility companies during the early 1890s in places like Highland Park, Joliet, and Streator. Efforts to bootstrap these localities up to big-city standards spurred the formation of central-station companies, although conditions indigenous to each place actually account for their origins. In the affluent suburb of Highland Park, an electric utility was started in 1890 by a resident, Bernard E. Sunny, who was the midwestern representative of the Thomson-Houston Company. In the satellite city of Joliet, a water-powered grist mill was easily and cheaply converted to run dynamos with an equivalent of 350 horsepower. And in Streator, the crossroads of the region's coal mining district, local businessmen started a company because fuel was readily available at very low cost.[9]

Unlike Chicago, however, services were usually restricted to Main Street where municipal contracts for street lighting afforded the utility enterprises some chance of financial success. Shopkeepers and home-owners nearby enjoyed modern lighting if they could afford it, but low densities precluded the extension of service lines far from the downtown generating stations. In addition, they operated only from dusk to midnight during the hours covered by the street lighting contracts.

Power to run motors and equipment was generally unavailable. The coming of the trolley and interurban in the 1890s would bring daylight service to some towns, but the high cost of distribution in areas of low density kept the electrical grid confined to downtown areas. In big cities and small places, the limits of technology suggested that electrical systems would follow the steam engine pattern of a widespread diffusion of self-contained equipment among individual users and small-scale operators rather than the gaslight model of central-station service.[10]

To forestall this outcome, Insull initially pursued the public utility operator's conventional strategy of monopoly. Before leaving New York, he secretly orchestrated a plan to buy up the stock of his most threatening challenger, the CALPC. Upon his arrival in 1893, the Edison executive immediately proceeded to buy out every other electric utility in the city. He also instructed his salesmen to offer secret discounts to potential customers who would be relatively large consumers and, therefore, most likely to purchase their own self-contained systems. Three years later, Insull's relentless pursuit of monopoly gave Chicago Edison control of all the central stations within the city. Equally important, Insull made sure that he secured the exclusive rights to sell all the various manufacturers' patented equipment. These patent licenses put Insull one step ahead of the "gray wolves" in the City Council, who attempted to blackmail him into buying their paper creation, the Commonwealth Electric Company. Instead, the aldermen sold it and its valuable fifty-year franchise at a bargain-basement price after discovering that they could not buy any generating equipment to carry out their threat of competition. In 1907, Insull would consolidate his city-wide monopoly under this favorable grant as the Commonwealth Edison Company.[11]

The lessons of the World's Fair pointed Insull toward a solution to the two major technological constraints on the supply of electrical energy. To reduce rates significantly, utility operators needed to attain economies of scale at the generating station and to cut distribution costs to the point where service could be made available throughout the city. In 1893, Westinghouse had met the diverse needs of the exposition by using a recent invention, the rotary convertor. It helped to create a universal system of distribution by changing alternating into direct current (AC into DC) or the reverse. Citywide service could be achieved at a reasonable price by coupling the rotary convertor to high-voltage transformers. Fortunately for Insull, Chicago contained a talented pool of electrical experts who quickly perceived the tremendous potential of the convertor device for the expansion of the distribution grid. Two college-trained engineers, Frederick Sargent and Louis Ferguson, were especially instrumental in applying the lessons of the World's Fair to Chicago.

Over the next five years, they helped Insull build a hierarchical system that could deliver energy anywhere in the city. The new configuration of central-station service contained a larger generator plant at Harrison Street, transmission lines, and several substations where high voltages were transformed and converted into a variety of currents for local distribution.[12] Yet technological advances alone were insufficient to put electric utilities on a competitive footing with self-contained systems or cheaper sources of artificial illumination. On the contrary, the resolution of the technological "battle of the [AC versus DC] systems" added pressure to solve the economic problem of rates and costs.

Insull's national and international contacts gave the Chicago executive original insights that led to a major breakthrough in understanding the unique economics of electrical technologies. Like other public utilities, electric companies were encumbered with extraordinary fixed costs for capital invested in plant facilities. These financial burdens were compounded by the peculiar requirement of utility operators to satisfy every period of peak demand. To calculate the generating capacity to meet this "readiness to serve," company managers had to sum up the maximum demands of all the customers. Everyone did not turn on all their appliances at the same time. During the early days, however, the system's load curve approximated the sum of every consumer's peak demand because electricity was used almost exclusively for evening lighting.

Insull and other utility operators faced the dilemma of how to structure rate schedules to induce both large and small consumers of energy to use central-station service. Rate schedules represented the single most powerful tool the electrical suppliers possessed to affect demand. The difficult problem was how to apportion fairly the fixed costs, as well as the ordinary operating expenses, among the various kinds of customers. From the start of commercial service, electric companies had imitated the traditional practice of gas companies. They simply charged a flat rate for a unit of energy. But it seemed essential for electric companies to offer special discounts to large users in order to lure them away from buying self-contained systems. At the same time, rate schedules could not shift the cost burdens too heavily onto the small consumers or they would continue to use cheaper alternatives.

With the aid of another recent invention, Insull created a novel, two-tier rate structure that solved the economic dilemma of central-station operators. While visiting England in 1894, the expatriate heard about a new device, a "Wright demand meter." It measured not only a customer's energy consumption but also the timing and the size of his peak demand on the system. On returning to Chicago, Insull dispatched his chief engineer, Ferguson, to study the device in Bristol, where its inventor, Arthur Wright, had already put it into widespread use. Insull's lieutenant reported that the meter allowed the Bristol company to charge one rate for maximum demand that corresponded to the utility's fixed charges and a second rate for energy use that corresponded to the company's operating expenses.

The two-tier approach met Insull's need for a highly flexible framework of rates. The demand meter allowed him to apportion the company's fixed costs equitably among the various kinds of consumers while offering them an incentive to use more energy. Customers would pay an equal amount for each kilowatt of demand but a lower and lower unit price for each kilowatt hour (kwh) of consumption based on a sliding schedule of rates for additional blocks of energy. Insull was among the first central-station operators in the United States to answer the question of how to calculate the interrelationship between the costs of this unique technology and the price of electricity. He seized upon the Wright demand meter as a fair and understandable way to establish a differential structure of rates.[13]

The demand meter also helped Insull to see the far-reaching implications of the new economics of electrical technology. Suddenly, the "engineering of selling" eclipsed the importance of improving the efficiency of production. The utility executive realized that utility companies needed to institute techniques of load management to build and spread the demand for energy over the course of the entire day. Making fuller use of expensive generating equipment would effectively reduce the fixed costs of a unit of electricity for every member of the community of energy consumers. In 1898, Insull argued that a self-perpetuating cycle of falling rates and rising consumption could be set into motion by aggressive marketing campaigns to attract a highly diverse range of customers, especially those who could use electricity during off-peak periods. His gospel of consumption led him immediately to offer special low rates for off-peak, evening use such as lighting drug stores, all-night restaurants, and advertising signs. He also began replacing the kilowatt hour meter with the demand meter and collecting data on his customers' patterns of energy consumption. The company's innovative statistical department studied these social and economic trends in order to target groups for special rates that could build and diversify the utility's load.[14]

Insull's announcement of a gospel of consumption at the 1898 meet-

By the mid-1930s, the design of the house and its surroundings for higher-income households presented paradoxical themes. Tudor and Georgian façades, along with Venetian wellheads and rustic bridges, suggested antiquity. Controlled plantings created the impression of regulated nature. But the interior had to be modern. The production of meals flowed in regular and predictable fashion from one machine to the next. Whether on the inside or out, invisible technology (modernity) in the form of gas, water, electric, and sewer lines allowed residents to shape their environments—perhaps holding back the city they were helping to create. The net result was a house that was emblematic of America's upper classes. Courtesy Western History Collection of the University of Missouri at Kansas City.

The finished kitchen of the "Little Magic House," Kansas City, 1926. Courtesy Western History Collection of the University of Missouri at Kansas City.

Smoke in the city helped make the suburbs attractive to householders. This photo was taken above the railroad yards in Kansas City, December 22, 1922. Courtesy Western History Collection of the University of Missouri at Kansas City.

A branch store of the Commonwealth Edison Company on State Street, Chicago, April 1929. Courtesy Commonwealth Edison Company Historical Archives.

Original waterpower site of the Joliet Economy Light and Power Company at Hyde's Grist Mill (*left*) and auxiliary steam power plant (*right*), September 1900. Overhead wires connected the two stations. Courtesy Commonwealth Edison Company Historical Archives.

June 1912 cover of *Electric City Magazine*, a free monthly distributed by Commonwealth Edison. (The ten-cent price on the cover refers to mailing costs.)

Most of the magazines were distributed through drugstores, and advertising space was given to the electric company in exchange for discounts on the store's utility bills. Courtesy Commonwealth Edison Company Historical Archives.

This advertisement for a luxurious "manor" in Andresy was published in a Parisian newspaper around 1930. It notes the various services available: piped-in water, sewers, electricity, and telephone.

ing of the industry's trade group, the National Electric Light Association, initiated a second stage of the electrification of the city. His understated conclusion that "low rates may mean good business" did not convince an industry built on monopoly practices. Nevertheless, Insull's unbroken record of success and leadership in engineering, finance, marketing, and public relations left few doubters among central-station operators a dozen years later.[15] Spearheaded by a sales campaign that became a major enterprise in its own right, Insull secured enough big customers to finance the construction of a second generation of truly large-scale generator stations. By 1911, these efficient power plants were connected to an expanding network of substations in the city and suburbs to form a unified distribution network on a metropolitan scale of 1,250 square miles. The steady removal of supply-side constraints translated into a 50 percent reduction in rates despite considerable inflationary pressures on the national economy (see Fig 12.1). Electrical service among commercial establishments became virtually universal in Chicago, while declining rates and better appliances began attracting significant numbers of residential and industrial customers.

The new rate structure allowed Insull to go after the single largest user of electricity, the street and elevated railways. Accounting for two-thirds to three-quarters of the city's demand for electricity, the railway companies offered a ready-made daytime load during the morning rush hour. Since the inauguration of rapid transit for the World's Fair of 1893, however, they had operated self-contained systems, with one exception. A decade later, Insull secured a contract with this line, the Chicago and Oak Park Railroad Company, by offering rates so low that they seemed to fall below the cost of production. For the central-station utility, the key provision of the power deal was a guarantee clause that required the transit firm to pay for a minimum average load, or "load factor," of 35 percent of maximum demand. This proviso meant that Insull could cover, if barely, the heavy costs of installing more generators while reaping handsome profits from their use during off-peak hours. Over the next four years, Insull signed up most of city's other transit companies on similar terms.[16]

The big power contracts encouraged Insull to take the high financial risks involved in being the first to build a second generation of large-scale central stations. At the Harrison Street station, Insull was already operating one of the world's biggest steam engines, which approached the technological limits for this type of reciprocating machine. Driving generators with an output of about 3,500 kilowatts, its pounding arms and huge mass tested the structural limits of steel parts and building foundations. An alternative type of prime mover, the smooth-spinning turbine, had long been used to tap the power of running water. But steam versions of the turbine engine for electrical applications had been limited to small models no larger than 1,500 kilowatts. After sending Ferguson and Sargent to study these machines in Europe, Insull persuaded company directors and General Electric (GE) to build a 5,000 kilowatt turbogenerator for a new, ultra-modern station at Fisk and 23rd Streets.[17]

On October 2, 1902, the successful test of the new technology virtually revolutionized the electrical utility industry. The steam turbogenerator opened up a new era in the achievement of economies of scale that continues down to the present. In the first eight years of commercial operation, the smooth-running machines had already cut the amount of fuel needed to generate a kilowatt hour of electricity from 7 to 3 pounds of coal. The new, less wieldy engines also slashed capital costs because they were more compact and easier to house. These savings were reflected in significant rate cuts, especially for large consumers but including a 40 percent reduction for average residential customers.[18]

The steady removal of supply-side constraints went hand in hand with the satisfaction of pent-up demand for more light and power awaiting a reasonable price. The economic logic of electrical systems gave Insull a strong incentive to "merchandise" energy. Building the company's load factor by filling up the daytime and evening valleys between the rush-hour peaks would keep otherwise idle equipment operating, reduce unit costs, and increase stockholder dividends. In 1901, Insull set up an advertising department to conduct a systematic sales campaign. This central office wrote copy for the newspapers and published a free monthly magazine entitled The Electric City. The new department also commanded a small army of door-to-door solicitors who periodically swept through the neighborhoods. Usually, they came armed with some kind of special inducement to persuade the prospective customer to install service, such as free wiring or irons for the householder and low-rental light fixtures or advertising signs for the shopkeeper.[19]

Next to the transit companies, the city's commercial enterprises made the greatest demands for better lighting and more power. The cultural linkage of electrical energy with notions of high technology, elite status, and urbanity made the demand for "modern" services almost universal among retailers, hotels, restaurants, clubs, and other places of popular amusement. Distribution lines spread rapidly along trolley routes, which were often located down the middle of neighborhood business strips. Shopkeepers of these inner-city "main streets" frequently banded together to petition Insull to install ornamental street lighting ahead of municipal plans. The rise in nightlife, fueled by a craze for five-cent movie theaters, the nickelodeons, brought the glamour of elite downtown establishments to working-class neighborhoods. And beginning in 1906, Insull's solicitors could offer more skeptical merchants improved tungsten filament bulbs that were three times more efficient than the old ones. Within five years, the new bulbs had virtually displaced gaslight in Chicago's business houses.[20]

A steady succession of new and improved appliances also fueled demand among businessmen for electric service. Once the small motor reached a point of practical perfection in the late 1890s, the application of electricity to mechanical devices was limited only by the imagination. The downtown hotel remained the leading edge in the use of innovative technologies to control interior space and to perform housekeeping chores. Built in 1909, the Blackstone Hotel, for example, had 9,000 incandescent lights, over 400 telephones, and several motors totaling 450 horsepower. These ran fans for forced-air heating of the lobby and restaurant areas. Separate exhaust fans were used in the kitchen. Other motors powered refrigeration machines for ice and food storage, as well as an air-cooling system for the dining rooms, cafes, and banquet halls. Electricity operated the water pumps, dumbwaiters, and elevators up the building's eighteen floors. The lobby, however, made the most conspicuous display of the consumption of energy. Automatic revolving doors spun around, and water fountains played under indirect lighting in the ceiling. Behind the scenes, the hotel's housekeepers relied on an expanding array of commercial-type appliances, including washing, drying, and ironing machines in the laundry room.[21] The widespread demand for electricity by business enterprises served to introduce Chicagoans to the advantages of central-station service. And the extension of distribution lines along neighborhood main streets made service increasingly available in residential areas of the city.

Most members of the middle class still could not afford to replace gas lighting with electricity. Shortages of the superior tungsten bulbs also acted as a significant supply-side constraint. Yet, electric lighting was no longer a luxury item enjoyed exclusively by the rich. By 1911, about 80,000 households, or 16 to 18 percent of the city's families, were central-station customers. Over the next five years, about 30,000 families a year would join the ranks of those with better lighting in their homes. Electric service was fast becoming a necessity of urban life among the middle class.[22]

In sharp contrast to the commercial sector, the introduction of electricity into the home was a slow process that did not upset traditional family routines. During this period, it was common for households to have both gas and electric lighting. The contemporary fashion of dual-purpose fixtures reflected a conservative attitude toward the use of novel technologies in the private preserve of the home. Electric lighting was usually restricted to the parlor or living room for a few hours after dinner. To overcome this resistance to change, Insull sent his sales force door to door with special incentives to encourage residents to become central-station customers. In 1908, he launched the first of a series of giveaways by persuading 10,000 housewives to accept a free GE iron in exchange for installing service. Another scheme offered landlords free wiring and discount rates for hall lighting if they would help persuade tenants to use electricity in their apartments.[23]

The same logic of load diversification that drove Insull to encourage home lighting during the evening also pointed him toward industrial applications during off-peak hours of the day. A differential rate structure allowed him to create special categories for big-power consumers who were given preferential treatment over residential and commercial customers. The contracts with the transit companies provided a model for the industrial sector of low rates and incentives to consume more energy. Nonetheless, industry proved to be the most difficult to convert to central-station service. Manufacturers were naturally reluctant to junk huge investments in factories organized around the steam engine and its elaborate transmission system of shafts, pulleys, and belts. At first, it was cheaper and easier to place self-contained generator equipment in the power house for illuminating purposes only. Factory owners were slow to recognize the numerous, albeit indirect, benefits of reorganizing work flows according to principles of scientific management by installing a separate motor for each machine.

The most powerful incentive for factory owners to become consumers of electric power was the extremely low efficiency of steam technologies, which delivered less than 30 percent of their energy to the machines. The poor performance of steam power created a pent-up demand for a substitute, especially among industries that already used lines of machines of the same type, such as printing, ready-made clothing, and machine shops. They led the way because conversion promised major savings and was relatively simple. Factory managers could use their existing shafts and belts to drive a group of machines with an electric motor. Electricity also satisfied long-thwarted demands for power applications that could not adopt steam technologies, including moving heavy materials in steelyards and stone quarries. Food processors also adopted electric power at an early point to run refrigeration machines in their candy, bakery, brewery, and meatpacking establishments. In addition, new manufacturing processes began to emerge from unique uses of electricity such as arc welding and electroplating.[24]

By 1911, big power contracts with transit and industrial firms ensured the fulfillment of Insull's plan to trigger a self-perpetuating cycle of rising consumption and falling rates. Since 1898, generator units had grown from 3,500 to 20,000 kilowatts, while transmission lines had increased from 2,300 to 12,000 volts. The company's forty-one substations in the city were interconnected with a similar, integrated network of power in suburban Cook and Lake counties. With the railways consuming about two-thirds of the 700 million kilowatt hours generated annually, the utility had boosted its output fifty times over 1898 levels. The resulting economies of scale and improved load factor meant that the average cost of residential service fell from 13.1 to 7.9 cents per kilowatt hour, or 40 percent. Declining rates and rising standards of living added to the momentum of consumer demand for a more intensive use of energy in everyday life.[25]

Rapid increases in the use of electricity after 1898 began to have cumulative impacts on city life and form. Reflecting consumption patterns, the trolley, elevated, and interurban lines probably produced the greatest changes by accelerating the process of population deconcentration to the urban fringe and beyond to the suburbs. The proliferation of "streetcar suburbs" and industrial "made-to-order" cities not only quickened the pace of metropolitan sprawl but also redefined the city's social geography. In the twentieth century, suburbanization has meant segregation by class, race, and ethnicity. Suburban subdividers quickly learned that the availability of electricity helped to persuade middle-class home buyers to make the move from better-served inner city districts to the periphery. For these families, the possession of electric lighting was fast becoming a necessity, especially as elite downtown institutions set new standards of urban modernity and conspicuous consumption. Whether white- or blue-collar, Chicagoans explored the new frontiers of nightlife opened by the Great White Ways, amusement parks, and popular entertainments along neighborhood main streets. Although greater use of energy was just beginning to have significant effects on daily life in 1911, the building momentum of consumer demand seemed to offer irrefutable proof of Insull's gospel of consumption.

The period from 1911 to 1919 marks a third stage of rapid diffusion of electricity in the residential areas of the Chicago region. The loosening of supply-side constraints helped reduce rates to a point where working-class households could afford central-station service. The number of residential customers skyrocketed to 375,000, or about 60 percent of Chicago's families. The figure would have been even higher if World War I had not intervened to delay the completion of home electrification until four years after the war ended.[26] Outside the city, similar technological, economic, and cultural forces were shifting the equation between supply and demand decisively onto the side of consumption.

Since the 1900s, Insull had been applying the principles of his gospel of consumption in the suburbs that he had used so successfully in the city. Beginning in suburban Cook and Lake counties, he put together a large and diverse community of energy consumers by a combination of system building, load management, transit contracts, and marketing campaigns. A new feature of Insull's aggressive drive to establish a monopoly on a metropolitan scale was the holding company that helped to finance his far-flung acquisition, modernization, and expansion programs. After constructing a metropolitan web of power around Chicago, he moved boldly to take advantage of the continuing migration of people and industry away from the central city. In 1911, he formed a new holding company, the Public Service Company of Northern Illinois (PSCNI), to consolidate his control of all the electric and gas utilities encompassing a 6,000-square-mile arc around Chicago.[27]

In 1910–1912, Insull opened a new marketing campaign to speed the electrification of the home. By this time, his demand meters had produced overwhelming statistical proof of the value of small commercial and residential consumers in reducing the unit cost of electricity. The time-of-use instruments also pinpointed the household appliance as a supplement to industrial power in load building during the slack daylight hours. Insull opened several neighborhood and suburban Electric Shops, department stores devoted to consumer products. His utilities provided easy installment plans to pay for home wiring and appliance purchases. He even helped to underwrite similar financial schemes offered by independent retailers and building contractors. Equally important, the tungsten filament bulb became generally available for home use, marking the death knell of the gaslight era. The improved lights produced three times as much light as the old carbon filaments while consuming the same amount of energy. Insull exploited the advantages of the new bulbs by extending special low-cost deals on home wiring and fixtures to working-class families.[28]

The healthy growth of residential and other uses of electrical energy gave Insull ample financial resources to continue constructing larger and larger central-station systems. Turbogenerators increased in size from 12,000 to 35,000 kilowatts, while transmission lines swelled in capacity from 12,000 to 22,000 volts.[29] This technological evolution of the scale of electric utilities went hand in hand with the growing dependence of a sprawling regional community on a more intensive use of energy to sustain daily life. The experiments along the North Shore had shown Insull that suburban patterns of energy use were significantly different from urban ones. In 1904, he began interconnecting distribution grids between Evanston and Waukegan to create a diverse community of transit commuters, shopkeepers, homeowners, local governments, and industries. Three years after laying this first suburban transmission line, he linked its network of consumers together with the city to take further advantage of diversities in the demand for energy. Insull took the logic of load building to its logical conclusion by showing that rural customers would add yet another beneficial dimension to Chicago's community of energy consumers.[30]

By 1911, he completed the integration of a metropolitan-scale system of 1,250 square miles. The economies of scale that Insull's unified network of power achieved inexorably undermined the viability of small-scale technologies. Until the formation of the PSCNI, shoestring utility operators and self-contained systems continued to proliferate in communities outside Chicago. Along the North Shore, however, the integration of several small grids into a single web of power cut operating costs in half. For example, the new turbogenerators at the Waukegan station consumed only 5.7 pounds of coal, compared to 12.1 pounds of fuel with the old equipment. Insull not only brought the rates down in the suburbs but he also put central-station service on a reliable, twenty-four-hour basis.[31]

In 1912, the PSCNI consolidated five gas and electric utilities serving 110 communities over a 4,300 square-mile expanse of territory around Chicago. The company usually junked the small-scale equipment of its predecessors, turning their generator plants into substations. Local distribution grids in the suburbs were gradually expanded until their crazy-quilt patterns overlapped to form a unified service territory. Large rural areas remained without modern amenities, however, because low densities and declining populations made the cost of stringing lines prohibitive for both the utility and the farmers. In contrast, the suburbs were gaining new residents at a rate 36 percent faster than the city itself. After the turn of the century, the growth rates of service industries, interurban railways, and heavy manufacturing in the suburbs and satellite cities were also impressive.[32] The economic advantages of Insull's system persuaded more and more of these large power users to opt for central-station service. New industrial facilities and public utilities were built without separate power plants. Simultaneously, those who already had small-scale equipment tended to abandon these self-contained systems as they wore out or needed to be replaced with larger units. Patterns of residential electrification followed an analogous two-part process.

Electricity helped to facilitate an exodus to the suburbs by supplying a level of modern convenience that the middle class had come to expect in the city. Home builders increasingly installed electrical wiring in new construction, while more and more owners of older structures had their dwellings retrofitted for central-station service. Subdividers had learned long before that extra amenities often enhanced the attractiveness of their properties, especially to middle-class buyers. They were becoming reluctant to move away from well-serviced neighborhoods in the city to fringe areas where a modern infrastructure was not already in place. The availability of electric service was often considered essential since gas lines were generally absent in these low-density communities. For old-time residents, the coming of the electric lines and poles also promised to improve daily life. At first, however, there was fear that just the opposite would result from the messy job of installing the new wires and fixtures. But as workmen gained experience, the advantages of modern urban services gradually overcame this source of consumer resistance. Insull's utilities wired 600 older homes in 1910, a number that grew annually to 10,200 in 1915 when the process of retrofitting the suburbs was about half finished.[33]

For the residents of both new and old houses, the desire for better lighting remained the primary source of demand for electrical energy. Ironically, leaders in the gas industry had been the first to notice how the incandescent bulb was sharply raising standards of interior lighting. Brilliant levels of illumination in downtown stores and public places had the effect of making people feel they were living in the dark.[34] Changing perceptions not only gave birth to the profession of illuminating engineer, but also a new style of domestic architecture. Frank Lloyd Wright best embodied the rising demand for more light in his prairie houses with their ribbons of art glass, huge picture windows, and indirect lighting fixtures. Although few could afford one of Wright's "city man's country house[s]," more and more Chicagoans were deciding that electric lighting in the home was a necessity of modern life. In 1912, the builders of working-class dwellings voiced this conclusion in appeals to Insull for an economy lighting package. He responded by offering to wire these houses for $12, plus $2.50 for each outlet and $1.75 to $6.00 for various fixtures. In comparison, the cost of wiring more affluent houses ranged from $100 to $300.[35]

During this period, household appliances had important but secondary impacts on the spread of electricity in the Chicago region. After 1911, manufacturers like GE and Westinghouse made rapid strides in creating mass consumer products. In a process repeated down to the present, the manufacturers miniaturized housekeeping and cooking technologies that had already proven successful in commercial establishments. Yet the fact that bulb-type outlets and sockets predominated into the late 1920s warns against placing too much emphasis on appliances as a principal agent of home electrification. Despite continuous advertising campaigns, only the iron and the vacuum cleaner would penetrate a majority of homes, with about three out of four families having them in Chicago. As late as 1927, the next two most-popular appliances—the washing machine and the toaster—would be found in far fewer homes, or 42 percent and 30 percent respectively.

In the home, Chicagoans clung tenaciously to traditional domestic routines. The emphasis of Insull's advertising and door-to-door solicitations suggest that household appliances were used mainly as something extra to persuade doubters to install electric lighting. Before World War I, ads usually depicted appliances in the hands of servants, which further indicates that a mass consumer market for these products had not yet been fully established. In 1923, for example, a survey of residential customers found that lighting comprised about 80 percent of their demand for electrical energy. The iron consumed three-quarters of the remaining demand. Even the iron, however, was adopted slowly by housewives. Appliance-repair records indicated that they used electric irons during the hot summer months but returned to the old "sadirons" during the winter season when stoves were in use for cooking and heating.[36] Although the use of electricity in the home was limited primarily to satisfying demands for better lighting before the late 1920s, the overall growth of energy consumption had profound impacts on the city and its inhabitants.

By the end of World War I, the emergence of a highly diverse community of energy consumers on a regional scale meant that electricity was influencing more and more aspects of daily life. In the decade from 1909 to 1919, electrical output per capita increased 171 percent (see Figure 12.1). Reflecting consumption patterns, a more intense use of energy to sustain urban life was most evident in the public and commercial sectors. Cheap forms of rapid transportation carried the masses

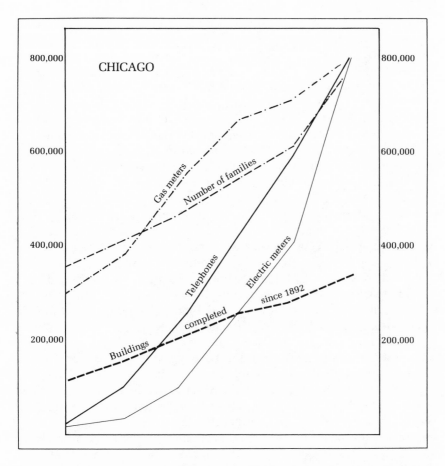

Figure 12.1 Comparative growth of Chicago's homes and its gas, telephone, and electric utilities. Electric utility outdistanced all others by 1925.
Source: "Statement Submitted by the Commonwealth Edison Company in Competition for the 1925 Charles Coffin Award," in Commonwealth Edison Company Historical Archives.

to homes, shops, and factories on the urban fringe. Insull's regional web of power now supplied Chicago's suburbs and satellite cities with the same modern amenities of light and power found previously only in the central city. Novel technologies seem to have had the least effect on the private realms of the home and family.

Nonetheless, electricity had subtle impacts that began to create an "invisible world" of energy, a world we take for granted today. Perhaps the best example of this systemic growth of energy dependence was the use of artificial refrigeration to supply the city with a wider selection of fresh foods. In the 1910s, the food industry applied artificial refrigeration to every step in the commercial food chain from transporting fresh meat and produce from more distant points to cold storage and processing plants to new-style supermarkets (self-service refrigeration units) and home iceboxes. The universal appeal of more fresh food, especially during the winter months, complemented the spreading popularity of gas ovens with thermostatic control. In this case, the use of greater amounts of energy quietly changed one of society's most basic necessities, its diet and nutritional standards.[37]

The evolutionary growth of a regional community of energy consumers probably would have continued without interruption into the 1920s, but World War I upset these trends in permanent ways. Major shifts in the national economy triggered the fourth and final stage of the electrification of Chicago. In contrast, the utility industry had been chiefly responsible for earlier turning points or transitions in urban energy supplies. At base, the war created unprecedented shortages in fuel, transportation, and labor that left each of these economic sectors with chronic problems of readjustment after the conflict.

Most important here, uncertain coal supplies turned industry in the Chicago region decisively away from self-contained power plants and toward central-station service. Moreover, war needs sharply accelerated the related processes of industrial deconcentration and modernization. Manufacturers moved to the urban fringe where they built one-story factories and instituted the latest innovations in the mechanization of production. In some cases, industries unable to make the transition were severely crippled and never recovered. In the natural ice business, for example, labor shortages and soaring freight costs combined to strike a death blow. The same pressing demand for workers also had long-term consequences for the domestic life of the middle and upper classes. The supply of immigrant white live-in servants dried up and was only partially replaced by southern black day helpers. Many housewives instead became servants to their home appliances.

The period from 1919 to 1932, then, witnessed the birth of an energy-intensive society. The industrial sector led a dramatic upsurge in the use of electricity in the Chicago region that had ever greater cumulative impacts on urban life and form. The rising momentum of demand for energy buttressed Insull's faith in a gospel of consumption. Placing new emphasis on coal conservation, he built a "superpower" ring around the city that boosted the size of generator units from 35,000 to 200,000 kilowatts and the capacity of transmission lines from 22,000 to 132,000 volts. In addition, average residential rates were cut by over 40 percent, adding further momentum to consumer demand. By the late 1920s, householders had become so dependent on electricity that even the onset of the Great Depression failed to curb their use of energy. Modern life was now unthinkable without central-station service; Chicagoans had created an "invisible world" of energy to maintain the quality of everyday life.

For most city dwellers, electrical energy became a ubiquitous influence at work and play as well as at home. Every major group in Chicago's community of energy consumers steadily increased its demands for central-station service in the 1920s, but the consumption of electricity by industry significantly outpaced the others. In 1919, the war economy had pushed the factories' usage to the point where the electric railways no longer accounted for over half of total consumption. Within five years, industrial power surged to the forefront in spite of a respectable growth in demand of 40 percent by the transportation sector. By 1930, the pattern of energy consumption was practically reversed because the factories now used about half the total supply compared to a quarter by the railways. In contrast, the mix of commercial and residential use remained basically unchanged. Home demand edged upward from 8 to 14 percent due to a gradual acceptance of appliances, especially after the appearance of commercial radio sets in the middle of the decade. During the postwar decade, Chicago increased the demand for electricity by 168 percent, which doubled the amount of energy consumed by the average city dweller.[38]

A closer look at particular businesses serves to illustrate the farranging implications of industry's wholesale conversion to central-station service. As noted above, the war served to kill off the natural ice business at the very time that the home icebox was coming into vogue. In addition, Chicago's meat packers started buying electrical energy from Insull during the war because coal supplies dried up while prices soared. In 1918, Swift and Company led the way, becoming Insull's single largest customer by 1925, with a demand equal to the city of Peoria. The meat packers used more and more energy in the form of heat and refrigeration to speed up their assembly lines. Together, the ice and food businesses used twice as much electricity as the next-largest industrial consumer, the steel and metal trades. The latter responded to the war by turning to the electric furnace, which produced a higher-quality product.[39]

During the 1920s, an intensive use of electrical power became an integral part of the production of a new group of mass consumer goods. New-style factories and assembly-line methods predominated in the growth industries of the decade, such as automobiles, chemicals, rubber, and electrical products. In each enterprise, increasing scale and mechanization spawned new work routines and workforces. New types of communities also formed in the neighborhoods around the plants. For instance, Cicero became a blue-collar suburb as thousands of workers moved closer to their jobs at Western Electric Company's massive Hawthorne Works. The transformation of manufacturing in the 1920s would set the stage for the rise of the Congress of Industrial Organizations (CIO) in the depression-racked 1930s.[40]

The fast-growing dependence of industry on electricity was paralleled by a swelling demand for more energy in the home. During the 1920s, consumption in the average household rose by 63 percent. Rate cuts held bills to a 20-percent increase, or from $24 to $30 per year. The war diverted resources from housing construction, setting the stage for Chicago's greatest building boom of the century. The new construction ranged from luxury high-rises along the Gold Coast to working-class bungalows and flats in the neighborhoods, and houses of every type in the suburbs. Almost all the new dwellings included electrical service, which helped to complete the electrification of the home by the middle of the decade.[41]

Although lighting remained the primary use of electrical energy in the home, appliances loomed in importance over the course of the period. Persistent advertising helped household technologies gain acceptance as cultural symbols of modern life with its emphasis on the home as the focus of consumption, leisure, and private comforts. For women, the iron and other cleaning devices greatly eased the physical drudgery of household chores. Nevertheless, recent studies have raised serious doubts about the promised benefits of modern technology in liberating women from their traditional domestic roles. For the working class, improvements in appliances and kitchen designs, found in the new-style bungalows and apartments, may have *mainly* speeded up the flow of women into full-time jobs, but without relieving their burden of household chores. For the middle class, the war-born revolution of domestic servants had analogous effects. Not only were housewives doing more of the work themselves but the standards of cleanliness and motherhood were being raised as well.[42]

By 1920, Insull's regional power grid had largely eliminated the differences in modern home service between the city and the suburbs.

The rich had long used servants to lift domestic life in the country up from rural squalor to a level of comfort on a par with the city. In large measure, the coming of electricity to the suburbs supplied a technological substitute for the corps of servants that the middle class could not otherwise afford. In addition, electricity could make up for a lack of other urban amenities by powering water pumps, cookstoves, and telephones. In sum, central-station service encouraged suburbanization by minimizing any sacrifices in the quality of middle-class life in the city while taking away none of the supposed advantages of having a home in the country.

In sharp contrast, the spreading use of energy in the home greatly exacerbated the conflict between urban and rural life. The war may have again been pivotal in the battle to "keep 'em down on the farm." Returning home was difficult for rural recruits who had experienced the modern conveniences of daily life on army posts, besides the sensuous enticements of city lights at nearby nightspots. For example, a 1919 study of farm life reported that the average family spent ten hours a week pumping and carrying water. Even so, it used far less water for cleaning, cooking, and personal hygiene in maintaining the quality of life than urban families. Another survey revealed that farm women found water chores to be their worst daily grind, followed by cleaning and trimming kerosene lamps. In fact, many commentators not only bemoaned the comparative decline of rural life but also saw electricity as its savior. These reformers would achieve success during the New Deal with the creation of the Rural Electrification Administration and the Tennessee Valley Authority.[43]

In the 1920s, however, the private utility companies made little progress in supplying rural areas, with their declining populations, at anything but a financial loss. Keeping pace with the rapid expansion of demand in the suburbs proved a much more profitable allocation of resources. In 1928, Insull finally began to underwrite the cost of extending lines and poles to the farms of northern Illinois. This plan and the advent of commercial radio marked the real beginnings of rural electrification in the Chicago region. Regardless of where one lived, the radio ended any doubts about the necessity of electricity in the home.[44]

The phenomenal success of the radio reinforces the conclusion that Chicagoans had become dependent on electricity to maintain a new, energy-intensive style of life. The links among the consumption of energy, technology, and leisure activities in the city had started forming long before with the lighting of downtown theaters and streets; the rise of the World's Fair Midway, White City, Riverview, and other amusement parks; and the spread of movie palaces throughout the metropolitan area. But the radio brought this new world of urban popular culture into the home. Perfected during the war, the radio became a mass consumer item in 1923 when commercial production began in earnest and the principles of national networks were established. Two years later, nighttime network shows were changing the rhythm of everyday life, causing Insull to remark that "the widespread use of radio receiving sets is a factor in the increased use of electricity for residential lighting." In 1930, the radio became the most popular home appliance, surpassing the iron in only seven years. A 1934 field survey confirmed that 94 percent of the company's customers had a set. It made life on the farm without electricity that much more isolated and unbearable.[45]

During the 1920s, the momentum of consumer demand for more electrical energy in the city and the suburbs had become so powerful that even the onset of the Great Depression failed to halt it. Although 1929 marked a peak year for total demand in the Chicago region, the average household continued to use greater amounts of energy for another four years. Furthermore, residential consumption never fell back to pre-depression levels, but started moving again in the opposite direction in 1934 and each year thereafter to set new records until the oil embargo of the mid-1970s. In 1930, the drop in the utilization of electricity by commerce and industry also reflected the pervasive influence of energy in the city and the nation. Economic activity and employment had become inseparably tied to mechanical technologies and electrical power. Whether in the home, shop, or factory, Chicagoans had created a world based on the intensive use of energy. By the end of the decade, farm families too would become part of this world, using an average of 75 percent more electricity than their counterparts in the city.[46]

In the Chicago region, the process of electrification occurred in a highly uneven manner. The shifting equation between supply and demand best explains these social and geographical patterns. The unique characteristics of electrical technologies that bound supply and demand in instantaneous balance created four distinct stages between the introduction of Edison's lighting system and the emergence of an ubiquitous world of energy consumption. Supply-side constraints characterized the first era of technological incubation. High rates and limited distribution areas were caused by the technical flaws of a complex, integrated system for the generation, distribution, and utilization of light and power. As engineers solved these problems, utility operators such as Samuel Insull worked out a structure of rates that corresponded to the costs of supplying electrical energy.

In 1898, a second era opened when Insull's gospel of consumption led to the construction of large-scale systems and the applications of load-management techniques. By 1911, the resulting reductions in the unit cost of electricity brought rates down to a point where Chicago's working-class families could afford home service. A third era of residential electrification ensued as Insull pursued the logic of his gospel of consumption to forge a community of energy consumers on a regional scale. The evolutionary growth of this network of power was permanently altered by World War I, which accelerated the conversion of industry to energy-intensive modes of production. From 1919 to 1929, the industrial sector underwent a revolution in the utilization of electricity to create highly mechanized work routines and production techniques. Even the Great Depression did little to slow the momentum of demand for the energy needed to maintain new styles of everyday life.

Notes

1. *Harper's Bazaar*, September 9, 1983, as quoted in Russell Lewis, "Everything under One Roof: World's Fairs and Department Stores in Paris and Chicago," *Chicago History* 12 (Fall 1983): 44. Also see *Electrical Engineering* 1 (1893); John P. Barrett, *Electricity at the Columbian Exposition* (Chicago, 1894); and David F. Burg, *Chicago's White City of 1893* (Lexington, Ky., 1976).

2. John F. Kasson, *Civilizing the Machine—Technology and Republican Values in America, 1776–1900* (New York, 1976); William R. Leach, "Transformations in a Culture of Consumption: Women and Department Stores, 1890–1925," *Journal of American History* 71 (September 1984): 319–342; T. J. Lears, *No Place for Grace—Antimodernism and the Transformation of American Culture, 1880–1920* (New York, 1981); David Noble, *The Progressive Mind, 1890–1917*, rev. ed. (Minneapolis, 1981), chaps. 1, 3.

3. For perspectives on the crisis of the late nineteenth century, see John Higham, "The Reorientation of American Culture in the 1890s," in *Writing American History—Essays on Modern Scholarship* (Bloomington, Ind., 1970), pp. 73–102; Robert H. Wiebe, *The Search for Order, 1877–1920* (New York, 1967). For Edison's symbolic role, see Wyn Wachhorst, *Thomas Alva Edison: An American Myth* (Cambridge, Mass., 1981). For early experiments in Chicago, see Commonwealth Edison Company Library, Historical Files, Files 5–6. (Hereafter cited as CEC-LF, F[ile] number-E[nvelope] number.) The files contain newspaper clippings, trade journal reprints, and reminiscences of several key participants. Also consider A. T. Andreas, *History of Chicago*, 3 vols., repr. ed. (New York, 1975), 3:596–600.

4. Forrest McDonald, *Insull* (Chicago, 1962), chaps. 1–2; Samuel Insull, "Memoirs of Samuel Insull," Samuel Insull Papers, Loyola University of Chicago.

5. CEC-LF, F5–6; "Files of Predecessor Companies," Commonwealth Edison Company, Historical Archives, Chicago, boxes 1–3, 5 (hereafter cited as CEC-HA, box number); W. C. Jenkins, *Chicago's Marvelous Electrical Development* (Boston, 1911).

6. Thomas P. Hughes, *Networks of Power, Electrification in Western Society, 1880–1930* (Baltimore, 1983); Harold C. Passer, *The Electrical Manufacturers, 1875–1900* (Cambridge, Mass., 1953); W. Paul Strassmann, *Risk and Technological Innovation* (Ithaca, 1956), pp. 158–183. For the development of electric lighting, see Arthur A. Bright, Jr., *The Electric-Lamp Industry* (New York, 1949); and Robert Friedel and Paul Israel with Bernard S. Finn, *Edison's Electric Light—Biography of an Invention* (New Brunswick, N.J., 1986). Indicative of the small-scale nature of the first generation technology, the Edison "Jumbo" dynamo had a capacity of 700 16-candle-power lamps. See Payson Jones, *A Power History of the Consolidated Edison System, 1878–1900* (New York, 1940), pp. 156–157.

7. See note 5; Andreas, *History of Chicago*, 3:596–597; and "Report of Chicago Edison Company to the Stockholders, January 8, 1889," CEC-HA, box 253, for a list of "isolated plants." On the introduction of electricity in the outlying neighborhoods of Chicago, see Ann Durkin Keating, "The Role of Real Estate Subdividers and Developers in the Nineteenth Century Service Provision: The Case of Chicago," paper presented at the meeting of the Social Science History Association, October, 1983.

8. On the development of housekeeping technologies, see Siegfried Giedion, *Mechanization Takes Command: A Contribution to Anonymous History* (New York, 1969 [1948]); Dolores Hayden, *The Grand Domestic Revolution: A History of Feminist Designs for American Homes, Neighborhoods, and Cities* (Cambridge, Mass., 1981); and Steven Davis, "'Of the Class Denominated Princely': The Tremont House Hotel," *Chicago History* 11 (Spring 1982): 26–36. On the importance of the Loop as an entertainment district, see Lewis A. Erenberg, "Ain't We Got Fun?" ibid., 14 (Winter 1985–1986): 4–21.

9. For an introduction to the electrification of the suburbs, see Imogene Whetstone, "Historical Factors in the Development of Northern Illinois and Its Utilities" (typescript, Chicago, 1928), in CEC-HA, box 9003. The records of the predecessor companies of the Public Service Company of Northern Illinois (PSCNI) comprise a large volume of material in the Commonwealth Edison Company Historical Archives. For an introduction to these materials, see PSCNI, "Important Papers Files," in CEC-HA, boxes 4205–4215. For perspectives on the suburbs, see Michael Ebner, "The Result of Honest Hard Work: Creating a Suburban Ethos for Evanston," *Chicago History* 13 (Summer 1984): 48–65; and Kenneth T. Jackson, *Crabgrass Frontier—The Suburbanization of the United States* (New York, 1985).

10. Whetstone, "Historical Factors."

11. Insull, "Memoirs," pp. 70–73; Insull, letter to C. A. Coffin, president of General Electric Company, August 22, 1892, "Chicago Arc Light and Power

Company (CALPC) Files," CEC-HA, box 2. In another memo on the acquisition of the downtown rival, Insull reveals that he wanted the "exclusive central station rights in Chicago." See ibid. Also see "CALPC Files," CEC-LF, F6–E1. The franchise of the Chicago Edison Company was restricted to the central business district. The Commonwealth Electric Company had a citywide franchise. See Bion J. Arnold and William Carroll, "Report on the Committee on Gas, Oil, and Electric Light to the City Council of Chicago," City of Chicago, *Proceedings of the City Council*, March 26, 1906, pp. 3225–3232. Also see McDonald, *Insull*, pp. 88–90.

12. Insull, "Memoirs," pp. 70–78; Hughes, *Networks of Power*, chap. 8; McDonald, *Insull*, pp. 66–74, 90–101; H. A. Seymour, "History of Commonwealth Edison [1934]," CEC-HA, box 9001; Sargent and Lundy, "The Sargent and Lundy Story," Chicago, 1960 (?) (ms. in author's possession). Cf. Sam Bass Warner, Jr. "Technology and Its Impact on Urban Culture 1889–1937," paper presented at the Lowell Conference on Industrial History, Lowell, Mass., 1983.

13. Arthur Wright, "Profitable Extensions of Electric Supply Stations," National Electric Light Association (NELA), *Proceedings*, 1897, pp. 159–189; "Debate on Rate Structures," ibid., 1898, pp. 67–91; Insull, "Presidential Address to the NELA, 1898," in William E. Keily, ed., *Central-Station Electric Service . . . The Public Addresses of Samuel Insull* (Chicago, 1915), pp. 48–53; Insull, "Address before the Eighteenth Annual Convention of the Association of Edison Illuminating Companies (AEIC)," in ibid., pp. 1–7; Insull, "Address at Purdue University," in ibid., pp. 8–33; Ernest F. Smith, "Development . . . of the Substations and Distributing Systems of the Chicago Edison Company and the Commonwealth Electric Company," *Journal of the Western Society of Engineers* 8 (March–April 1903): 209–222. For the cut in rates, see George E. McKena, "Significance of Statistics," NELA, *Proceedings* 2 (1910): 293.

14. Samuel Insull, "Address to the Y.M.C.A.," in Keily, ed., *Central-Station Service*, p. 452; Wright, "Profitable Extensions, p. 162; McKena, "Statistics," pp. 291–305; H. B. Gear, "Diversity Factor in the Distribution of Electric Light and Power," *Journal of the Western Society of Engineers* 15 (September–October 1910): 572–586. Also consider Thomas P. Hughes, "The Electrification of America: The System Builders," *Technology and Culture* 20 (January 1979): 139–153; Seymour, "History of Commonwealth Edison," p. 321. For the quotation, see Insull, "Address before the Western Society of Engineers," in William E. Keily, ed., *Public Utilities in Modern Life—Selected Speeches by Samuel Insull* (Chicago, 1924), p. 383. Also see the memoir of the chief statistician, E. J. Fowler, "The History of the Statistical Department," in CEC-LF, F23–E2.

15. Insull, "Address to the AEIC," in Keily, ed., *Central-Station Service*, p. 81, for the quotation; ibid., pp. 73–83; Insull, "Presidential Address," in ibid., pp. 48–53. Also consider the speeches of Insull's chief salesman, John F. Gilchrist, *Public Utility Subjects*, 2 vols. (Chicago, 1940) (available in the CEC-Library). On the leadership of Insull's companies, see "The Lighting System of the Chicago Edison Company," *Electrical Review* 40 (January 1902): 42–51;

"Electrical Service in Chicago," *Electrical World* 61 (May 1913): 1137–1145; "Electrical Service in Chicago Suburbs," ibid., pp. 1243–1254; Jenkins, *Electrical Development*; Hughes, *Networks of Power*, chap. 8.

16. Seymour, "History of Commonwealth Edison," pp. 77–80; Meetings of the Board of Directors, Commonwealth Edison Company, CEC-HA, box 300. A review of these meetings helps clarify the process of decision making. Insull secured the power contracts first as a prerequisite to financing and actually building additional generator capacity. On the politics of transit policy, see Paul Barrett, *The Automobile and Urban Transit—The Formation of Public Policy in Chicago 1900–1930* (Philadelphia, 1983), pp. 91–99.

17. Insull, "Memoirs," pp. 74–78; Seymour, "History of Commonwealth Edison," pp. 340–344; "Sargent and Lundy Story," chap. 2; Hughes, *Networks of Power*, pp. 209–214.

18. "The Commonwealth Edison Company," *Electrical Review* 52 (January 11, 1908): 65–69; "The Commonwealth Edison Company," ibid., 52 (May 16, 1908): 756–762; "New Features of Central-Station Service in Chicago," ibid., 54 (May 29, 1909): 968–976.

19. See John Gilchrist, "Campaigning for Business," paper presented to the meeting of the AEIC, 1902, in Gilchrist, *Public Utility Subjects*, 1: 1–20; E. W. Lloyd, *Notes on the Sale of Electric Power* (Chicago, 1906), in CEC-LF, F4–E2. A complete set of *The Electric City* is found in CEC-LF.

20. On the day-to-day process of electrification, the best sources are the trade journals and periodicals, such as *The Electric City* (note 19) and *Electric World*. The latter was published in Chicago and represents a virtual newspaper of electrical developments in the area. On the impacts of the tungsten bulb on the commercial sector, see Seymour, "History of Commonwealth Edison," pp. 86–90; the recollections of special lighting agent Oliver R. Hogue, "Store Lighting," *The Electric City* 10 (January 1913); 8–10; *Chicago Tribune*, May 2, 1907, and May 5, 1909; E. W. Lloyd, "Compilation of Load Factors," NELA, *Proceedings* 2 (1909): 586–601; Insull, "Address to the Y.M.C.A.," pp. 445–475; Keating, *Lamps*, pp. 61–73; and John Winthrop Hammond, *Men and Volts: The Story of General Electric* (New York, 1941), pp. 333–336, who concludes that "it [the tungsten filament] was the greatest step made in incandescent electric lamps" (p. 336). For perspectives on the nickelodeon craze, see Jane Addams, *The Spirit of Youth and the City Streets* (New York, 1909); Robert Sklar, *Movie-Made America: A Cultural History of American Movies* (New York, 1975); and Kathleen D. McCarthy, "Nickel Vice and Nickel Virtue: Movie Censorship in Chicago, 1907–1915," *Journal of Popular Film and Television* 5 (1976): 37–55.

21. "Electrical and Mechanical Equipment of the Blackstone Hotel, Chicago," *Electrical World* 54 (November 1909): 1099–1101; for the Palmer House and Plaza Hotel, see ibid., 54 (September 1909): 723.

22. For a detailed statistical study of residential electrification, see Commonwealth Edison Company, Departmental Correspondence, E. Fowler to J. Gilchrist, December 4, 1916, in CEC-LF, F23–E4. In 1908, Chicagoans living in

apartments paid an average of $1.59 per month for service, while homeowners paid an average of $3.02 monthly. The difference in bills reflected a larger number of rooms in houses compared to apartments, not greater daily use. Also see Insull, "Address to the Y.M.C.A.," pp. 445–475.

23. On the limited use of electricity in the home, see Henry B. Fuller, *The Cliff Dwellers* (Ridgewood, N. J., 1968 [1893]), p. 134; Anna Leach, "Science in the Model Kitchen," *Cosmopolitan* 27 (May 1899): 96; Gear, "Diversity Factor," p. 582. For modern perspectives, see Charles A. Thrall, "The Conservative Use of Modern Household Technology," *Technology and Culture* 23 (April 1982): 175–194; and Joann Vanek, "Household Technology and Social Status: Rising Living Standards and Status and Residential Differences in Housework," ibid., 19 (July 1978): 361–375. On the sales campaign, see *The Electric City* 1 (April 1903): 5; ibid., 3 (October 1903): 3; Commonwealth Edison Company, "Reports of the Advisory Committee," October 19, 1908, CEC-HA, box 34; and NELA, *The Electrical Solicitor's Handbook* (New York, 1909).

24. Richard DuBoff, "The Introduction of Electric Power in American Manufacturing," *Economic History Review* 20 (December 1967): 510–513; Warren D. Devine, "From Shafts to Wires: Historical Perspectives on Electrification," *Journal of Economic History* 43 (June 1983): 347–372. For a list of Insull's industrial customers, see "Electric Service in Chicago," pp. 1144–1145.

25. Seymour, "History of Commonwealth Edison," appendix, pp. 170–178; Commonwealth Edison Company, "Report of the Statistical Department," March 10, 1926, CEC-LF, F23–E3.

26. See note 25; Gwendolyn Wright, *Building the Dream: A Social History of Housing in America* (New York, 1981), pp. 153–214; Keating, "Real Estate Subdividers"; Ann Durkin Keating, "From City to Metropolis: Infrastructure and Residential Growth in Chicago," *Essays in Public Works History* 14 (December 1985): 3–27; Marc A. Weiss, "Planning Subdivisions: Community Builders and Urban Planners in the Early Twentieth Century," ibid., 15 (September 1987): 21–37.

27. PSCNI, *Important Papers of the North Shore Electric Company*, 3 vols. (Chicago, 1902–1911); PSCNI, *Important Papers of the PSCNI*, vol. 1 (Chicago, 1916), CEC-HA, box 4045; Whetstone, "Historical Factors"; "Electrical Service in Chicago Suburbs," pp. 1243–1254.

28. Gear, "Diversity Factor," pp. 572–586; "Electric Shops," CEC-LF, F4; Ernest A. Edkins, "An Idea Journey through a Modern Electric Shop," *Electrical Merchandise* 16 (July 1916): 9–13.

29. Seymour, "History of Commonwealth Edison," appendix, p. 171; [J. F. Rice], *History of Commonwealth Edison* (Chicago, 1954), pp. 72–77.

30. North Shore Electric Company, "Corporate Record Books," CEC-HA, box 4211; North Shore Electric Company, "Balance Sheets," ibid., box 4222. On the experiments in rural electrification, see Insull, "The Production and Distribution of Energy," paper presented at the Franklin Institute, Philadelphia, 1913, in Keily, ed., *Central-Station Service*, pp. 357–389.

31. See note 30.

32. PSCNI, *Annual Report* (1912); "Electric Service in Chicago Suburbs," pp. 1243–1254; Whetstone, "Historical Factors," pp. 141–174; U.S., Department of Commerce, Bureau of the Census, *Fourteenth Census of the United States, 1920: Population* (Washington, D.C., 1921), 1:394.

33. John C. Learned, "10,211 Wiring Contracts," *Electrical Merchandise* 15 (March 1916): 87–88. Also consider Fred H. Scheel, "Retailing Electric Washers—By the Carload," ibid., 13 (October 1914): 259–260; "Furnishing Power for Home Conveniences," newspaper advertisement in 1924 Scrapbook of the PSCNI, CEC-HA, box 4182.

34. The trade association of the gaslight industry is an invaluable source of information on the energy systems of cities and the public utility industry. See, for example, Thomas Turner, "Presidential Address," American Gas-Light Association, *Proceedings*, 8 (1888): 201–208; Alpheus B. Slater, "Presidential Address," ibid., 9 (1889): 16–33; Emerson McMillin, "Presidential Address," ibid. (1890): 239–246.

35. Robert C. Twombly, "Saving the Family: Middle Class Attraction to Wright's Prairie House, 1901–1909," *American Quarterly* 27 (March 1975): 57–72; Harold Allen Brooks, *The Prairie School: Frank Lloyd Wright and His Midwest Contemporaries* (Toronto, 1975); Gwendolyn Wright, *Moralism and the Model Home* (Chicago, 1980). For the lighting deal, see Commonwealth Edison Company, "Reports of the Advisory Committee," September–October 1912, CEC-HA, box 34; and compare with Learned, "10,211 Contracts," pp. 87–88.

36. On the use of appliances, see H. B. Gear, "A Survey of Electrical Appliances Used in Chicago Residences," paper presented at the annual meeting of the Illinois State Electric Association (1923), in CEC-LF, F4A–E2; "Survey of Residence Electrical Installations," report presented at the annual meeting of the AEIC, *Proceedings* (1923):609–637; Commonwealth Edison Company, *Yearbook* (1927), p. 22. On the persistence of the bulb-type socket, see Fred E. H. Schroeder, "More 'Small Things Forgotten': Domestic Electric Plugs and Receptacles, 1881–1931," *Technology and Culture* 27 (July 1986): 525–543.

37. Commonwealth Edison Company, Statistical Department, "Kilowatt Hours Sold per Capita [1896–1926]," February 10, 1927, CEC-LF, F23–E3; "Refrigeration," CEC-LF, F4–E2; Oscar E. Anderson, *Refrigeration in America: A History of a New Technology and Its Impact* (Princeton, N.J., 1953); Mary Y. Kujovich, "The Refrigerator Car and the Growth of the American Dressed Beef Industry," *Business History Review* 44 (Winter 1970): 460–482; Georg Borgstrom, "Food Processing and Packaging," in *Technology in Western Civilization*, ed. Melvin Kranzberg and Carroll W. Pursell, Jr., 2 vols. (New York, 1967), 2:386–402. For the best insight on the changing American diet, see Robert S. Lynd and Helen Merrell Lynd, *Middletown: A Study of Modern American Culture* (New York, 1929), pp. 153–157.

38. Seymour, "History of Commonwealth Edison," appendix, pp. 82, 171, 176.

39. "Symposium on Merchandise and Service of the Commonwealth Edison Company," paper presented at the Illinois State Electric Association, Chicago, 1919, p. 6, CEC-LF, F4B–E4; "The Cry for Electric Steel—and Why," *The Electric City* 15 (July 1917): 7–9, 20; "Wholesale Power Load Shows Rapid Growth," *Edison Round Table* 9 (August 31, 1925), in CEC-Library. The best guide to the growth of the use of industrial power in Chicago during the 1920s is Commonwealth Edison Company, *Yearbook* (1922–1930), in CEC-HA, box 113.

40. Dr. Stephen Meyers, University of Wisconsin, Parkside, deserves credit for highlighting the relationship between industrial methods and community change outside the factory. Also see David A. Houndshell, *From the American System to Mass Production* (Baltimore, 1984). For contemporary perspectives on the links between the factory and working-class culture, see American Academy of Political and Social Sciences, *Annals* 117 (March 1925).

41. See note 38; Chicago Plan Commission, *Residential Chicago*, 2 vols. (Chicago, 1943); Carl W. Condit, *Chicago 1910–1929* (Chicago, 1973); Celia Hilliard, " 'Rent Reasonable to Right Parties': Gold Coast Apartment Buildings, 1902–1929," *Chicago History* 8 (Summer 1979): 66–77; Daniel J. Prosser, "Chicago and the Bungalow Boom of the 1920s," ibid., 10 (Summer 1981): 86–95.

42. Ruth S. Cowan, *More Work For Mother: The Ironies of Household Technology from the Open Hearth to the Microwave* (New York: 1983); David M. Katzman, *Seven Days a Week: Women and Domestic Service in Industrializing America* (New York: 1978).

43. A. M. Daniels, "Electric Light and Power in the Farm House," U.S., Department of Agriculture, *Yearbook* (1919) (Washington, D.C. 1920), pp. 223, 223–238; D. Clayton Brown, *Electricity for Rural America: The Fight for the REA* (Westport, Conn., 1980), p. 7; Richard A. Pence, *The Next Greatest Thing* (Washington, D.C., 1984). For insight on the urban–rural conflict of the 1920s, see Don S. Kirschner, *City and Country: Rural Responses to Urbanization in the 1920s* (Westport, Conn., 1970).

44. PSCNI, *Yearbook* (1928), pp. 21–22; ibid., 1929, pp. 12–14.

45. Samuel Insull, as quoted in Commonwealth Edison Company, *Annual Report* (1925), p. 9; *Public Opinion and the Peoples Gas Light and Coke Company and the Commonwealth Edison Company* (Chicago, 1935), in CEC-HA, box 9000. See Lynd and Lynd, *Middletown*, chap. 18; Kirschner, *City and Country*, pp. 70–74, 244–250; and James E. Evans, *Prairie Farmer and WLS: The Burridge D. Butler Years* (Urbana, Ill., 1969).

46. Seymour, "History of Commonwealth Edison," appendix, p. 81; Commonwealth Edison Company, *Annual Report* (1940), p. 14; Commonwealth Edison Company, Comptroller Department, "Comparative Statement of Residential Electric Service Sales for the Year Ending December 1941, and December 1940" (photocopy in author's possession).

13 Municipalities as Managers: Heat Networks in Germany

Roselyne Messager

THE most prominent networks in the field of energy are the gas and electricity supply networks. This is logical, as these are nationally developed networks and transport the most important energy supplies—electricity and gas—consumed within European society. But there are also networks, known as district heating systems, for the distribution of heat that were constructed in the past and that have been revitalized since the first oil crisis. The operation of these systems is completely different from that of electricity and gas. Heat networks exist only at the city level; they have no national interconnections, and the municipality is in charge of both the investment and management of the operation.

This study focuses on the evolution of urban heating in Germany, although it also provides some comparisons with the French experience. The development of urban heating networks has been more successful in Germany than in France, primarily because the administrative and institutional environment in German cities has allowed them to develop systems of *cogeneration* (combined generation of electricity and heat) more readily than in French cities. In Germany, cogeneration is the dominating technology, but aside from two exceptions (Paris and Grenoble), it is not utilized in France. This chapter explores the institutional milieu in Germany as a means of explaining this difference.

The Technology and Characteristics of Heat Networks

The technology of urban heat networks is similar to that of central heating, but on the scale of a city rather than that of a building. The various parts of the system include a heat producer, a distribution system, and consumers. Heat producers vary considerably. They include traditional boilers burning various fuels, incinerator plants for urban wastes, geothermal sources, or a combination of electrical and heat production (cogeneration). Cogeneration is the use of the heat normally lost during the process of electrical production for urban heating purposes. These systems usually use either gas turbines, combustion engines, or steam engines for heat generation. The latter include counterpressure turbines and extraction turbines.

In December 1982, there were 116 plants having a thermal power of 29,320 MJs or thermal MW in Germany; in France (where data are more imprecise), network power was estimated at 16,400 MW, or barely half as much. Paris has the largest network (the CPCU) of any city in Europe, but, in general, German networks are larger, with the twenty largest German cities all equipped with at least one heat network and occasionally two or three. Nineteen German towns of under 50,000 inhabitants have heat networks, whereas in France this population size is considered the threshold below which it is no longer considered profitable to install a network.

In Germany, 73 percent of the heat delivered in urban networks is derived from cogeneration plants and only 27 percent from traditional boilers used for heat alone. In France, only two companies use cogeneration: the company operated by the city of Metz and the CPCU through the TIRU in Paris. The explanation for this critical difference lies in the different institutional environment.

German Institutional and Administrative Structures

The primary characteristics of the institutional structure of the German energy sector are its decentralized character and its relationship to the local community. Considerable control by German municipalities over municipal functions and services explains the rapid progress in the development of urban heat networks in that country.

The German electrical sector—in contrast to that of France, where one national company provides electricity—has a multitude of companies of all forms and sizes. At the highest level, the large companies mainly produce electricity and supply it to other companies in the network for distribution. At a second level, regional companies supply industries directly or communal companies have their own delivery networks. Finally, a very large number of local companies produce and distribute their own power or buy power from other suppliers and then redistribute it.

A survey of German urban heating companies shows that 80 percent of them are owned by the public and operated at the municipal rather than the Bund (federal state) or Länder (federated states) level.[1] Many municipalities operate systems themselves, producing electricity and heat. In other situations, single municipalities or groups of municipalities own shares in energy companies and influence them in this manner. There are also situations where municipalities purchase their heat from privately owned firms.

For instance, the largest urban heating system in the country, located in Hamburg, is operated by an electrical company (HEW) of the first or primary generating level. It produces all its heat by cogeneration. Hamburg owns 73 percent of the assets of this company. The nation's second largest urban heating company, BEWAG, which supplies Berlin, is also a large electricity-producing company. In 1977 the Länd of Berlin controlled 58.3 percent of the assets of this firm.[2]

Another exception is in Dortmund, where the heat network is operated by the second largest electrical company, VEW (Vereinigte Elektrizitatwerke Westfalen). The assets of this company, however, are mainly held by diverse local collectivities rather than private investors.[3] In contrast, in Essen, where the fourth largest urban heat network is located, the electrical company that supplies the heat is controlled by private industry. And in Gelsenkirchen, with the twentieth largest heat network, the heat is supplied by VEBA Krafte Ruhg, a large regional electrical company.

These pecularities, especially the large amount of municipal control of cogeneration and heat networks, is explained largely by the ideological climate present in German municipalities at the beginning of the twentieth century rather than by any shortage of energy. France, as well as Germany, lacked energy, but France did not develop this form of energy production. The key questions that remain to be answered, therefore, are why municipalities are so involved in the production of power and heat in Germany, and why private electrical corporations are involved in heat production?

The Formation of the German Electric Industry and the Role of Local Communities

The earliest applications of discoveries concerning the production of electricity were made at the end of the nineteenth century in a strictly industrial context: Companies whose production processes required a supply of electricity produced it themselves on the premises. Later, small private firms involved in the production of electrical equipment, wishing to find a market, invested in the production and distribution of electrical energy. This was the case, for example, with Siemens and Halske in 1879 and in 1884 in Berlin with Edison, which in 1912 became the AEG (Allgemeine Elektrizitats Gesellschaft) corporation.[4]

Some of these small companies are the ancestors of the gigantic and structurally complex corporations that dominate the electric industry today: corporations whose high-tension and ultra-high-tension distribution networks are tied together to form the interconnected network Verbund, which in turn is connected with national networks of neighboring countries.[5]

Thus, from the end of the nineteenth century to the early twentieth century, many local thermal power plants were created on the initiative of private capital. Thanks to rapid progress in the long-distance conduction of electricity, the industry expanded through the takeover of small companies by larger ones, establishing power plants at the extraction sites of the most profitable fuels and increasing the supply area.

German authorities opposed this monopolizing tendency on the part of private capital, and the early twentieth century witnessed the transfer to civil communities of the property of many industrial corporations.

Community motivation in becoming involved in electrical production, and the manner in which communities proceeded, allows us to understand why heat distribution later became a secondary activity. This essay describes the ideological context, the association between public and private capital in the electric industry, and the structure of the fledgling electric industry at the beginning of the twentieth century.

In the years preceding World War I, German economic life was far more marked than that of other countries by the influence of community economics, which considerably surpassed the scope of the socialist parties.[6]

Historically, municipalities had always owned community enterprises; that is, municipal institutions supplied the inhabitants with the amenities of life: sanitation and hygienic services, public security, educational services, etc. The end of the nineteenth and the beginning of the twentieth centuries brought considerable extension of these institutions. Municipalities then created "economic enterprises" that had their own legal characteristics and intervened in activities that had originally been the concern of private capital: the distribution of water, gas, and electricity, and public transportation.[7]

Such activities provoked vigorous debate between partisans of direct community management of economic concerns of public interest (production and distribution of gas, electricity, etc.) and partisans of private enterprise.[8] The legal and institutional problems generated by these debates centered on two questions. First, should the development of industries involved with public goods and the public interest, as well as catalysts to economic activity (such as gas and electricity), be left in the hands of private capital, risking the formation of cartels, or should they be placed in the hands of public authorities through municipal takeover in order to assure an equitable rate structure?

The second question involved the pricing policies that the community-run industrial services should follow. More precisely, since the community was becoming tacitly involved in selling services at prices lower than or equal to those of private enterprise, should the management of those utilities be based on principles prevailing in community sanitary authorities, which permitted deficits, or should it be a means of assuring revenue for the city?

For historical reasons as well as reasons of purely internal politics, the "municipal socialism" movement was particularly strong in Germany, and relations between the public and private economy were somewhat different than in France.

The strength of local ideology, made concrete by the adoption of legal opinions facilitating the entrance of municipalities into industrial activity, strongly influenced the organization of the German electrical industry. Thus, early in the twentieth century, besides the industrial firms, many strictly local developments were created. But particular to Germany was a mixed form of private–public development. In the light of the ideological context presented here, it is possible to understand the process of municipal intervention in the electric industry, the form of this intervention, and its consequences.

The first example of important municipal intervention in the electric industry was that of the Rheinische Westfalisches Elektrizitatswerke (RWE). That agreement created a sensation in the business world and later was widely imitated, so it would seem particularly interesting to retrace the history of the motivation for association between public capital and private capital within RWE.[9]

In 1898, the electric firm of W. Lahmeyer signed a contract with the city of Essen whereby the company was committed to build an electric power plant within the community; for its part, the city committed itself to concede to the company the right to use the public highways for forty years. A few months later, Lahmeyer and several bankers founded RWE, which took over the contract. The *Oberburgermeister* (equivalent to a mayor) of Essen and the industrial giant of Mulheim-Ruhr, Hugo Stinnes, among others, joined the supervisory council of this new corporation.

The power plant was built in 1900, in direct conjunction with a coal mine owned by Stinnes. Lahmeyer and his group then sold their shares to H. Stinnes and A. Thyssen. The new supervisory council, consisting of these two great industrialists from the Ruhr and the mayor of Essen, envisaged a great expansion of the company; specifically, to supply the entire Rhenish–Westphalin industrial region with electric power.

Accomplishing these great plans required a considerable sum of capital, which was found by issuing bonds and negotiating long-term loans from certain kreisen (administrative territories)[10] and communities. In this way, the RWE saw an explosive growth through 1910 by buying shares of electric plants in neighboring localities.

Then, as rival firms (particularly AEG and Siemens) and communities felt threatened, public opinion and the press were unleashed against this "electricity monopoly in the hands of Stinnes." Agreements were made to limit the respective local supply areas of the partners.

After 1910, RWE continued its expansion at a more moderate pace, including the local communities in its development. At that time, the company and neighboring communities signed a contract typical of the new economic situation, wherein local governments associated themselves with ordinary citizens from a position of quasi equality. The participation of the kreisen in the financing of some proposed projects was mentioned in the contract.

Numerous other communities joined with private companies and, following the example of RWE, founded mixed-economy corporations. Their motivation was to avoid placing an activity considered by the communities to be a public utility into the hands of private interests, with their increasing propensity for merger. Such organizations seemed an excellent means for communities to impose their demands on the general policies of the company while preserving the competence and efficiency of the kind of management found in private enterprise.

For their part, private companies were quite satisfied with this new arrangement for two reasons: (1) It greatly facilitated the financing of their expansion or initial construction, since the communities, kreisen, or other public entity could both assure loans and purchase shares in the firm much more easily than private individuals; (2) the industrialists often acquired greater influence in the management of the corporation than they would have had with only an investment of capital. This fact seemed to be a drawback neither to the community nor to the industrialists because, in the minds of both parties, "the interplay of private initiatives eased the burden that would have resulted from overly cautious management by administrative organs and of their insufficient adaptation to commercial methods."[11]

TABLE 13.1.
Electric Companies in Germany, by Type of Corporation
and Composition of Capital, 1913

	Power	Percentage of Total Power
Plants entirely owned by communities	673 MW	47.3
Plants operating under a mixed economy	380 MW	26.7
Plants entirely owned by private companies	370 MW	26.0

Thus, whether they built power plants themselves or became associated with private firms, German communities strongly dominated the electric industry at the beginning of the twentieth century. The extent of community participation in the genesis of the electric industry is shown in Table 13.1.

The formation of the German electric industry is characterized, then, by the fact that, from its outset, local communities have participated in the rapid development and management of electric companies. It is also interesting to note that the legal form of this cooperation did not change the kind of management because the companies retained their corporate status and operated as such. Moreover, when the electric companies belonged entirely to the community, a legal framework based on the principles of industrial management was established in order to permit efficient operation.

Once the local authorities were established in the electric industry, the idea of producing and distributing heat from the municipal plants seemed completely natural.

Expansion of the Role of Municipalities in the Distribution of Heat

The first heat networks were supplied by steam recovered from thermal power plants. It is, in fact, easy to understand the idea of using steam wasted in the production of electricity to heat public buildings. Urban heating largely owed its genesis to an ideological context that prompted local authorities to intervene in every economic and industrial domain of a public utility. Although it is not easy to define exactly what a "public utility" is, the distribution of heat exhibited all the characteristics of one, as did the distribution of water, gas, and electricity.

Moreover, the argument used by communities establishing themselves in this endeavor was that they would save large amounts of fuel. This was a decisive argument at a time (during and after World War I) when cities were very much preoccupied with supplying their inhabitants with wood or coal for heat.

But beyond this, some saw heating as an operation that allowed municipalities to resist the phenomenon of consolidation that characterized the electric industry at the time.[12] For reasons of economy, large power plants were replacing or downgrading smaller ones. Anxious to stay in business, firms in which communities had an interest saw the production and distribution of heat as a means of preserving their usefulness and remaining established in the community.

Statistical confirmation of these hypotheses is not available for the first half of the century, but it is confirmed by data from the post–World War II period. The statistics illustrate the dominance of electric plants in the field of urban heating: In 1950, the twenty companies concerned with the production and distribution of heat operated thirty-one production units, twenty-two of which produced both electricity and heat and only nine of which were solely for the production of heat.[13]

The statistics clearly show that whether out of concern for energy savings or a deeper fear of losing control on the local level, communities already involved in the production and distribution of electricity expanded their conception of a public utility in order to include heat production. When they did not themselves manage these enterprises, they formed ties with industrial entrepreneurs in mixed-economy corporations in order to smooth the way for the new heat-production process.

This explanation is much more satisfactory for the evolution of urban heating up to 1958–1959 than for the period that followed; indeed, other reasons must be given for the growth of urban heating after that time. Paradoxically, during the years of inexpensive energy resulting from the massive importation of oil and gas, urban heating maintained its pace of development. The motivation here, especially after 1958, was the state's desire to support the coal industry during the crisis created by competition from new and cheaper sources of energy. One method of support was to provide large financial subsidies for the construction of coal-burning district heating plants.

The long-distance distribution of heat, therefore, appeared in the first half of the twentieth century as a community venture and was carried out primarily by municipal electric plants; then, over time, the method of producing heat diversified to include plants producing only heat, fueled primarily by coal. The pattern of urban heating operations in Germany today is still highly influenced by these origins.

This analysis of the genesis of the German electric industry allows us to propose a hypothesis that suggests reasons why heat and electricity production techniques became established. That is, the choice of a technology cannot be analyzed outside its institutional context. As for the combined production, it seems clear that its existence resulted from the perception of municipal enterprises of their mission to provide heat as a public service and to the general economic goal of providing the land with energy. In the end, the power accorded to the communities by the state explains the relative importance of the heat energy sector.

Municipal Power and Its Consequences for Urban Heating

In this section, the long-distance production and distribution of heat are placed in an institutional context, that is, into the body of regulations and structures that determine the scope of energy-related operations.

The administrative organization of the German energy industry, modeled after the political organization, is such that it creates the means to smooth out considerably the "technicopolitical" obstacles linked with the operation of a heat-power plant. In a more general way, these arrangements also facilitate the status of heat power with respect to competing sectors. The following analysis attempts to show that the Energy Supply Enterprise (*Energieversogungsunternehmen*, or EVU), the organization at the community level that provides the foundation of the country's supply system, is a structure that offers particularly interesting possibilities for the further development of heating networks.

To compensate for the financial hardship of communities during the economic depression between the world wars, a series of legislative measures were taken that had fundamental consequences for community economic development. In 1935, laws for the Regulation of German Communities were approved, covering management principles of community enterprises. In addition, the Energy Law (*Energie virtschafts-gesetz*) of December 13, 1935, was passed, defining the basic principles relating to the economic and administrative organization of the energy industry. These principles are still valid today.

The Energy Law was a response to the question whether energy operations, given the ever increasing role they occupied in economic activity, should be left to the operations of the marketplace or should be put under the supervision of public authorities. The law established the principle that energy distribution is the responsibility of community authorities. It delegated to the *Länder* (regional states) administrative supervisory powers over the communities in their production and distribution of gas and electricity. The law, however, related only to gas and electricity production and did not mention urban heat production. The community was therefore left completely free in its decisions and actions concerning heat.

During the reconstruction period following World War II, Germany's administrative structure was reorganized,[14] resulting in considerable decentralization. The formation of the *Länder* and the transfer of administrative responsibility into the hands of the municipalities seemed to be effective means of generating economic revitalization on a local basis. This same logic, which emphasized efficiency and rationalization, justified the reinforcement of the decentralized concept of energy distribution made explicit by the 1935 law.

Since 1974, the federal government has emphasized the importance of community responsibility in the field of energy. The "federal energy program" (second revision) clearly shows the role the government wants community authorities to play in supplying energy to the nation.[15] It encourages communities to develop a local conception of energy supply in order to secure a rational cooperation between electricity, gas, and the economic potential of cogeneration from both electrical-generating plants and industrial boilers.

The communities must therefore define their local conception of energy needs and supply (*lokal Energieversorgungkonzept*) and set their own goals in the context of the general goals of federal energy policies and the specific requirements of community development. In regard to heat energy, the fact that communities are charged with managing the expansion of gas, electricity, and heating networks is a major element. The primary obstacle to the development of urban heating networks is their lack of immediate profitability.

Because of their capital intensity, these investments will be profitable only in districts with a high density of heat requirement and few competing networks using other sources of energy. The existence in Germany of a municipal policy that reserves a "place of favor" for urban heating in the most advantageous districts is undoubtedly an essential element in explaining its success.

How can the community coordinate the development of different, competing forms of energy? The EVU provides the management structure to facilitate this mission. H. Eitenmeyer defines the EVU in the following manner: "By Energy Supply Enterprise [EVU] is meant enterprises providing the public with electricity, gas, and heat over a long distance; that is, enterprises that, independently of their legal form and independently of the extent of participation of public capital, in response to the Energy Law of December 13, 1935, fulfill their duty by providing energy to the population and the economy."[16] This definition reflects the fact that a great variety of firms can fit it, the only requirement being to deliver gas, heat, or electricity—that is, an energy network. The only common ties among the firms are that they are public utilities and that their concern is with distribution (but not necessarily only with distribution) to the eventual consumer. But, diverse as the firms are, the legal power of community enterprises always allows the community to make its wishes heard on the general policy of the enterprise.[17]

These firms generally function according to a particular type of organization, called Querverbund. (This type of organization does not exist in France.) In the Querverbund, various utilities are grouped under the same administration, although they retain a certain amount of independence. The utilities involved are: water supply distribution, public transportation, gas distribution, electricity distribution, and heat distribution.

A characteristic of the Querverbund is that it offers the possibility of compensation for losses in certain branches of the operation by surpluses realized in others. This fact is important in regard to heating networks because of their limited initial profitability. In this type of enterprise, it becomes possible for the general administration to let the constraint of profitability carry less weight in its decision than if it had to be considered separately. Thus, the Querverbund frees the community seeking to establish a rational policy of energy supply from making decisions solely on the basis of sector-based profitability. Even if the immediate profitability of a heating network is not assured, the community can order its construction in locations that it considers most advantageous.

It also becomes possible for the community to respond to other needs directly tied to energy operations, such as the treatment of household refuse, the abolition of individual heating systems that are polluting, and the recovery of waste industrial heat. Thus the Querverbund offers to a community enterprise the possibility of adjusting the rate structure of related services in order to offer heat to customers at a price that will assure a satisfactory rate of connection into the system, eventually making possible its profitability.

Conclusion

The study of German heating networks illustrates the concept that a technology should not be studied outside the institutional context in which it was established. A historical approach allows one to note the aspects peculiar to the development of urban heating in Germany. Paradoxically, the community's domination over electric networks facilitated the appearance of heating networks; heat-power technology (or cogeneration) is thus at the crossroads of the supply mission incumbent on municipalities. In fact, the key to understanding the German model of development of heating networks involves two points:

1. For ideological reasons, German cities found themselves involved in the production and distribution of energy, and created an industrial type of operation providing for gas and electrical networks. But the tendency toward a communal economy conflicted with the tendency of the electrical industry toward concentration, causing difficulties for small central communities.

2. To reinforce municipal independence, and, to some degree, to resist the tendency toward concentration, some towns extended their functions and produced heat from the municipal power stations. This could be done because the administrative framework of the supply companies made it possible to use capital appropriated for one activity to support another.

Moreover, this extension of municipal activities to include heat lies completely within the realm of functions delegated to towns in 1935 in order to promote a relationship between federal policy and local development. In this way, the energy supply companies are essential factors in the development of urban heating in Germany.

Notes

1. H. Stumpf and E. Orth, "Stand der Fernwarmeversorgung in der BRD, unter dem Gesichtpunkt der Marktgegebenheiten und der Eigentumverhaltnisse," *Fernwarme International*, no. 3, (1974).

2. C. Profrock-Vagneron, *l'Industrie électrique en RFA. Structure et croissance* (Grenoble, 1979).

3. Ibid.

4. J. P. Angelier and E. Lalanne, *Histoire des relations entre l'Etat et l'industrie électrique en Europe* (Grenoble, 1979).

5. It was in this way that W. Lahmeyer and Co., a private corporation, was created in Frankfort in 1898. It rapidly became the RWA (Rheinische West-falisches Elektrizitatswerke), the largest electric corporation of its day, based on the quantity of electricity produced. In the same way, in 1925, a small company was founded in Dortmund; its evolution led the VEW (Vereinigte Elektrizitats-werke Westfalen) to rank second in the same classification.

6. On this subject, see the several articles in the *Annales de la régie directe*, nos. 103–104 (1918), especially "La Construction de l'economie collective."

7. Centre d'études sur les Entreprises Publiques (CEEP), *Les Entreprises publiques dans la CEE* (Paris, 1967).

8. These debates took place during the International Congress on Cities, held in Seville in 1929, and later in the same year in Amsterdam. On this subject, cf. Oskar Mulert, "L'Activote economique des communes allemades," *Annales de l'economie collective*, nos. 252–253 (June–December 1930).

9. All the factors relating to this contract can be found in R. Kaeppelin, "Le Système dit d'économie mixte dans les entreprises publique en Allemagne," *Revue d'économie politique* 34 (1920).

10. *Kreisen* is an administrative territory, larger than the community.

11. Renaud, *Les Entreprises électriques et les collectivites locales* (Bordeaux, 1931).

12. Energiewirtschaftlisches Institut von Koln, *Fernwarme* (March 1980).

13. "Grundung der AGFW bei der VDEW," *Fernwarme Interntional*, no. 1 (1973).

14. A. Grosser, *l'Allemagne de notre temps* (Fayard, 1978).

15. Economic Ministry, Energy Program of the BRD, 2nd revision (Bonn, 1977).

16. Helmut Eitenmeyer, "Marketing in der Energieversorgung," *Fernwarme International*, no. 11, (1975).

17. In the case of corporations, "The manner in which the community is rep-resented at the general meeting of members is determined by the arrangements of the various community regulations, which vary considerably. In general, it is expressly stipulated that community representatives on the main board of the corporation must obey the orders of the community. In this way it is possible for the community to direct these firms, particularly to appoint members of the board and of management and the chairmanship at its convenience."

In the case of autonomous companies controlled or owned by the state, the influence of the community is maintained "as far as is necessary to realize the objectives of community policy."

In the case of local government-controlled enterprises where a small com-munity is involved, the representatives of the community have decisive power over the affairs of the enterprise.

14 Utility Networks and Territory in the Paris Region: The Case of Andresy

Gabriel Dupuy

Most of the research carried out until now on the development of utility networks has been sector based. Researchers have been interested in the distribution of drinking water, in tramway networks, or in electrical utilities. This can be explained by the fact that the development of each kind of network is distinct, the decision makers are often different, and the documentary sources are specialized. Whatever the reason, this approach to infrastructure networks complicates our understanding of the territorial meaning of their expansion. Although any one network is capable of having an effect on urbanization, it is the totality, the combination of several different networks, that corresponds to the new forms of spatial development, redefining the territory and bringing about the transformation of local society.

From this global point of view, I have chosen to study Andresy. For a long time in this community in the Paris region, one concessionaire was responsible for the construction of water, gas, and electric utility networks, extending them gradually to numerous neighboring towns.

The study of the development of these utility networks, simplified by the existence of a single concession, has been carried out using collections of community archives, general historical documents, and, keeping in mind the time period in question, interviews with persons who participated or helped in the development of the utility network.[1]

The limits of such a study are evident: There are geographical limits with respect to studies covering the totality of the region, and documentary limits with respect to approaches using the historical method. Nevertheless, the scale retained permits an in-depth understanding, and the results stress the territorial meaning of utility networks, a factor that has rarely come up in previous studies.

Andresy

The case I have selected to study is Andresy, a community in the Department of Seine and Oise, located approximately 30 kilometers from Paris. A rural town[2] until the middle of the nineteenth century, and relatively isolated, Andresy underwent enormous changes in the next fifty years, changes precipitated by the double influence of inland water navigation and the railroad.

Around 1850, relations between Andresy and Paris were very limited. The Seine and Oise rivers presented formidable obstacles to communication with the capital.[3] But, gradually, links began to form. Starting in 1850, the canalization of the Seine and the Oise meant that their confluence (where Andresy and Conflans-Sainte-Honorine are located) became an important navigation port, linked closely to Paris by the river. Andresy was a stopping point for river traffic going from Paris to the lower Seine and to the north of France; along with Conflans, Andresy constituted a kind of advance port for Paris.[4] Diverse construction and ship-repairing activities, along with supply and administrative functions, began to develop at the same time that a large maritime population, transient or stationary, settled in the town.

In 1892, the Paris–Conflans–Mantes railway reached Andresy, then Triel, thanks to the Eiffel Bridge over the Oise, assuring a good rail connection with the capital. This new railway would have many effects, as we shall see.

The first effects were felt in agriculture: Fruits and early vegetables cultivated in the fertile zones of the loop of the Oise could now be brought more easily and rapidly to Parisian tables. Going in the other direction, "night soil"—a compost of garbage and waste matter—could be evacuated from Paris toward the railway stations of Andresy and its surrounding region to be used as fertilizer, thus further encouraging the development of truck farming.

Finally, the train created a link between Andresy and the communities on the banks of the Seine for Parisian "tourists" who came there to spend Sundays or vacations. Starting at the turn of the century, these towns began to enjoy the development of a tourist industry in the form of hotels, restaurants, rented vacation houses, and especially the construction of second homes for well-to-do Parisians.[5]

Around 1890, another link—an underground one this time—was set up between Carrières-sur-Seine (a town bordering on Andresy) and Paris. A sewage pipeline, completing the one at Achères, continuously discharged wastewater from Paris onto the alluvial plain of Carrières, to be used for the sewage farming of fields. The negative side to this contribution was that the underground water table in certain sections of Carrières and Andresy became contaminated, making it difficult or even impossible to use wells that had ordinarily, until then, supplied drinking water. It was as if the problem of water for Paris, which the authorities had solved by piped-in water and sewer systems, had been put out of sight and out of mind via the pipeline and sent to Carrières, 30 kilometers to the west, forcing Andresy—for better or worse—to enter into the problems of the Parisian ecological system.

These new connections were not the only elements to affect Andresy's relationship with Paris. Naturally, we must also add improvements to the major trunk roads beginning in the early years of the twentieth century; the establishment of a telephone line with Paris (via Saint-Germain-en-Laye) around 1900; the construction, in 1912, of another railroad by the departmental Seine and Oise Railroad, connecting the communities along the banks of the Seine and Oise rivers with the Halles (the marketplace) of Paris[6] (with the same kind of freight traffic as on the Conflans line: "night soil" in one direction, fresh fruits and vegetables in the other); and so on.

Thus, by the end of the nineteenth century, Andresy was no longer a rather autarchic rural community. It was now another element in a much larger totality—the Paris region. This explains its new urban, rather than rural, character: problems of finding suitable drinking water; a permanent population of rural and river inhabitants, but one that was affected by its working relationships with the capital; the seasonal presence of a well-to-do Parisian population wanting to find in Andresy not only vacation pleasures but also all the comforts to which they were accustomed at home.

It must, of course, be stated that such transformations were not unique to Andresy. Nearby towns, located along the banks of the Seine and the Oise, linked to Paris by the Western Railway or by major roads, would be affected by the same changes in their relations with the capital. All these towns would capitalize on the Parisians' need for relaxation (fishing, second homes, etc.) by emphasizing the advantages of the situation of the Hantil, a hill covered with lush vegetation overlooking the Seine and the Oise, and the natural advantages of bodies of water created by the canalization of the river. Projects for new construction of secondary residences habitually encountered the problem of drinking water created by the pollution of the underground water table caused by the practice of using night soil to fertilize the surrounding truck farms.

The Mallet Enterprise and Its Utility Networks

In 1894, "the community of Andresy conceded to Mr. Mallet the rights and exclusive privileges, for sixty years, to set up, keep, and maintain in all the roads and other places located in the territory of said community, the pipes necessary to conduct all the water required for public and private uses."[7]

Mallet then built a pumping plant on the Oise River to serve Andresy and, soon after, in 1896, the neighboring community of Conflans-Sainte-Honorine. The same year, the concessionaire agreed, at the request of the municipality, to take on the task of public lighting in Andresy; at the time, lighting was provided by outdoor oil lamps. In 1897, Mallet began the changeover from oil lamps to gaslights, which he supplied with a new utility network.

The public water supply network and the gas system had originated together in the same place: the pumping plant on the right bank of the Oise. They, in fact, complemented one another. A part of the natural gas produced by the combustion of the coal was sent through the utility pipes, and the rest was used to operate the water-extraction pumps.

During the years that followed, and by means of a modification of conditions outlined in the original concession, the Mallet utility networks would be extended into neighboring communities as the Andresy networks were finished (see Figure 14.1). In the area of public lighting, oil was abandoned in favor of natural gas. In addition, gas lines supplied the fuel for domestic use (e.g., lighting, cooking). The water supply network was more and more in demand; the underground water table was now totally contaminated, and wells were abandoned.

In 1924, after a delay caused by World War I, a decision was made to bring electricity to Andresy and the surrounding region. Two years later, Mallet obtained this new concession and set up a large electricity network, connected to the Parisian electric company Nord-Lumière.

A little later, with increasing pollution, the quality of the Oise had deteriorated to such an extent that treating the water[8] no longer had any effect, and so in 1932 Mallet proposed to the community a supply of artesian well water to be pumped from very deep wells. The distribution of this artesian well water would also be extended to many neighboring communities.

Thus, in just under forty years, numerous public utility networks were installed in Andresy and the surrounding region: gas lighting, pressurized river water, household gas, artesian well water, and electricity (for public lighting, industry, and home utilization).

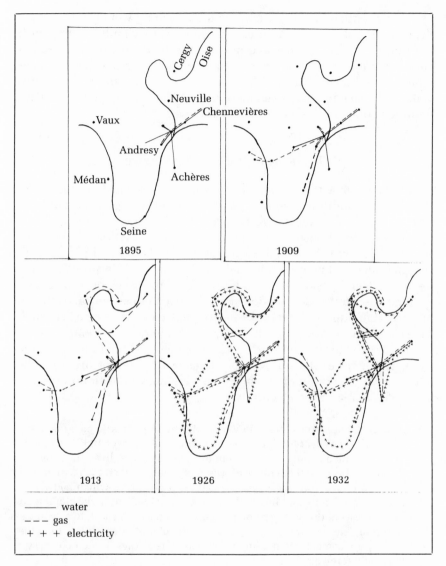

Figure 14.1 Extensions of the Mallet factory network to supply water, gas, and electricity, 1895–1932

For more than half a century, the Mallet family would preside over the destinies of Andresy utilities. The network they built, starting with the pumping plant constructed on the Oise in 1895, would become a veritable public utilities network providing natural gas to sixteen communities and nearly 30,000 inhabitants in the Seine and Oise valleys, water to five communities and more than 14,000 inhabitants, electricity to fifteen communities and 24,000 inhabitants.

The Role of the Concessionaire

When, in 1894, the municipality of Andresy conceded to Paul Mallet the rights to build a pumping plant for water and gas, the city was dealing with a large business enterprise of a dynastic nature.

The founder of the dynasty was a chemist named Alfred Mallet, born in 1813. He invented a process for purifying the natural gas used in lighting, and he knew how to make his patent yield a profit. On his death in 1885, he bequeathed to his son Paul the basis for what would become the Mallet Corporation. Paul Mallet built chemical factories and gas refineries and networks in different parts of France. Through his wife's family, the Cozes, he was tied to another gas-manufacturing corporation; at his death in 1923, he would leave behind him a veritable industrial empire. His son Marcel Mallet would further diversify the activities of the corporation while consolidating its position in the key areas of water, gas, and electricity.

> Among the numerous affairs in which Marcel Mallet participated, none was perhaps as dear to him as the small group at Number 10 Rue de Milan in Paris, which sprang directly from his own impulses, lived the same rhythm as his own thoughts, and was completely marked by his own personality. First among these was the natural gas business which had originated in the family and in which he took special delight in developing; in the interests of this group he dedicated himself, as to a sport, to a "concession hunt," in which considerations of pure interest counted less than the joy of pure combat, often grim, but in which he always wanted courtesy to remain foremost. [9]

The least that can be said is that, with such a person acting as agent for the concession, the size of the little municipality of Andresy hardly mattered. The sixty-year concession gave Andresy an entrepreneur who, with a remarkable sense of continuity, turned his pumping plant and utility networks, into which he had invested his own capital, into a bridgehead for the development of the entire region. The expansion of the Mallet networks would not come to an end until the accidental death of Marcel Mallet in 1944, followed by the nationalization of gas and electricity in 1946.

What primarily interested Paul Mallet was not providing services for the little community of Andresy, but the potential for development he perceived in a region that was slowly transforming itself into a Paris satellite. One episode is quite revealing with respect to this attitude. Less than ten years after obtaining the concession, Paul Mallet began negotiations with the city to obtain the rights to construct gas conduits destined to serve the neighboring towns. Even though this was not one of the provisions figuring in the original contract, he was eventually able to win the case by promising to provide electricity for the community in the future (it would not come about until 1926).

The municipality of Andresy, and very likely other local collectivities in the region, maintained relationships with the Mallet enterprise that were rather complex. The minutes of the Municipal Council mention frequent skirmishes over the price of gas, water, and lighting, as well as the quality of service in one neighborhood or another. But over and above these official, rather firm stands taken by city councillors, one senses a tacit agreement not to abrogate the concession. Even more, the technico-economical "status" of the Mallet enterprise was such that the concessionaire seemed to inspire the deepest respect among city authorities: "The agent for the gas and water concession in Andresy is a hardworking individual in whom one can place the greatest confidence and whose enterprise unites all the possible guarantees of stability and high performance." [10]

For this reason, the enterprise was eventually entrusted with the construction of the other networks. There was a sort of understood delegation of municipal authority in everything having to do with technology; it is understandable that the local government hardly had the means to oppose Mallet in this area. In addition, local resources were never sufficient to allow for large investments. Mallet frequently reminded the mayor of Andresy in letters to him that he had borne, at his own expense, costs that normally should have fallen to the municipality.[11]

The concessionaire knew equally well how to respond to various aspects of local demand. It was not his policy to look for ready-made solutions. Rather, he perfected techniques that were adapted to the particularities of the place in question, techniques that would guarantee him at one and the same time a comfortable profit, acceptance by the local users, and professional prestige. He accomplished his goals with such success that the plant at Andresy rapidly became known as a pilot factory of the Mallet group. Thus, for example, he installed high-powered gas-engine-driven pumps so that the first water distribution network could supply houses in Andresy located at an unusual altitude with respect to the level of the Oise. To cite another example, the pumping plant at Andresy made the most of the coal distillation and natural gas treatment by immediately exploiting the by-products: coke, benzol, road-surfacing material—all of which helped in balancing the finances of the station and at the same time justifying the local basis of the concessionaire, who became a supplier of carbochemical products. Later on, Paul Mallet and his engineers perfected high-pressure gas feeders, one of which was 7 kilometers long, thus allowing the extension of the network into the valleys of the Seine and the Oise. In 1932, the drilling of an artesian well at a depth of 500 meters, which gave new impetus to the development of the water network of the Mallet enterprise, was considered a technological feat. Examples could be multiplied to show that the installation of the utility networks in Andresy and its region was not done in a standardized manner but by adaptations on the part of the concessionaire, in his own best interests, to ensure the incorporation of his enterprise into the local milieu.

The dynastic continuity of the Mallets, the economic power of the enterprise, the length of the concession, and finally their technical competence allowed the Mallets to acquire a kind of local "mystique." Through the decades, the enterprise became a quasi institution, more powerful than a local government lacking in resources, political power, competence, and sufficient stability.

The New Territory of the Utility Networks

The power given to the Mallet factory and its utility networks created new physical interdependencies,[12] and at the same time defined a new territory. In his study on the geography of power, Claude Raffestin has noted: "From the State on down to the individual, by way of all the large or small organizations possible, one finds syntagmatic actors[13] which 'produce' territory"; and "every network is an image of power, or more exactly, of the power of the dominating decision maker(s)."[14]

In the municipality of Andresy, as in the neighboring towns, a power had been set up in the form of a utility network, that would "produce" a territory distinct and separate from the communal territories. In fact, from the point of view of basic services, the installation of water and gas pipes and electrical cables around Andresy brought a number of communities into a relationship of solidarity with one another; up to that point, they had not had much in common. From 1895 on, the inhabitants of Chanteloup, Andresy, and Maurecourt could drink the same water. From 1909 on, the inhabitants of Triel, Vernouillet, Verneuil, as well as those of Andresy and Conflans-Sainte-Honorine, would light their gaslights with gas from the same refinery; after the 1930s, they would use the electrical current provided by the same network (see Figure 14.1). Moreover, they all participated, again in solidarity with one another, in the financial equilibrium of the Mallet enterprise, while their elected representatives together guaranteed its territorial power.[15]

But can we speak of "inhabitants"? Who were these utility consumers, these occupants of the new territory of the utility networks? For the community of Andresy, as for the surrounding communities, the installation of the totality of utility networks was one factor in the city's spatial and social transformation. To quote from the Municipal Council: "[We may consider] that the community of Andresy, by conceding, for a long period of time, the exclusive monopoly for the installation of gas for lighting and for water for its territory, [did so] in order to give new impetus to its prosperity and to its future, to increase the well-being of the inhabitants, and finally to attract the largest number of inhabitants who, for their vacations, would specifically choose a place where life was comfortable and agreeable." In another place: "It is . . . to work towards prosperity and local expansion to render habitable those plots of land which are fortunate enough to be located close to the railroad and which, up until now, have been almost without value due to lack of water."[16] Even when the well-being of the resident population is mentioned, it clearly appears—and the ulterior development of the utility networks in Andresy would confirm it—that concessions were made, and networks built, in the interests of a new population.

The villas of the well-to-do Parisians could be built on the banks of the Seine and along the slopes of the Hantil, on plots sold for good prices by farmers and winegrowers. It was, then, a question of making available for this construction all the urban services already available in Paris, in other words, to "urbanize" Andresy and its region.

It was in relationship to this project, shared by other municipalities in the region, that the Mallet utility networks were extended over and above particular concessions and communal boundaries.

The logic of the firm's expansion was thus up against a project concerning the transformation of social space. The territory that resulted from this encounter would no longer resemble what had previously existed. Aside from infringing on the municipal zoning of the communities themselves, the new territory would no longer conform to the preceding lines. Historically, links between Andresy and its environs had been established with two towns to the south, Poissy and Saint-Germain-en-Laye; now the territory of technical networks would regroup the riverside communities of the Seine or the Oise (chiefly to the west and north) around Andresy and would set up new real estate opportunities.

This new territory would also be defined in its relationship with Paris. We have seen how the development of relations with the capital, notably the existence of good railway service, affected the construction of new utility networks in the region of Andresy. But the relationship of the new territory with Paris would also be marked by the presence of new inhabitants. The Parisians had, to some extent, annexed these towns. The less wealthy would flock out on Sundays, crowding the dance halls and taking over the boats for hire. The well-to-do were more demanding; in addition to well-preserved countryside and pastoral landscapes, they required the elements of comfort to which they were accustomed in Paris. Running water, gaslight, artesian well water, and electricity meant that the new territory was an extension of Parisian space to be used for the benefit of those who wished to, and could, annex it for their own pleasure.

The ultimate evolution would affect the Mallet enterprise, the networks, and the territory. The nationalization of energy after World War II, the development of natural gas and electricity distributed by nationalized enterprises, would mean that the organization of the utility networks in the region of Andresy would be called into question. The improvement of highway and rail transportation services, the industrial development of the Seine River Valley, the massive construction of tenements in Parisian suburbs, in particular in the new town of Cergy Pontoise—all these factors would modify the space of this zone by transforming certain communities into bedroom suburbs, by creating new phenomena of social segregation, and so on. As a result, even though the Mallet utility networks still exist physically, they no longer have any territorial significance whatsoever.

Conclusion

What is to be learned from this study of the period from 1896 to 1946? First, the construction and extension of different utility networks around Andresy were not separate phenomena, irreducible one from the other. Aside from the different technologies involved, two global factors appear that concern the totality of the utility networks.

The first, an economic factor, is the strategy of the Mallet enterprise, which invested simultaneously or successively in different networks. This strategy aimed at extending the networks and reaching new markets of consumers in order to pay off the major equipment of the networks.

It is clear that in pursuing its strategy, the concessionaire leaned on networks that had already been conceded in order to obtain the concession for new networks or the extension of ones that already existed: The water concession brought about that of gas, and that of electricity, in the same way that the concession for a particular town brought about a concession in a neighboring town.

The other factor is social. The entrance of Andresy and its region into the Paris orbit created a need to provide services of the urban type, specifically linked to hygiene and comfort, needs that the local authorities would try to satisfy.

These two separate sets of reasons—that of the entrepreneur and that of the local authorities—would come together and define a territory designated by the distribution networks of the Mallet plant. The new territory was that of the entrepreneur, of his plant, techniques, concessions, capital, and consumers. But at the same time, this was equally the territory of second homes, of tourists and leisure activities, of those who made their living from it, those who improved and sold their land or their houses—the territory, in short, of the municipalities in which all these interests were located. In this sense, the new territory becomes an annex, an extension of the territory of Paris.

While the new territory thus presents a double face, would it not be because the techniques of distribution, in the form of a network of basic services, ensures a kind of equalization between the economics of their production and the characteristics of a local society in the process of change? Would it not be in this direction, in the specificity of a relationship between city and technology, that we must look for the unity of various urban infrastructure networks?

Notes

1. This research was carried out in collaboration with Philippe Mustar of the Institute of Urbanism in Paris, who is currently completing a doctoral dissertation on the genealogy of utility networks.

2. There were 1,265 inhabitants in 1896.

3. Only in 1836 was a modest suspension toll bridge constructed over the Oise. Traveling to Paris by means of this waterway was an extremely long trip due to the meanderings of the Seine.

4. Until 1860, animal traction was the principal means of energy in river navigation. From 1860 on, warpage tugs were progressively installed, including one in Andresy going in the direction of the lower Seine. In 1880, the steam tugboat made its first appearance on the river; this began a veritable revolution in river navigation.

5. Cf. a chapter in the autobiographical novel of Julien Green, *Partir avant le Jour* (Paris, 1972), in which the author tells of his childhood vacations spent in Andresy.

6. G. Dupuy, *Histoire du chemin de fer Pontoise–Andresy–Poissy* (Club Historique d'Andresy, 1977).

7. Cahiers des charges de la Concession Mallet, November 11, 1894.

8. The treatment consisted of filtering the water through a bed of sand.

9. J. L. Vaudoyer, *Le Souvenir de Marcel Mallet*, (1946).

10. Deliberations of the Municipal Council of Andresy, January 13, 1898.

11. For example, the construction of a special telephone line to warn the pumping station of Andresy in case of a direct lightning hit on the transformer of Puiseux, which supplied the electrical network from the Nord-Lumière network.

12. *Solidarity* is used here in its original etymological sense of "whole," "solid." I have translated it as "interdependencies."

13. That is, boosters of a project.

14. Claude Raffestin, *Pour une géographie du pouvoir* (Litec, 1981).

15. Concerning the electrical utility network, the communities formed an Intercommunal District, which borrowed from the public in order to finance the purchase of equipment; Mallet Enterprise was responsible to the district for the actual building and operation of the utility network.

16. Declaration of the Municipal Council of Andresy, January 13, 1898. Italics added.

PART V

Communication

OF the major technologies that constitute the infrastructure of the networked city, the least attention has been paid to
those involving communication. And yet urban scholars generally agree
that cities evolved in order to facilitate communication and that technologies such as the telegraph and telephone profoundly affected urban
structure and functioning. The two chapters in this part are unusual in
that they discuss communication history and attempt to link the history
with current telecommunication issues. Seymour J. Mandelbaum is particularly concerned with the idea of the city as a "deep community," a
concept that would have required municipal concern with communication infrastructure. Such a policy, he argues, was repressed by liberal
fears of a restrictive polity and conservative fears of too open a polity.
New telecommunication technologies offer the possibility of enhancing
a "deep community," but Mandelbaum is skeptical that urban polity
will produce such an outcome. Chantal de Gournay's chapter focuses
on the pattern of initial telephone diffusion and use in France and Great
Britain, and compares developments at the beginning of the century
with telecommunication policy today. Great Britain's telephone system,
she maintains, was better adapted to the nation's economic and regional
realities than that of France, which did not allow for business needs
or cross-regional communication. These basic sociopolitical factors, she
argues, continue to shape telecommunication policy today, with the
British providing a much more flexible and less political approach than
the French.

15   Cities and Communication: The Limits of Community

Seymour J. Mandelbaum

I⊤ has long been fashionable to consider the impact on cities of great international shifts in the flow of information. Rarely, however, is local communication treated in the way we are accustomed to think about waste treatment or the distribution of water. The political theorists who helped shape the conception of U.S. city government in the first years of this century barely spoke of communication.[1] The guides to local planning and public works published by the International City Managers Association since the 1930s deal only briefly with the problems of overhead or underground wires and the use of various media to convey information about city plans. They do not imagine an urban communication infrastructure as an object of design, policy, or politics.[2] The recent Urban Institute studies of the "capital plants" of U.S. cities analyze the flow of people, water, waste, and energy, but not the movement of disembodied information.[3]

I was a member of the first generation of academic enthusiasts for the broad social promise of cable television.[4] In small circles (bolstered by the sense that we owned the future), a few people dreamed of multipurpose broadband networks as the central technical element of a synthetic conception of urban communication. When I first sat down to plan what I would say at the Paris conference on urban technologies, I thought to talk about the failure of cable television to realize or even to test seriously those early dreams. That might, indeed, have been an appropriate topic, since the French are in the midst of a great wave of cable fantasies.[5]

That focus appeared to me, however, to minimize the lessons of the

Reprinted with permission and with minor changes from *Telecommunications Policy* 10, no. 2 (June 1986): 132–140. I am grateful to Mitchell Moss for his generous comments before and after the drafting of this essay.

last two decades. The poverty of the cable experience may be easily laid to limitations of the flesh: power, capital, and authority. The more interesting failure is one of spirit or idea. How is it—why is it—that the synthetic idea of an urban communication infrastructure and policy agenda has fared so badly in the United States? Is it possible now to invest it with political meaning?[6]

This exploratory inquiry into these questions rests on a broad assumption: The idea of an urban communication infrastructure would have entered deeply into our thinking during the last twenty years if participants in city politics had defined a communication policy agenda and fought for influence over the generation and flow of information. The failure to create such an agenda should not, however, be attributed to the openness of communication systems. All the infrastructural systems discussed so far in this volume are open and expansive, stretching well beyond the spatial and political range of limited city polities. Communication is different, not because it cannot be contained by city polities, but because of what it does to them.

This assumption subtly shifts the focus of inquiry away from the analytic problems of an idea and toward the institutional difficulties of realizing it in practice. (Flawed ideas often come into good currency if they fit). It also shifts it away from the last twenty years and toward a broader historical frame. The image of urban communication built around broadband networks was squeezed between stable but conflicting values that preceded the shift in distribution modes.

Specialized urban polities, such as sewer and water districts, have long been organized around components of the physical infrastructure. General-purpose governments in a variety of very profound ways have treated infrastructural decisions—which neighborhoods are developed, who gets a new highway, what sort and where sewers should be built—as important elements of intergroup relations. Fragmentary beginnings that treated components as distinct have, over time, been linked (symbolically, professionally, and organizationally) under common rubrics, such as *city, environmental planning, urban design,* and, indeed, *infrastructure.*

Not so in the communication field! City polities have been deeply concerned with the flow of information. Schools—institutions for intergenerational communication—dominate local budgets in the United States and a great deal of city politics. The history of motion picture distribution and of libraries has been marked by repeated city attempts to control access to messages. New waves of immigrants reinvigorate old local disputes over the dominance of English in civic discourse.

Beyond this, general-purpose urban governments or their delegates have sometimes accepted explicit authority over a broadcasting frequency. Public broadcasting stations assigned to nonprofit "community" groups may sensibly be seen as specialized communication polities. Virtually all city governments operate substantial telecommunication facilities for the use of police and fire departments.

Indeed, if there was not so strong and varied a tradition of local communication activity, my speculations about the absence of a broad view would be tendentious and undisciplined. It is, however, precisely the density of the local communication policy field that leads me to wonder about the absence of a synthetic image and politics. Why weren't debates surrounding the design of schools generalized to shape urban radio and television? Why weren't persistent discussions of intergroup communication transformed into a program for urban telephony? Why didn't the tradition of analytic comment on the implications of changes in the metropolitan press generate a significant local political agenda?

Barriers to an Urban Communication Policy

The questions asked in the previous section, however provocative, would be ahistorical if there were impassable barriers between the local policy field as it emerged and the notion of an urban communication infrastructure. There are three possible candidates as barriers. None of them are quite as broad on close inspection as at first appearance.

Institutional Settings

A great many policy ideas are never articulated fully because there is no forum in which they can be discussed, no arena in which they can be resolved. As a result, they suffer from underinvestment. The contemporary group of urban communication advocates have certainly often felt that there was no one in city government with whom they could readily speak, that the professional staffs of municipal and regional planning agencies might talk glibly about the "information society" of the future, but would not undertake serious efforts to guide its development.

The impatient sense that there was no one home to communication must be tempered by a historical perspective. Advocates of new urban political functions during the nineteenth and early twentieth centuries repeatedly complained of gaps in the design of institutions: no one to take responsibility for neglected children, polluted streams, noisy streets, or congested houses. Repeatedly, however, the gaps were filled in one or another of several different ways. Most recently, the federal government mandated and (in part) financed local institutional development. In older patterns, policy communities succeeded in enlarging the scope of an established agency or investing a voluntary association with public authority.

Many of the conditions that encouraged institutional innovations in housing, education, social services, and all the conventional elements of infrastructure were available in communication. Telegraphy, telephony, electronic broadcast communication, and motion pictures each developed in a context of emergent engineering professions, technical and trade journals, and widespread public speculation about the impact of new technologies on urban form, social order, and personal behavior. The regulation of telephone companies by state public utility commissions provided a focal setting in which a distinct urban policy concerned with the design and policies of point-to-point electronic communication might have been articulated. Even more tellingly, the federal government in the 1920s created the basis of special-purpose communication districts. Herbert Hoover, then secretary of Commerce, described an ambitious version of these districts in 1925. He suggested that the federal government assign frequencies to communities and monitor interference, "leaving to each community a large voice" in the actual assignment of wavelengths. He recognized that these radio districts could not be circumscribed by city or state lines, but it was possible, he insisted, "to establish zones which will at least roughly approximate the service areas of stations and to a very considerable extent to entrust to them the settlement of their local problems."[7]

The initial form of these districts was much more modest. The major communication acts of 1927 and 1934 provided for the assignment of frequencies to spatially contiguous market areas—characteristically designated by the name of the central cities. These market areas were not given political authority, but licensed broadcasters were required to ascertain the "needs" of their "communities" and demonstrate that their programming met those needs.

The politics of telephony and broadcast did not, however, grow from the rudiments of utility regulation and community ascertainment into stable elements of the process of urban governance. Market areas did not become special communication districts controlling the frequencies assigned to them. The concept of an urban communication infrastructure was not expressed in local institutions comparable to those that realized public concerns with education, environmental quality, housing, or transportation.

The First Amendment

The First Amendment to the United States Constitution prohibits Congress from "abridging the freedom of speech, or of the press." Many states have similar provisions in their constitutions, and—at any rate since the 1920s—the U.S. Supreme Court has consistently acted to apply the First Amendment prohibition to all political jurisdictions.

The prohibition has inhibited the development of certain substantive communication policies, such as prior restraint and restrictive licensing of print. It has not, however, stood as a wall blocking communication polities or policies at any scale of the federal system. The copyright and post office provisions of the Constitution describe developmental and regulatory roles for the national government, roles that have blossomed over time. Washington has prescribed rules for common carriers, licensed broadcasting in a way that would not be tolerated for print, and invested massive funds in the creation of the largest publishing house in the country. It has been the principal mover in the development of worldwide satellite networks and what is now sometimes called "informatics."[8]

The federal government has sometimes preempted local and state regulatory authority. Even when administrative and judicial limits on regulatory discretion rested on the First Amendment, they did not preclude developmental initiatives. Cities might have adopted aggressive developmental roles by building distribution networks, managing and programming broadcast stations, and subsidizing print without violating constitutional strictures. The little bits and pieces of such activity that mark the last 150 years only highlight the failure to create a local communication policy field.

Conceptual Deficits

Recent discussions of a communication infrastructure have been encouraged by an intellectual revolution that eroded the boundary between information processing and transmission, imagined broadband networks that could carry a great variety of codes, and described in general terms the properties of communication systems and the role of information in human affairs. If this "revolution" is very recent—if it is, for example, epitomized by John von Neumann, Claude Shannon, and Norbert Weiner—then it may have been impossible to have imagined a general communication infrastructure (under the diversity of channels, codes, and social relationships) before 1950. In effect, there was a conceptual deficit that inhibited urban communication policymaking.

There was no such deficit. The recent expansion of high-capacity wired networks and the confluence of computation and communication have certainly encouraged a sense that a single integrated system may transmit a full range of media. Major elements of that integrated image appeared repeatedly, however, in earlier discussion of telegraphy, telephony, and radio. In the 1840s, the printing telegraph suggested to knowledgeable contemporaries the possibilities of a Telex network and what we would now call "electronic mail."[9] Telephony suppressed technical innovation in teletypewriting, but only reinforced the sense of integration. The economic, political, and legal battles between Bell and Western Union and then between AT&T and RCA were all driven by an understanding among engineers and business leaders of the flexibility of the media: music, pictures, and data could be conveyed by wire or through the atmosphere; radio and ultimately television signals could serve as a common carrier of targeted (if not necessarily private) messages.

The image of an infrastructure was a simple complement of this conception of flexibility. Alexander Graham Bell, in a letter to British investors in 1878, explicitly compared the proposed telephone system to the subterranean network of gas and water mains:

> At the present time we have a perfect network of gas pipes and water pipes throughout our large cities. We have main pipes laid under the streets communicating by side pipes with the various dwellings, enabling the members to draw their supplies of gas and water from a common source.
>
> In a similar manner it is conceivable that cables of telephone wires would be laid under ground, or suspended overhead, communicating by branch wires with private dwellings, counting houses, shops, manufactories, etc., uniting them through the main cable with a central office where the wire could be connected as desired, establishing direct communication between any two places in the city. Such a plan as this, though impracticable at the present moment, will, I firmly believe, be the outcome of the introduction of the telephone to the public. Not only so, but I believe in the future wires will unite the head offices of telephone companies in different cities, and a man in one part of the country may communicate by word of mouth with another in a distant place.[10]

The same graphic image appeared in the 1910 annual report of the American Telephone and Telegraph Company:

> One system with a common policy, common purpose and common action; comprehensive, universal, interdependent, intercommunicating like the

highway system of the country, extending from every door to every other door, affording *electrical communication of every kind*, from every one at every place to every one at every other place.[11]

The agreement that created the National Broadcasting Company in 1926 was launched with similar imagery: a system combining telephone long lines with broadcast transmission to provide what the opening advertisement described anomalously as "machinery" to distribute "high quality" radio messages nationally.[12]

Finally, and most important, a broad public understood that the new means of communication were not discrete phenomena long before 1950. I suspect that it was the movies rather then the telephone or radio that shaped this consciousness. The movie system—as an industrial organization and as a set of visible local artifacts—was more tangible than the golden webs of telephony or broadcasting. Movies were not simply a new way of delivering public drama and pageants, a shift from one transmission mode to another. In Muncie, Indiana (whose voices speak through the *Middletown* volume), and across the country critics and commentators argued that films fundamentally threatened to remove children from the socializing control of their parents and the institutions of family, church, and neighborhood. That perception —repeated over and over again—exposed the processes of social communication to inspection and suggested that it was possible (indeed imperative) to design not simply the media but society as a communication network.[13]

Communication and the Concept of Cities in the United States

The politics of cities in the United States has always been dominated by the openness of the polity. Aggrieved parties frequently employ the threat of exit if their wishes are not met. "If you are unwilling to tolerate our noxious emissions," the owner of a factory may insist, "we'll move our jobs elsewhere." "If you alter the schools in ways we don't like," parents announce, "we'll take our families to other districts."

In the ordinary course of city politics, these threats are critically assessed: Some bluffs are called, some warnings heeded. The plasticity of the federal systems also allows for an extraordinary form of politics in which the context of the game of threat and response is altered. The possibility of creating new polities as havens increases the cogency of exit threats. Metropolitanwide government or the assumption of responsibilities by the national center has the opposite effect. The creation of a uniform field from which it is difficult to escape enhances the voice of relatively immobile factors of production and groups.

I do not mean to imply that urban politics in the United States is played in an atomistic state of nature. General-purpose urban polities —many of which have long histories—engender a political rhetoric in which claims of obligation and loyalty, rather than the threat of exit, are the coins of influence. Civic institutions and rituals cultivate the sense of a corporate entity whose members are bound by a commitment to rules and to each other in a way that tempers short-term calculations of interest. The language of Daniel Burnham's 1909 Plan of Chicago may now seem very old-fashioned: His designs were to engender and symbolize "the feeling of loyalty to and pride in the city"; "good citizenship" was the "prime object of good city planning."[14] There is not, however, much difference between Burnham or the civic pageants of the early twentieth century, on the one hand, and the contemporary tribal celebrations of sports victories and community goals, on the other. Indeed, much of professional urban history and system analysis is written to enhance the urban polity as a community of mutual obligation. Once you understand the linkages of a dense urban field and the ubiquitous externalities of private action, then you must act with a sense of communal responsibility.

The creation of deep communities of mutual obligation depends on the richness of dialogic processes and their intimate connection to action. Diminish the richness, limit mutual exchange, or weaken the link between talk and deed, and obligation falters. Efforts to establish obligations in the absence of communication—in effect, equating the claims of strangers and neighbors, families and aliens—always seem shrill and partial or to depend heavily on coercion.[15]

What was at stake in the expansive politicization of urban communication was the possibility of developing city polities as deep communities. That possibility has been very attractive to many participants in urban political debates over the last 150 years and (probably beyond its salience in the historical record) to professional analysts. The resistance to the creation of deep bonds of mutual obligation has, however, been very powerful and, at least until now, has prevented the emergence of the idea of an urban communication infrastructure and its institutional complements.

The enemies of the city as a deep community have been both liberal and conservative. Liberal critics have insisted that competent adults should be free to choose from a great and expanding array of messages. Public policy should positively encourage the openness of communication networks and, even more stringently, avoid restrictions. Censorship of library books, films, or video in the name of "community values" has been anathema; individual competence and choice and not social solidarity the desiderata. Some issues have troubled and divided liberals—notably pornography and the protection of children—but they have uniformly resisted the ideas that communication flows should be subject to local majorities or that the well-being of social entities, rather than the preferences of individuals, should be the measure of communication improvement.

Conservative critics, in contrast, have argued for the moral salience of social groups, for the quality of community life rather than individual choice as the measure of value. The conservative attitude is sometimes only a dream of a deep community yet to be formed: the nation, the city, the working class. It has been most powerful when it expresses the shared values of a real group in which sustained interaction creates a web of mutual obligation. Broad individual choice and communication without direct interpersonal contacts threaten such webs, suggesting that talk and action may be separated, that meaning is in the message and not in the social relationship. These threatening lessons generate in turn powerful counterattacks to protect neighborhood and parochial schools, to challenge the autonomy of cosmopolitan professionals, to discipline libraries, and to assert group values against the ubiquitous presence of the metropolitan press and broadcasting systems.

The communities that conservatives sought to protect were never diverse urban polities. Large cities were frequently forums in which groups debated and arenas in which at least some communication conflicts were resolved. Neither conservatives nor liberals, however, conceded that they were the primary communities of obligation, the dominant arenas for communication policymaking or the final courts to legitimate public action. Even in towns and suburbs where groups and polity converged, the development of motion pictures, radio, and television shattered the illusion that the community was natural rather than historic and vulnerable. Control required complex alliances outside the boundaries of the local polity.[16] Reactive measures mirrored the forms they were intended to counteract. Radio and television ministries, for example, imitate conventional broadcast programming and —necessarily—distort the dynamics of religious organizations in their communities.[17]

The idea of the city as a deep community that must attend to its communication arrangements was squeezed between liberal apprehensions that it was too restrictive and conservative fears that it was too expansive. As a result, neither urban theorists nor practitioners developed a synthetic notion of an urban communication infrastructure. Nor did they build professions and institutions embedding that notion within the political culture of American cities.

Urban Broadband Networks

In the 1950s, the first wires were strung in American cities to improve local television reception and import signals from distant transmitters. Cable had little impact on television viewing into the 1970s, but, as an idea and as a possibility, it excited a great deal of interest in small circles in the late 1960s. Since then, grand hopes have often been dashed, but technological innovations and waves of new recruits who do not remember the first dreams have constantly refreshed enthusiasm for the promise of urban broadband networks. As cable penetration levels have increased, franchises granted in the largest cities and the first fiber optic loops constructed, the circle of dreamers has grown. Smart telephones and personal computers seem to resolve the difficult problems of developing low-cost terminals that beset the early experiments of the 1970s. The first trials of videotext systems have been discouraging, but the fantasy of a universal encyclopedia that responds quickly to discrete questions is not easily dashed.[18]

The idea of cable television (as distinct from the particular pattern of current practice or technology) is at the heart of the current talk of an urban communication infrastructure. Broadcast market areas never developed into special communication districts, but it was clear virtually everywhere that local polities would have either to string wires themselves or enfranchise private companies. Few chose or even seriously considered the former option—although in the same period, most cities assumed ownership of mass transit lines—but the process of granting franchises raised communication to the symbolic domain of an urban infrastructure.

The political requisite was supported by potential technical capacities of cable transmission. Cable systems could be designed so that they:

• would allow for an audience response through a return loop.

• could address messages to individual viewers or subsets of the entire population of subscribers.

• provided an abundance of channels far beyond those available in the broadcast spectrum or in the early systems.

• provided production facilities available to large populations.

• integrated information transmission, processing, and storage in a variety of modes.

These potential attributes would support rich dialogic processes intimately connected to action—the essential requirements of a deep community. Both national institutions and major local actors have, however, been very reluctant to realize these potentials. In the 1960s, the Federal Communications Commission (FCC) asserted its authority to set national cable policy and to regulate franchise procedures and provisions despite the absence of any specific authorization in the 1934 Communications Act. Cable, in the eyes of the commission, would have such a profound impact on broadcasting that to allow it to grow freely would put national communication goals (and broadcasting stations) in jeopardy.

Since 1968, the commission and the courts (in a sometimes uneasy alliance) have acted to protect broadcasters against cable companies, cable companies against both cities and the telephone system, and cities against themselves. Even in the most recent wave of deregulation, in which many federally mandated franchising rules were rescinded, the FCC has never acted as if city polities might decide whether they wanted cable companies to act as common carriers or not, to inhibit or allow the integration of cable and telephone systems or the cross-ownership of newspapers, broadcast stations, and cable companies. Indeed, several recent court decisions and legislative actions surrounding both cable and satellite transmission suggest that the local franchising process (even in its modest current versions) may be at risk of diminution or extinction.[19]

Without significant authority, cities have not developed an expansive image of communication tasks and opportunities. Even the technical promises that excited the first dreams have not been realized. The first enthusiasm for public access and the multiplication of local programming was encouraged by the gap between the number of potential channels and the limited programming available through the importation of signals from conventional broadcasting stations in distant cities. The development of commercial recordings for home video, computer systems, and the efflorescence of satellite programming services has closed the gap. The differentiation of audiences competes, in effect, with efforts to erase the essential distinction between audience and producer.

The two-way capacity and interinstitutional links that excited early enthusiasts have also not been realized in ways that integrate urban polities. Specialized national and international common carriers and AT&T (rather than cable companies) have created interactive teleconferencing and data-sharing networks outside the framework of local control and, often, even local understanding. These viewers increase the mobility of information and further detach central business districts from their immediate environs. They make it more difficult to imagine cities as deep communities or to define even a limited domain in which they may act as corporate entities.[20]

The game is not yet over. Urban polities may yet attempt to discipline communication networks and deepen the bonds of mutual obligation. Certainly, retrenchment and the decline of intergovernmental payments have revived a rhetoric of community and shared fates among working politicians, business leaders, and political theorists. I suspect, however, that the very pressures that encourage city polities to seek consensual developmental policies[21] will continue to inhibit the assumption of deep mutual obligations. The idea and practice of urban communication politics will continue to be marked by a skittish irresolution.

Notes

1. W. B. Munro, *Municipal Administration* (New York, 1934); and F. Goodnow, *Municipal Government* (New York, 1909).

2. F. S. So et al., eds., *The Practice of Local Government Planning* (Washington, D.C., 1979).

3. G. E. Peterson, ed., *America's Capital Stock* (Washington, D.C., 1979–1981).

4. S. J. Mandelbaum, *Community and Communications* (New York, 1972).

5. P. Flichy and G. Pineau, *Images pour le cable. Programmes et services des reseaux de videocommunication* (Paris, 1983).

6. M. Moss, *The Changing Telecommunications Infrastructure in New York City* (New York, 1983); idem, "The New Urban Telecommunications Infrastructure," *Computer/Law Journal* 6 (1985): 323–331.

7. U.S. Senate, *Hearings Before the Committee on Interstate Commerce, Radio Control, S.1 and S.1754*, 69th Cong., 1st sess. (Washington, D.C., 1926), pp. 57–58.

8. I. de S. Pool, *Technologies of Freedom* (Cambridge, Mass., 1983).

9. I. de S. Pool, *Forecasting the Telephone: A Retrospective Technology Assessment* (Norwood, N.J., 1983), p. 30.

10. Ibid., p. 21.

11. Ibid., p. 22; italics added.

12. E. Barnouw, *A Tower in Babel. A History of Broadcasting in the United States*, vol. 1, . . . *to 1933* (New York, 1966), p. 87.

13. R. S. Lynd and H. M. Lynd, *Middletown: A Study in American Culture* (New York, 1929), pp. 251–271.

14. P. Boyer, *Urban Masses and Moral Order in America, 1820–1920* (Cambridge, Mass., 1978), pp. 270–276.

15. H. Arkes, *The Philosopher in the City: The Moral Dimensions of Urban Politics* (Princeton, N.J., 1981); R. M. Cover, "Nomos and Narrative," *Harvard Law Review* 97 (1983): 4–68; C. Pateman, *The Problem of Political Obligation: A Critical Analysis of Liberal Theory* (Chichester, England, 1979); and M. J. Sandel, *Liberalism and the Limits of Justice* (New York, 1982).

16. G. Jowett, *Film: The Democratic Art* (Boston, 1976).

17. G. T. Goethals, *The TV Ritual: Worship at the Video Altar* (Boston, 1980).

18. J. Carey and M. Moss, *New Telecommunications Technologies and Public Broadcasting* (New York, 1983).

19. E. N. Noam, "Local Distribution Monopolies in Cable Television and Telephone Service: The Scope for Competition," in *Telecommunications Regulation Today and Tomorrow*, ed. E. N. Noam (New York, 1983).

20. S. J. Mandelbaum, "What Is Philadelphia? The City as Polity," *Cities* 1 (1984): 274–285. In a volume in progress on the poetics of policy and planning arguments, I develop the contrasting idea of "open moral communities."

21. P. E. Peterson, *City Limits* (Chicago, 1981); S. L. Elkin, *City and Regime in the American Republic* (Chicago, 1987).

16 Telephone Networks in France and Great Britain

Chantal de Gournay

Until the 1970s, the telephone system in France was considerably behind the systems in Great Britain and Germany. In Britain, the telephone followed a normal growth pattern, which nonetheless merits study because of the importance of private industry's role in its development, an exceptional situation in Europe, where communication systems were traditionally under public control. During the telephone's first three decades, the British government participated little in the construction of urban (local) networks. Installations set up by the National Telephone Company (NTC), which had a concession from the government to operate urban telephone systems, while the state maintained a monopoly over long-distance lines, became public property only in 1912; in France, the telephone was nationalized in 1889.

Different rationales governed telephone development in the two countries. This study assays these differences in order to determine if economic and industrial needs are not sometimes restrained by subtle factors connected to the interplay of political forces or to social elites' attitudes toward innovation.

Differences between French and British Equipment at the Turn of the Century

The beginnings of telephony in France, from 1876 to 1890, were too hesitant and uncoordinated to merit a lengthy analysis here. Suffice it to say that a private company, the Société Générale du Téléphone (SGT), was given the concession to operate the telephone system until 1889. But beginning in 1884, SGT became involved in disputes with the public administration; aware of the precarious nature of their concession, they stopped investing in telephony and constructing new networks. Progress halted until 1889, a period of inactivity that set France back seriously in comparison to its European neighbors.

Nevertheless, the superiority of the English telephone system over the French system lay not so much in the amount of equipment (France invested as much, and sometimes more, on equipment than Great Britain did) as in the adaptation of the British telephone system to the nation's economic and regional realities. The installation of networks in France followed a rationale that did not allow for business needs (or at best only to a small degree) or for communication between various local communities within a given region, and that catered least of all to the needs of private individuals. As Yves Stourdzé has put it, France had "a telephone for local elites" made out of disconnected, fragmentary networks. A plethora of small rural networks existed, but very few long-distance connections (except for those that reinforced traditional centralization in Paris), and there was an almost total lack of interregional connections and even connections between large cities and their suburbs (see Figure 16.1). An exception was the fairly integrated telephone network connecting northern textile industry cities; it corresponded quite well to real economic relations between different regional industrial centers.

This explains why, if one uses only the criterion of equipment, such as the number of telephone exchanges or the length of aerial or underground cables, the differences between early telephony in France and England are insignificant. For example, the number of telephone exchanges in Great Britain was often lower than that in France. In contrast, if optimal utilization of these installations by the greatest possible number of users is considered, that is, if one takes into account the democratization of telephone use, by connecting the maximum number of people at the risk of saturating exchange capacity, Great Britain was ahead of France.

This section selects three criteria for evaluating equipment level in each country: (1) the number of exchanges; (2) the number of public telephone booths; and (3) the number of private telephones. A comparison of these criteria over several years can be seen in Table 16.1.

As the table makes clear, the number of exchanges (*stations centrales*) in France grew constantly after 1899 and was much larger than the number in Great Britain. In 1903, for example, the number of exchanges in France was twice that of Great Britain, for about one-third the number of private telephones. Telephone installations in France, much like post offices, were spread out over the whole territory, including the smallest towns, with no concern for differentiating service by

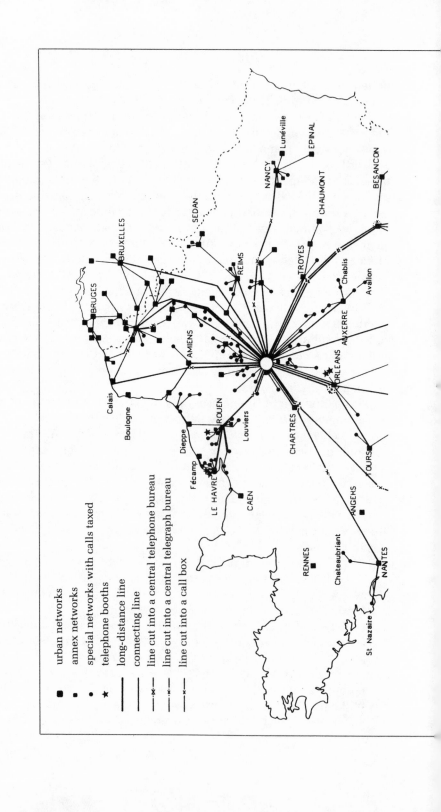

urban networks

annex networks

special networks with calls taxed

telephone booths

long-distance line

connecting line

line cut into a central telephone bureau

line cut into a central telegraph bureau

line cut into a call box

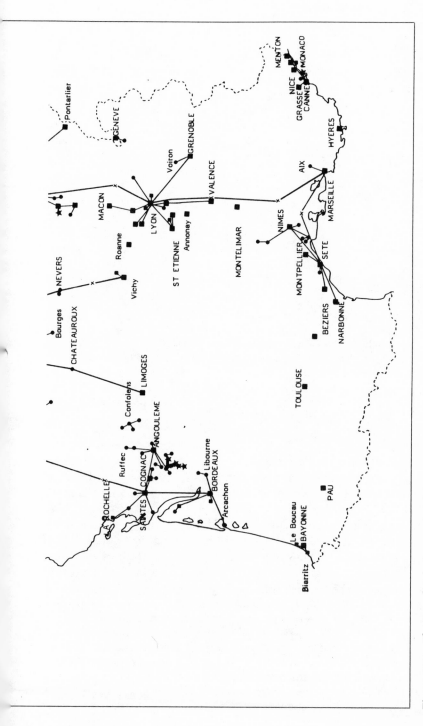

Figure 16.1 The French telephone network in 1891

TABLE 16.1.

Telephone Equipment, France, Great Britain, and Ireland, 1899–1908

Year	Equipment	France State	Great Britain and Ireland State	NTC	Total
1899	Exchanges	975		960	
	Public telephone booths	1,463		2,304	
	Private telephones	60,729		169,925	
1901	Exchanges	1,558		1,019	
	Public telephone booths	2,369		2,610	
	Private telephones	79,536		219,849	
1903	Exchanges	3,227	511	1,072	1,583
	Public telephone booths	5,129	655	3,571	4,226
	Private telephones	108,946	22,506	277,589	300,095
1906	Exchanges	5,379	720	1,285	2,005
	Public telephone booths	9,442	1,835	6,740	8,575
	Private telephones	152,072	68,098	407,736	475,834
1908	Exchanges	6,904	959	1,561	2,520
	Public telephone booths	11,956	2,489	8,692	11,181
	Private telephones	182,204	84,255	489,747	574,002

Sources: For networks operated by the state, International Bureau of the Telegraph Union. For urban networks in Great Britain, figures are available only from 1903. Figures for 1899 and 1901 are from the London Post Office archives.

size of population or different locality. The result was overequipped small towns and an underequipped nation as a whole. To summarize, one might say that there was territorial egalitarianism accompanied by a social elitism in telephone use. This apparently paradoxical phenomenon was in keeping with French sociopolitical realities, which were characterized by the importance of the local landowning bourgeoisie.

The statistics cited in the table confirm Catherine Bertho's analysis:

> The national telephone system is characterized by the proliferation of small-scale networks; every town and almost every village has its own. France, according to the Minister, is far ahead of other nations in the number of its local networks (almost 14,000 compared to 7,000 in Germany). More than 8,000 townships out of the some 88,000 in the country are equipped. We have a hypertrophy of rural telephone systems with thousands of small circuits covering very small distances (an average of 15 kilometers while the German average is 30 kilometers). Together, they form a close-knit national system of very small networks that can-

not be economical since it serves a small number of users and transmits relatively few calls. While small networks that serve basically two local dignitaries, the country doctor and the chateau owner, have proliferated, large interurban [long-distance] connections have been sacrificed.[1]

The lack of long-distance connections was also denounced in 1920 by the deputy Pierre Robert in the *Official Journal*: "In this huge cobweb . . . there are many lines going out from the center, but transverse lines are few: We have neither interurban connections nor connections between regions."

The conclusions of my comparative study of telephone networks at the beginning of the twentieth century add nothing to the pronouncements of technical and administrative directors of the state Postes, Téléphone, et Télégraphe (PTT) in 1910 and later. Their reports, available in PTT archives and in the *Official Journal*, clearly show that the problems of the French telephone system were perceptible even at its beginnings.

The illogical nature of the French telephone system's structure, characterized by a proliferation of local networks and a lack of national interconnections, is often ascribed to the financing system set up for telephone installations in 1890, called the system of "reimbursable advances." The system worked this way: Any town that wanted telephone equipment gave the money necessary for the installation of its telephone network to the state in the form of an advance. Ownership of the network, along with its profits, belonged to the state, which reimbursed the town's advance (without interest) as soon as there were enough profits to do so. Long-distance connections were financed by the same system. If two towns wanted to be connected, each had to advance one-half the cost of the link. The system accentuated local prejudices. Each town was quite aware of the advantages of having its own network, but was hesitant to finance equipment that served national, rather than local, interests.

Although this financing system accounts for the unsoundness of the national French telephone system to some extent, it cannot account for the system's inadequacy with regard to economic needs. Why was industry not strong enough to impose reforms capable of compelling telephone systems to conform to the needs of business centers and install exchanges that reached far beyond the local level? How did France dispense with a rational economic ranking of networks, a ranking that other European nations respected, and manage to maintain a dispersed telephone structure that, according to Bertho's analysis, was closer to the rural than the industrial world?

The next section attempts to show how the British telephone system had a fundamentally rational development *despite* the slow growth of long-distance installations, which has led some specialists to claim that the British telephone system was as underdeveloped as the French at the turn of the century. This relative underdevelopment was made up for by a *spatial rationality* that governed the development of the British telephone by giving it a structure based on regional economic realities rather than artificial administrative divisions.

Spatial Configurations of French and British Telephone Networks

The Predominance of the Industrial Factor in Britain

The superiority of the British system was based, above all, on urban networks: Many more telephone users were connected to the system in Great Britain than in France (169,925 in Britain and Ireland. 60,729 in France in 1899). This illustrates the efficiency of British private enterprise, since the Post Office (the state body in charge of telephone networks) was not in competition with the private sector at the time, except in a few cities such as London and Glasgow.

Nevertheless, although the urban phone system grew more rapidly in England than in France, long-distance connections, which were under state control, did not develop quickly. Until 1889, London had only one long-distance connection, a line linked to Brighton, which opened in December 1884. During the same period, Paris would not have tolerated being cut off from large provincial cities. The following list shows French long-distance connections in chronological order of installation:

Rouen–Le Havre, January 1885	Paris–Rouen, June 1887
Paris–Rennes, December 1885	Paris–Lille, 1888
Paris–Bruxelles, February 1887	Paris–Marseille, 1888
Paris–Le Havre, March 1887	Paris–Bordeaux, 1889–1890

Bertho points out that these connections "reflect patterns of telegraph connections, which favored the ports on the lower Seine and the northern border region."[2]

As for London, its first major long-distance line began to function in July 1890; it was a line 130 miles long connecting Birmingham and later Liverpool and Manchester (with two circuits in Birmingham, one in Liverpool, and one in Manchester). But this does not mean that long-distance connections were nonexistent until then. Violating the state monopoly, the Lancashire Telephone Company had set up a system that was almost a perfectly integrated regional network. As of December 22, 1880, the company had established trunk lines between exchanges in Manchester, Liverpool, Oldham, Ashton, Stockport, and Bolton. The Post Office even sent a warning to the company on March 25, 1881, stating that the company was suspected of planning to establish a direct line between Liverpool and Manchester. Until that time, interurban lines that the company had installed in the region had been more or less tolerated because they were relatively short. In March 1976, *Electronics and Power* published maps showing that, between 1880 and 1890, this system of connections was the one in Great Britain that most closely resembled modern networks.[3]

The British experience offers a particularly rich terrain for urban studies in that the distinction between private and public operating zones required clearly defined spatial criteria for identifying strictly urban areas. When they limited the field of private enterprise to urban zones, public authorities must have felt that the borders between zones were clear, obvious, and even "natural," between the entity that is a city and the neighboring countryside. Where does a city stop? This question, which was easily answered in the case of Paris, a city surrounded by fortifications until 1913, became problematic when applied to fast-growing industrial cities. When granting operating licenses, therefore, the British Post Office prudently stipulated a limited geographical zone in each contract, and this "operating zone" was negotiated for each site and each business; the Post Office thereby avoided having to set up a rigid norm that could apply to the whole nation. Lancashire is a particularly interesting example because it was a highly industrialized region, and because it was the first region to develop a telephone system. Moreover, Lancashire is typical of a continuous urban area within which boundary lines cannot be drawn, a form of urbanization well known to geographers under the name *conurbation*. Towns proliferated close to one another, and communication between them consequently intensified.

The Lancashire and Cheshire Telephone Company obtained an operating license for Liverpool and its surroundings after eliminating the Edison firm from competition in about 1880. At that time, the company's sphere of operations was limited to a circle with a radius of 5 miles from the city center in order to prevent any attempt to establish long-distance connections. For other towns in the region, the operating radius was smaller: between 3 and 4 miles. Such arbitrary limits must have been so far from regional urban realities that the Post Office, after many negotiations, decided to lift the geographical restriction in 1884 when the operating license was renewed. From then on, there was no longer a limit on the operating radius; the company was allowed to operate within urban zones whose limits were ill defined. The result was an immediate extension of the telephone network into a real web that linked towns to one another as soon as they were large enough to have a sufficient number of users. Towns around Liverpool, such as Wigan, St. Helen's, and Runcorn, whose exchanges went into operation in 1882, were rapidly connected to Liverpool. Well before 1890, which was the date of the first long-distance connection between London and this region, there were already the beginnings of a homogeneous telephone network designed to link various manufacturing centers. The following is a list of towns that had telephone exchanges; an asterisk marks those that had long-distance connections.

Liverpool district—Liverpool (includes Birkenhead, Bootle, Garston, Gateacre, Old Swan, Waterloo, Wavertree, and West Derby), Southport, Chester, St. Helen,* Widnes and Runcorn,* Wigan.*

Manchester district—Manchester,* Bolton,* Warrington,* Todmorden, Altrincham and Bowdon, Ashton-under-Lyne,* Rochdale,* Middleton,* Heywood,* Bury,* Radcliffe,* Oldham, Staleybridge,* Stockport.*

North Lancashire district—Blackburn,* Blackpool,* Preston,* Barrow, Burley,* Accrington,* Ulverston, Darwen,* Lancaster, Fleetwood, Windermer, Chorley,* Dalton.

North Wales district—Wrexham, Bangor, Holyhead, Welshpool, Newton, Flint, Carnarvon.

Between 1880 and 1890, the only long-distance connection that was *official*, that is, tolerated by the authorities—and the English archives do not specify whether it was constructed by the Post Office or by the Lancashire Company—was a circuit connecting Liverpool and Manchester. The 40-mile line, built along a railway track, started operating on January 1, 1885.[4] Yet, as the list above shows, a dense network of small circuits satisfied the need for exchanges between manufacturing and commercial centers. This growth of the telephone—technically illegal, since it violated the government's monopoly on long-distance connections—illustrates the exceptional dynamism of telephone communication in industrial zones.

This semilegal development of private long-distance lines was not restricted to Lancashire (see Figure 16.2). Everywhere in the nation, and wherever there was industrial or commercial dynamism, homogeneous small-scale networks appeared and formed regional telephone systems. What is striking about this phenomenon is the autonomy of these systems, their integration with economic reality, and their genuinely *local* identity, which was true even in distant regions in the north.

The map of the French national network (Figure 16.1) reveals the latter's centralization; there was a veritable "telephone desert" in Brittany and all the zones running from north to south between the Creuse and the Pyrenees including the Creuse, the Puy-de-Dôme, the Corrèze, the Cantal, the Dordogne, the Lot, Lot-et-Garonne, Tarn-et-Garonne, the Tarn, the Landes, the Averyron, the Ardeche, the Lozère, the Gers, the Haute-Garonne (with the exception of Toulouse), the Ariège, the Aude, and all of the Pyrénées.

How can we account for the regional integration of telephone networks in Great Britain? I feel that it was primarily the result of great flexibility on the part of the government in defining the areas covered by private concessions. In France, the concessions granted to the Société Générale des Téléphones until 1889 were strictly limited to city centers; in Great Britain, the restriction to a radius of 5 miles from city centers was abandoned as early as 1884, allowing private companies to operate on a regional scale. Thus the first English long-distance lines were set up within a given region in order to connect networks that were part of the same concession, while the state took charge of the construction of relatively unprofitable long-distance connections, for example between Dublin and Belfast and the rest of Britain. Interregional communication was also stimulated by a system of unequal rates. Rates were different for calls within a given concession and calls between different concessions. In the 1890s, the rate for a three-minute long-distance conversation outside a concession was twice that within the same concession.

A French contemporary observer, M. Depelley, pointed out the advantages of the British regional telephone system to the administration of the French Postes et Télégraphe in a report delivered to its advisory commission on May 9, 1889.

> We have just seen that the organization of the French telephone system has not yet enabled us to *link many suburban areas with urban networks, nor have we been able to group different telephone networks within the same region.* Abroad, some countries are ahead of us in this area. The principal reason for this is that their telephone concessions are not strictly limited to the city proper. Generally, the concession covers a rather large zone which includes neighboring small towns and the suburbs, in other words, the whole agglomeration surrounding the center of the concession; sometimes, it is even spread out over a whole region, in which case all the regional networks are part of the same concession.[5]

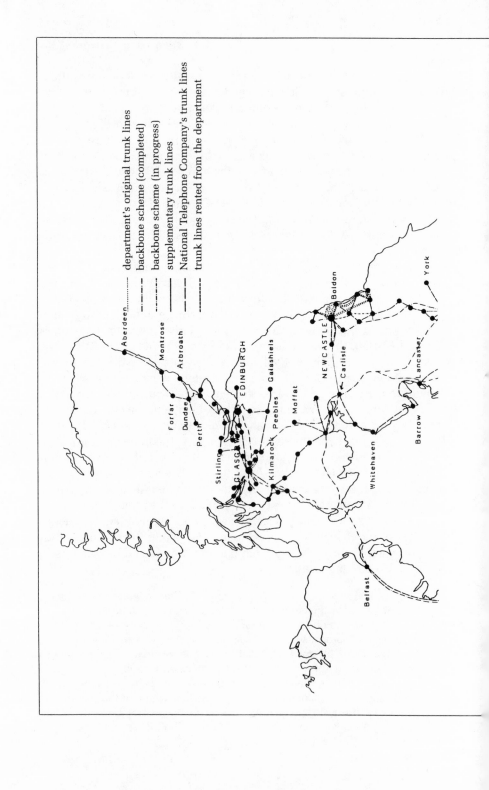

department's original trunk lines
backbone scheme (completed)
backbone scheme (in progress)
supplementary trunk lines
National Telephone Company's trunk lines
trunk lines rented from the department

Aberdeen

Montrose
Arbroath
Forfar
Dundee
Perth
Stirling
GLASGOW
EDINBURGH
Galashiels
Kilmarock Peebles
Moffat
NEWCASTLE
Carlisle
Boldon
Whitehaven
Lancaster
Barrow
York
Belfast

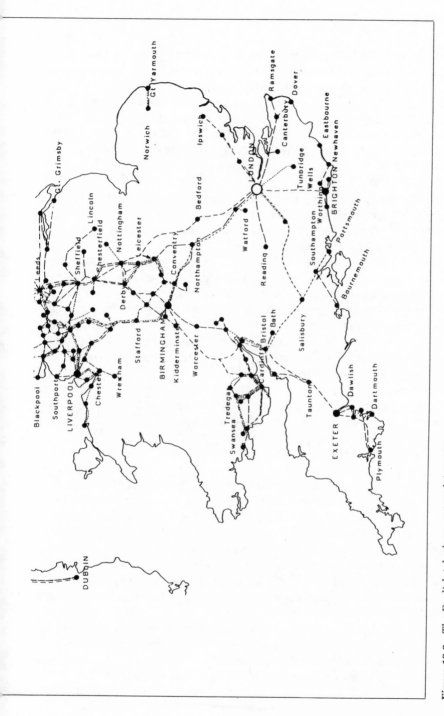

Figure 16.2 The English telephone network in 1892

Depelley's report highlights two shortcomings in the *spatial* organization of the French telephone system: difficulties in calling suburban areas from urban networks, and an absence of links between urban centers situated in the same region. I examine this question in greater detail later in this essay, because the same thing may recur in the future development of cable networks, the present idea of a "local network" being as restrictive as the idea of the early telephone network.

Regional Characteristics of the French Telephone Network, 1892[6]

In this section we are discussing a rural telephone system serving the local bourgeoisie and linked to the wine business; it was also a telephone system installed for leisure pursuits and associated with the development of tourist centers.

I have chosen this period of early telephone development for a specific reason: The telephone was too new to have become an object of directed planning. In both Great Britain and France, the government was hesitant to launch a technological program on a large scale and preferred to let private initiative explore the new terrain until a time when practical experience could be exploited. I shall limit the study to a period when the state monopoly was not yet firmly established. Within this context, two factors played a decisive role in the telephone's development: an economic factor (the telephone offered advantages to industry and commerce), and a social factor (the telephone offered the bourgeoisie an opportunity to improve domestic comfort and relations with the outside world).

Industrial needs seem to have had almost no influence on the telephone's development in France. They are perceptible only in the north (Lille and Roubaix-Tourcoing, Le Havre and Rouen). The east with its steel industry and the region around Marseille had almost no telephone service in comparison to English industrial regions. In contrast, the region around Lyon had a developed telephone service, and in nearby Dijon the winemaking region had equally good service. An industrial town such as Le Creusot did not even have a public telephone booth, although Pommard (a smaller town in a winemaking area) did.

Aside from the industrialized areas around Le Havre and Rouen and the whole northern region, there were a few places with dense telephone networks whose development cannot be accounted for by industry. The most surprising example is the Angoulême region bordered by La Rochelle, Saintes, Angoulême, Libourne, Bordeaux, and Arcachon. The whole Charente and Charente Maritime regions were in fact part of this phenomenon, including, to the north of Angoulême, the networks of Confolens, Ruffec, Champagne, Chabanais, Saint Claud, Mansle, and Saint Armand-de-Boixe. Still more impressive was the close-knit chain of networks and public telephone booths in the wine-growing region: Cognac, Jarnac, Salles-d'Angles, Hiersac, Chateauneuf, Barbezieux, Baignes, Brossac, Blanzac, Montmoreau (all special networks where calls were taxed), with, in addition, a series of public telephone booths in Saint Vallier, Sauvignac, Chatignac, Sainte Soulme, and Saint Félix.

The fact that phone calls on these networks were taxed indicates that new lines were viewed as private-interest lines; this meant that users had to pay, not only for the telephone, but also for the construction of a line outside the town, as well as fixed rates to cover maintenance and user costs based on distance and therefore proportional to the length of the section used beyond the urban network. The cost of an urban telephone varied between 150 and 200 francs for main urban networks within the city proper. The cost was 50 francs on special networks where calls were taxed, plus the rates based on the length of conversations as well as supplementary charges for line maintenance. These charges differed on state networks and those constructed by the Société Générale de Téléphone; the state began to construct its own networks as of 1888.

Clearly, not everyone could afford a telephone on the special networks, and only considerable pressure from private citizens could bring the state to lay out the large sums necessary for installation. (Most of the special networks were built by the state.) In many instances, the special networks had a small number of users. The one in Arcachon, for example, opened on December 16, 1891, had only 6 users; Bordeaux's urban network, installed much earlier, in 1881, by the Société Générale de Téléphone, had 637 subscribers in 1891 (and 280 as early as 1883). Occasionally the state was prepared to undertake construction in order to connect a single user: Special networks in Trouville in Calvados, Etretat in the Lower Seine, and Hautvillers in the Marne, all of which began to function in 1891, each served only one user. No figures are available to estimate the sizes of other special networks, but it is fairly certain that they rarely had more than 6 users. In 1891–1892, there were 93 urban networks, 125 special networks with a tax on calls, and 20 ancillary networks (i.e., networks connected to urban networks that had the same fixed-fee system—in fact, suburban networks). This approximate count gives an idea of the relative number of special networks within the whole of the system constructed by the state. It reflects the fact that social demand for the telephone was not restricted to towns but emanated as much from small, isolated groups spread over the whole French countryside—with the exception of the north where industry was undoubtedly the main stimulus in telephone development.

The first networks created by the state Postes et Télégraphes after 1888, which were added to the urban networks installed the previous decade, were primarily in wine-producing regions and coastal resort areas, the remaining being installed in fairly rich agricultural areas. (I am, of course, ignoring the north and the region around Paris for the moment.)

More specifically, in wine-producing regions, there were, among others, the following networks: those of Charente (mentioned above) with an impressive number of small-scale networks around Saintes, Cognac, and Angoulême, as well as around Confolens; networks on the Côte d'Or and in the Saône-et-Loire département[7] between Dijon and Mâcon; also, in Bourgogne, there were some networks around Chablis and Auxerre; finally, there were networks in the Hérault winemaking region.

In holiday areas, as one might expect, the Côte d'Azur had networks from Cannes to Menton, and Normandy had networks from Trouville to Fécamp. Normandy had a relatively well developed system, owing perhaps not only to the industrial dynamism of Le Havre and Rouen but also to the region's agricultural wealth.

The spatial structuring of the telephone system would lead to this interpretation: The telephone's institutional use (its appropriation by influential local figures who wanted to stay in constant touch with political and business centers) was not as predominant as one might think. Another hypothesis may well account for the telephone's development: Even at an early stage, a social use for the new technology emerged, and was adapted to the needs and way of life of a particular sector of the bourgeoisie, the landowners of huge vineyards who had broad economic influence and were attached to traditional values. This group's attachment to a very closed way of life, restricted to the family, and to a puritanical morality (in opposition to the cosmopolitan existence of the Parisian bourgeoisie) may have predisposed this group to use the telephone earlier than others, as a means of communication that protected their isolation while diminishing the disadvantages of it.

In addition to the development of the telephone in wine-producing regions, another phenomenon was the rapid installation of telephone equipment in vacation areas, the coast of Normandy and the Côte d'Azur being the most striking examples. Networks on the Norman coast were probably installed to serve the bourgeoisie of Paris and Rouen, who had established country homes in the region. The presence of English tourists accounts for the development of international connections. The area around Cognac, where the telephone developed very early, had many contacts with England. The proliferation of secondary networks in France (as opposed to the main urban networks) in holiday areas and wine-producing regions is difficult to account for without deeper and more detailed research into geographic factors than is possible here.

From Telephones to Video: Continuity in Reasoning

What conclusions can be drawn from this study of telephone network development that might help us analyze the present development of television networks and video communication?

Despite the essentially decentralized nature of the British telephone system at the turn of the century, local governments played a minor role in network conception and installation; private enterprise and economic factors gave Great Britain a telephone system that was integrated on a regional level and relatively independent of political decision-making centers.

In France, the weight of political considerations and social differentiation seems to have outweighed economic rationality or functional rationality. The nation's centralized administrative framework is not incompatible with an affirmation of local powers, which from the beginning set up a barrier between the telephone and the majority of the population.[8] More than in other nations, the concept of local territorial jurisdiction underlies political cohesion, setting up tangible limits to the spatial extension of networks. It is therefore no surprise that the administrative reform carried out under Gérard Théry[9] during the telephone boom of the 1970s was above all an attempt to impose a telecommunication management structure that did not correspond geographically to local administrations. The Directions Régionales des Télécommunications, or DRT (Regional Telecommunications Management Offices) and the Directions Opérationnelles des Télécommunications, or DOT (Operational Telecommunications Management Offices) were laid out in zones that purposely did not coincide with local administrative divisions. It was claimed that this was for technical reasons, but in fact its effect was to prevent local administrations from controlling network management. Through its "extraterritoriality" the telecommunication management thought it would be able to escape the local political control that had traditionally blocked "transnational" circulation of information.

The sociopolitical factors that shaped telephone systems at the beginning of this century are so powerful that they could, in addition to new economical and technological trends, continue to influence the development of telecommunication systems in Great Britain and France.

Notes

1. Catherine Bertho, *Télégraphes et téléphones. De Valmy au micropocesseur* (Paris, 1981).

2. Ibid., p. 205.

3. From Stan Roberts, *The First Hundred Years of the Liverpool Telephone Area*, 1979.

4. Ibid.

5. M. Depelley, dossier 161 of Commission Consultative des Postes et des Télégraphes entitled "Organisation d'un système de communications téléphoniques intermédiaires entre les réseaux urbains et les réseaux interurbains," Paris, May 1889.

6. I have chosen this particular date for several reasons: (1) As of 1890, the technical problems of connections between networks had been solved. Before that date, long-distance communication was possible only from the two endpoints of the same line. On the circuit between Paris and Le Havre, only users within one of the two cities could be connected. After 1890, long-distance networks were set up to allow communication between satellite networks, that is, all the towns surrounding Le Havre and Paris, or situated between the two main cities, could communicate with each other. Only if there is a general interconnection is there truly a regional network or a regional telephone system. (2) Before 1889, there was no information available on the whole telephone network. Only after the nationalization of the telephone in 1889 was such information gathered. The only available map of the national network in 1891 is Ortoli's map. (3) Finally, this date has been chosen to facilitate comparison with British networks. The difficulty lies in finding descriptions of national networks in the two countries that were drawn up at the same date in order to compare the two systems at different stages of technical development. Fortunately, the National Telephone Company established an inventory of long-distance lines after the British government's decision to buy up private long-distance lines in 1892. A map drawn up of all long-distance connections, both state and private, is sufficiently accurate to allow comparison with Ortoli's map of the French network.

7. The *département* is a French territorial administrative division, the whole of continental France being divided into 93 "*départements*." [Translator's note.]

8. See Armand Mattelart and Yves Stourdzé, *Technologie, culture, et communication: rapport remis à Jean-Pierre Chevènement, ministre d'Etat, ministre de la Recherche et Industrie*, 2 vols. (Paris, 1982–83).

9. Head of the Direction Générale des Télécommunications during Valéry Giscard d'Estaing's presidency.

The Contributors

Letty Anderson is a member of the faculty, Atkinson College, York University, Toronto.

Gabriel Dupuy is a member of the faculties of the Institut d'Urbanisme de Paris and the Ecole Nationale des Ponts et Chaussées, Paris.

Jean-Pierre Goubert is a member of the faculty, Ecole des Hautes Etudes en Sciences Sociales, Paris.

Chantal de Gournay is associated with the Centre National d'études des telecommunications, Paris.

André Guillerme is a member of the faculty, Ecole Nationale des Travaux Publics de l'Etat, Paris.

Georges Knaebel is a member of the faculty, Institut d'Urbanisme de Paris.

Dominique Larroque is associated with the Centre de Documentation d'Histoire des techniques, Paris.

Seymour J. Mandelbaum is a member of the faculty, University of Pennsylvania, Philadelphia.

John P. McKay is a member of the faculty, University of Illinois, Champaign–Urbana.

Clay McShane is a member of the faculty, Northeastern University, Boston, Massachusetts.

Martin V. Melosi is a member of the faculty, University of Houston, Texas.

Roselyne Messager is associated with Institut Economique et Juridique de l'Energie, Paris.

Harold L. Platt is a member of the faculty, Loyola University, Chicago, Illinois.

Mark H. Rose is a member of the faculty, Michigan Technological University, Houghton, Michigan.

Anthony Sutcliffe is a member of the faculty, University of Sheffield, England.

Joel A. Tarr is a member of the faculty, Carnegie Mellon University, Pittsburgh, Pennsylvania.